Synthesis Lectures on Mathematics & Statistics

Series Editor

Steven G. Krantz, Department of Mathematics, Washington University, Saint Louis, USA

This series includes titles in applied mathematics and statistics for cross-disciplinary STEM professionals, educators, researchers, and students. The series focuses on new and traditional techniques to develop mathematical knowledge and skills, an understanding of core mathematical reasoning, and the ability to utilize data in specific applications.

Sumati Kumari Panda ·
Velusamy Vijayakumar · Ravi P. Agarwal

Fixed Point Theory from Early Foundations to Contemporary Challenges

Springer

Sumati Kumari Panda
Department of Mathematics
GMR Institute of Technology
Rajam, Andhra Pradesh, India

Velusamy Vijayakumar
Department of Mathematics
School of Advanced Sciences, Vellore Institute of Technology University
Vellore, Tamil Nadu, India

Ravi P. Agarwal
Department of Mathematics and Systems Engineering
Florida Institute of Technology
Melbourne, FL, USA

ISSN 1938-1743 ISSN 1938-1751 (electronic)
Synthesis Lectures on Mathematics & Statistics
ISBN 978-3-032-03584-4 ISBN 978-3-032-03585-1 (eBook)
https://doi.org/10.1007/978-3-032-03585-1

This Springer imprint is published by the registered company Springer Nature Switzerland AG
The registered company address is: Gewerbestrasse 11, 6330 Cham, Switzerland

To my husband Anand Prabhakar and our beloved son Keerthan.

—Sumati Kumari Panda

To my wife Devikarani.

—Velusamy Vijayakumar

To my wife Sadhna.

—Ravi P. Agarwal

Preface

Fixed point theory is an active domain with applications in non-linear analysis, computation, discrete equations, complex systems, and various fields of engineering. Fixed points are fundamental, but fixed point theory also encompasses computing fixed points, algorithms, and approximation approaches.

The formulation of a fixed point hypothesis is as outlined below:

- Let X be a set and $\mathbb{T} : X \to X$ be a self map. A point $x \in X$ is known as a *fixed point* of $\mathbb{T}$ if $\mathbb{T}(x) = x$.

An assertion that presents conditions on X and $\mathbb{T}$ that ensure that $\mathbb{T}$ has a fixed point in X is known as a *fixed point theorem.*

While there are three main categories of fixed point theory, metric fixed point theory has been applied in much wider scientific and practical contexts than other categories like discrete and topological fixed point theory. Still, metric fixed point theory has played a significant role in these fields. Beginning with the traditional theorems of *Banach, Schauder*, and *Krasnosel'skiĭ*, it explores how their groundbreaking theorems laid the foundation for one of the most powerful areas of nonlinear analysis. These early contributions paved the way for a rich tapestry of developments, fostering new perspectives and solutions to problems that were once considered insurmountable.

Nowadays, *fixed point theorems* continue to address the most complex challenges through a good number of disciplines. From discretization in data science to synchronization analysis in neural networks, and even in the mathematical models of infectious disease, theory's applications are as profound as they are.

Here, our book demonstrates the significance, applicability, and widespread nature of fixed point theorems in contexts outside of mathematics. Moreover, this book seeks to follow the outstanding journey of *fixed point theory* from its early, fundamental principles to the modern research with applications. Our book not only highlights these modern hurdles but also explores how the field has accommodated and grown in response to them.

Moreover, this book provides a comprehensive exploration of both the classical underpinnings and the cutting-edge advancements in *fixed point theory*. It is designed to serve as both a guide for those entering the field and a resource for seasoned researchers looking to deepen their understanding of its modern applications.

There are two parts to the book: Part I, we discuss the historical notes in fixed point theory which consists essential characteristics of the traditional metric space along with the primary fixed point theorems. We examine several metric space generalizations and related fixed point theorems. This part also explores the developments in the fixed point theory. In the second part, we have covered modern applications in the fixed point theory in science and engineering.

There are eight chapters in the book. Preliminary information and historical notes concerning distinct types of metric spaces and fixed point theorems are presented in the first four chapters. The first four chapters provide introductory information, nontrivial examples, convergence axioms, and historical notes about different types of metric spaces and fixed point theorems.

Using the contraction mapping theorem, a recursive algorithm related to metric notions is provided in Chap. 5 for adaptive systems to guarantee the existence of the solution.

In Chap. 6, the fractional-order neural network is chosen as the focal point. The primary objective of this chapter is to provide a comprehensive overview and analysis of the latest advancements in fractional-order neural network, utilizing fixed point theory from both computational and theoretical perspective. We summarise the vast amount of research on the existence and uniqueness of solutions for the different mathematical models in Chap. 7. We have presented the existence and the uniqueness of the model the fractal-fractional system of the computer virus problem using the Atangana–Baleanu operator under applying the nonlinear F-contraction.

In Chap. 8, we demonstrate theoretically and practically how an admissible perturbation of fixed point mappings might improve ant-based algorithms for medical imaging processes.

To provoke additional research, each chapters ends with *open questions*, except Chap. 1, which is introduction.

Due to analogy, we refrain from proving certain theorems and corollaries.

Rajam, India — Sumati Kumari Panda
Vellore, India — Velusamy Vijayakumar
Melbourne, USA — Ravi P. Agarwal

Acknowledgements The researchers who contributed substantially to *Fixed Point Theory and its Applications* deserve our *gratitude*.

Contents

Part I Historical Notes on Fixed Point Theory: The Growth and Impact of Fixed Point Thought

1 Introductory Perspectives on Fixed Points 3
1.1 What Could Be the Origin of the Fixed Point? 4
1.1.1 Generalization of Fixed Point 6
1.2 How the Study of Fixed Points Evolved Into Fixed Point Theory? 6

2 Generalized Spaces: A Metric Evolution 11
2.1 Beyond Traditional Metric Spaces: A Comprehensive Overview 16
2.1.1 Classification of Generalized Metric Spaces 17
2.2 Symmetric Spaces: Convergence Axioms 17
2.2.1 Implications and Non-implications Among the Axioms 20
2.2.2 New Non-implications Among the Axioms 24
2.3 b-Metric Spaces: Topology and Some Inequalities 26
2.3.1 Implications Among Metric, b-Metric, Generalized b-Metric, and Strong b-Metric Spaces 29
2.3.2 Topology of b-Metric Space 32
2.3.3 Discussion on Certain Inequality-Related Results 41
2.4 θ-Metric Spaces: Understanding Their Structure and Properties 45
2.4.1 Some Key Generalizations of θ-Metric Space 46
Open Questions 50
References 50

3 Renowned Theorems in Fixed Point Theory: Topological and Discrete Perspectives 55
3.1 Some Standard Fixed Point Theorems 55
Open Questions 68
References 68

4 Foundational Extensions of the Contraction Mapping 71
4.1 The Banach Fixed Point Theorem 71
4.1.1 A Simple Proof of Banach Fixed Point Theorem According to *Richard S. Palais* [4] 74
4.1.2 Short Notes on Converse of the Banach Fixed Point Theorem ... 76
4.2 Some Definitions of Contractive Mappings 77
4.2.1 Some Special Cases Among Contractive Mappings 82
4.2.2 Pair of Mappings 83
4.2.3 Some Well-Known Fixed Point Theorems for Various Contractive Mappings 84
4.3 Selected Key Contractive Mappings in Modern Era 95
4.3.1 Nonlinear Contractive Mappings: A Few Non-implications 98

Part II Fixed-Point Theory for Modern Challenges

5 Control Process 107
5.1 Introduction to Control Systems 107
5.2 Adaptive Control Systems 109
5.3 Utilization of Contraction Mapping Principle in Control Process 111
5.4 $\mathcal{F}$-Contraction Mapping in Automatic Flight-Path Generator 115
5.4.1 Automatic Flight Path Generation for Terrain-Following Aircraft Through $\mathcal{F}$-Contraction 116
Open Questions 120
References 121

6 Equilibrium Points, Dynamics and Synchronization of Neural Networks ... 123
6.1 Dynamics of Fractional-Order Neural Networks 125
6.1.1 Existence, Uniqueness, and the Uniform Stability of the Solution 128
6.1.2 Finite-Time Stability Results Under Caputo-Katugampola Derivative 143
6.2 Dynamics of Fractional-Order Complex-Valued Neural Networks 149
6.2.1 Various Stability Results Under Caputo Derivative 150
6.2.2 Uniform Stability Results Under Caputo-Katugampola Derivative 158
6.3 Discrete Fractional-Order Complex-Valued Neural Networks 162
6.3.1 General Class 163
6.3.2 L_p-Sense 169
6.3.3 Synchronization 171
Open Questions 181
References 181

7 Modeling and Simulation 185
7.1 Computer Virus Modeling 186
7.2 Infectious Disease Modeling 194
7.2.1 COVID-19 Mathematical Models 198
7.2.2 HIV/AIDS Disease Modeling 204
7.2.3 Tuberculosis (TB) Disease Modeling 206
7.3 Existence of Solution for the Fractal-Fractional Model of the Computer Virus Problem Using (φ, F)-Contraction 208
Open Questions 213
References 213

8 Medical Image Processing 221
8.1 Ant-Based Algorithms in Medical Image Edge Detection 223
8.1.1 Key Principles Applied to Edge Detection 223
8.1.2 Algorithm Steps in Medical Edge Detection 224
8.2 The Role of Fixed Points in Image Process 227
8.2.1 Digital Image Process 227
8.2.2 Medical Image Process 230
Open Questions 242
References 247

Part I

Historical Notes on Fixed Point Theory: The Growth and Impact of Fixed Point Thought

Introductory Perspectives on Fixed Points

1

In high school, we were introduced to *Venn diagrams* as a tool to represent sets and their relationships visually. At the same time, we encountered basic functions like the **identity function**—a function that maps every element of a set to itself. For example, if we have:

$$\mathbb{T}(4) = 4,$$

this is a simple case of the identity function, where each input equals its output.

A **fixed point** of a function is any element x that remains unchanged when the function is applied, i.e., $\mathbb{T}(x) = x$. In the case of the identity function, every element in the set is a fixed point because for every x, the output is the same as the input: $\mathbb{T}(x) = x$. This realization shows how the identity function is a basic yet perfect example of **fixed point**.

Thus, even though we might not have explicitly recognized the idea of fixed points in high school, we were already working with it through familiar concepts like identity mapping. As we explore more advanced mathematics, the concept of fixed points becomes essential in a broader range of contexts, but its roots trace back to these fundamental ideas we first encountered with Venn diagrams and simple functions.

A basic definition of fixed point is as follows:

- A **fixed point** of a function is a point that remains unchanged when the function is applied to it. Formally, given a function $\mathbb{T} : X \to X$, a point $x \in X$ is called a **fixed point** if:

$$\mathbb{T}(x) = x.$$

S. K. Panda et al., *Fixed Point Theory from Early Foundations to Contemporary Challenges*, Synthesis Lectures on Mathematics & Statistics,
https://doi.org/10.1007/978-3-032-03585-1_1

In other words, when we apply the function $\mathbb{T}$ to x, the result is x itself. The study of fixed points is important in various branches of mathematics, particularly in areas such as analysis, topology, and applied mathematics.

Example 1.1 For the function $\mathbb{T}(x) = x^2$, the fixed points are those values of x where $\mathbb{T}(x) = x$, i.e., $x^2 = x$. Solving this equation:

$$x^2 - x = 0 \quad \text{implies} \quad x(x-1) = 0.$$

Thus, the fixed points are $x = 0$ and $x = 1$.

Example 1.2 Consider the function $\mathbb{T}(x) = x^2 - 2$. We want to find the fixed points of this function, which are the points where $\mathbb{T}(x) = x$. To do this, we solve the equation:

$$\mathbb{T}(x) = x \quad \Rightarrow \quad x^2 - 2 = x.$$

Rearranging the terms, we have:

$$x^2 - x - 2 = 0.$$

Now, solving this quadratic equation using the quadratic formula:

$$x = \frac{-(-1) \pm \sqrt{(-1)^2 - 4(1)(-2)}}{2(1)}$$

$$\Rightarrow x = \frac{1 \pm \sqrt{1+8}}{2} = \frac{1 \pm 3}{2}.$$

The two solutions are:

$$x = \frac{1+3}{2} = 2 \quad \text{and} \quad x = \frac{1-3}{2} = -1.$$

Thus, the fixed points of the function $\mathbb{T}(x) = x^2 - 2$ are $x = 2$ and $x = -1$.

Below given Fig. 1.1 is a plot of the function $\mathbb{T}(x) = x^2 - 2$ and the line $y = x$, with the fixed points indicated by red dots.

The points of intersection between the curve $\mathbb{T}(x) = x^2 - 2$ and the line $y = x$ are the fixed points, $x = 2$ and $x = -1$.

1.1 What Could Be the Origin of the Fixed Point?

The origin of a fixed point refers to the underlying mathematical or functional structure that leads to the existence of a point where $\mathbb{T}(x) = x$. The fixed point arises from certain properties or conditions of the function itself.

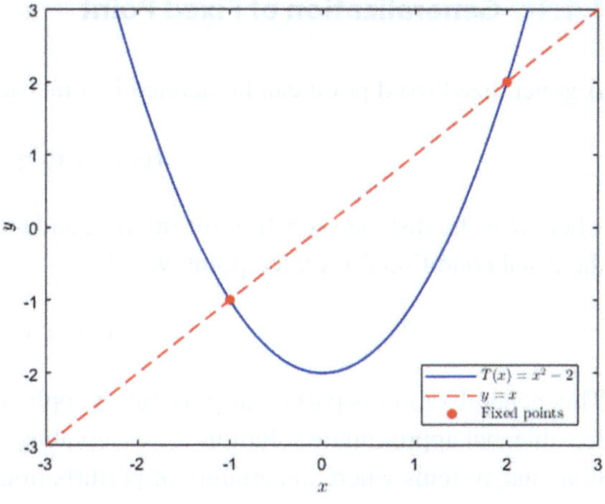

Fig. 1.1 Visualization of fixed points for $\mathbb{T}(x) = x^2 - 2$

- Fixed points can originate from the **symmetry** or **invariance** of a system. For example, in a dynamical system or physical process, certain points may remain unchanged under the action of the function or transformation. In such cases, the fixed point can represent an **equilibrium** or **steady state**.
- In topological contexts, fixed points often arise due to general conditions imposed by the space itself. For example, the **Brouwer Fixed Point Theorem** states that any continuous function mapping a convex, compact set to itself (e.g., a closed disc in $\mathbb{R}^2$) has at least one fixed point. This fixed point exists due to the structure of the space rather than any specific property of the function.
- Fixed points can also be observed when a function is **iterated** (applied repeatedly). If repeated application of the function brings points closer to a particular value, that value may be a fixed point. For example, when applying Newton's method to find roots of a function, convergence often occurs near a fixed point.
- In physics and engineering, fixed points often represent **steady states** or **equilibria**. For example, in a dynamical system governed by differential equations, fixed points can represent states where the system is in equilibrium (i.e., no change occurs over time). Such points can be stable (the system returns to the fixed point after a small perturbation) or unstable (the system diverges from the fixed point under perturbation).
- A function that has some form of **self-similarity** may naturally possess fixed points. For instance, the function $\mathbb{T}(x) = x$ has every point as a fixed point, and functions like $\mathbb{T}(x) = \cos(x)$ tend to stabilize at a fixed point due to the balancing nature of the cosine function.

1.1.1 Generalization of Fixed Point

A generalized fixed point can be defined in a metric space using the following condition:

$$d(\mathbb{T}(x), x) \leq \epsilon,$$

where d is the distance function (or metric), and ϵ is a small positive number. This relaxes the usual condition for a fixed point, which is:

$$\mathbb{T}(x) = x.$$

This generalization is particularly useful in applications where exact solutions may not be feasible, but approximate solutions are acceptable, such as in numerical methods, optimization, and systems where uncertainty or perturbations are present.

However, in the literature, we can find another strong generalization of fixed point notion as follows:

Let $\mathbb{T} : X \to X$ be a self-mapping of the set X. A point $x \in X$ is defined as a **periodic point** of $\mathbb{T}$ if there exists a positive integer m such that

$$\mathbb{T}^m(x) = x.$$

Specifically, we require that $\mathbb{T}^m(x) = x$, where $\mathbb{T}^m$ denotes the m-fold composition of $\mathbb{T}$ with itself. Importantly, the integer m, known as the period of x, represents the smallest number of iterations needed for x to return to its original position. If $m = 1$, then x is a fixed point of $\mathbb{T}$. However, for $m > 1$, x exhibits cyclical behavior under repeated applications of $\mathbb{T}$. Such periodic points are critical in the study of dynamical systems, as they often signify regular or repeating behavior in the system's evolution.

1.2 How the Study of Fixed Points Evolved Into Fixed Point Theory?

The transition from the concept of fixed points to the broader field of **fixed point theory** involves studying fixed points within various mathematical contexts and frameworks, yielding significant results in analysis, topology, and applied mathematics. Here's how this transition unfolds:

- **Interpreting Fixed Points**
  - At its core, the concept of a fixed point arises from a function $\mathbb{T} : X \to X$ where there exists an element $x \in X$ such that $\mathbb{T}(x) = x$. This idea serves as the foundation for exploring solutions to equations and understanding dynamic behavior in iterative processes.

- **Advancement of Fixed Point Results**
When mathematicians delved into exploring fixed points characteristics they developed theorems, on fixed points that lay out the conditions, for fixed point existence. Some noteworthy instances are:

  - **Contraction Mapping Theorem**: For a contraction mapping $\mathbb{T}$ on a complete metric space (X, d), there exists a unique fixed point $x^\star \in X$ such that $\mathbb{T}(x^\star) = x^\star$. Formally, if $d(\mathbb{T}(x), \mathbb{T}(y)) \leq c\, d(x, y)$ for some $0 < c < 1$ and all $x, y \in X$, then a unique fixed point exists.
  - **Brouwer's Fixed Point Theorem**: Any continuous function $\mathbb{T} : \mathscr{A} \to \mathscr{A}$, where $\mathscr{A}$ is a convex compact subset of $\mathbb{R}^n$, has at least one fixed point, i.e., there exists $x \in \mathscr{A}$ such that $\mathbb{T}(x) = x$.

- **Expansion into Various Mathematical Sub-Fields**
The exploration of fixed points has evolved from functions to different sub-fields of mathematics such as:

  - **General and Applied Topology**: Investigating fixed points in the context of continuity, compactness, and homotopy. For example, in topological fixed point theory, we explore continuous mappings $\mathbb{T} : X \to X$ where X is a topological space.
  - **Operator Theory**: Examining fixed points of nonlinear mappings, particularly those that may not satisfy the contraction condition or may have multiple fixed points. This is common in solving nonlinear operator equations.
  - **Game Theoretical Analysis**: Analyzing Nash equilibria through fixed points of best-response mappings in strategy spaces.

- **Applications in Real-World Problems**
Fixed point theory has significant applications in various fields such as:

  - **Differential and Integral Equations**: Fixed point methods, such as applying the *Schauder Fixed Point Theorem*, are often used to prove the existence of solutions for both differential and integral equations.
  - **Optimization Methods**: Fixed point theorems are a popular tool for solving optimization and approximation problems in nonlinear analysis. Here are some fixed point theorems and their applications:

    Ekeland's Variational Principle: This is a powerful tool for nonlinear analysis with applications in optimization theory, fixed point theory, and global analysis. One of the main problems in contemporary communication networks is the network bandwidth allocation problem, which is a convex optimization problem for

a strictly convex objective function covering the fixed point set of a nonexpansive mapping.
It is common practice to use fixed point assumptions for determining equilibrium points in dynamics of stability and/or Nash-equillibria in decision-making theory. As an instance, Kakutani's fixed point theorem has demonstrated the existence of Nash-equilibria.
Fixed point hypotheses are regularly utilized to find economic strategies. Used to address the question of whether and under what circumstances a market can be in equilibrium.

- **Computational Applied Mathematics**: In fixed point semantics, the denotation of a recursive function is constructed by approximation. This is done by repeatedly plugging the current approximation of the denotation into the function's definition. The process continues until the complete denotation is obtained.

  Kleene's Fixed-Point Theorem: This theorem states that for any continuous function on a complete lattice, there exists a fixed point. It is important in computer science and mathematical logic because it helps define recursive functions and prove program properties.

- **Chaos Theory**: In chaotic attractors, 'unstable fixed points', along with periodic orbits, help to define the attractor's topology. They serve as "skeletons" of the attractor, around which chaotic trajectories are organized. These points can give insights into the attractor's properties by revealing invariant structures within the system.

  For example, in the **Lorenz attractor**, fixed points lie at the center of the two "wings" of the butterfly-like structure, and trajectories spiral around these points in a chaotic manner as in below Fig. 1.2.

• **Interdisciplinary Links**

  - The implications of fixed point theory extend into various areas of mathematics, such as topology, functional analysis, and computational mathematics. For instance, the existence of fixed points can provide insights into the stability, convergence, and behavior of dynamical systems. The iterative process $x_{n+1} = \mathbb{T}(x_n)$, where $x_n \to x^\star$ as $n \to \infty$, represents a common approach to finding fixed points.

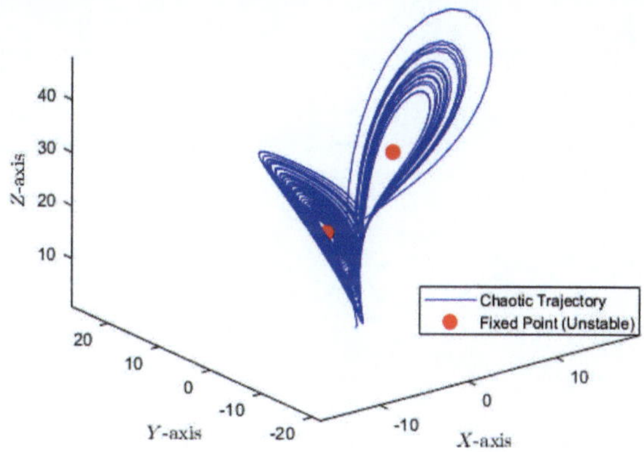

Fig. 1.2 Lorenz attractor with fixed points

Modern mathematics heavily relies on ***fixed point theory*** as a tool that connects many fields, such as science, engineering, and medicine and contributes to a deeper understanding of 'dynamic systems' and /or 'dynamic behaviour of complex systems'.

2 Generalized Spaces: A Metric Evolution

The concept of a **metric** extends the classical idea of distance between two points, which originated in the systematic geometry of Euclid. In modern mathematical foundations, the notion of a metric space was first axiomatized by Maurice René Fréchet [1], who referred to it as an "L-space". Fréchet's work laid the groundwork for formalizing distance in more abstract settings, beyond just the Euclidean plane. Later, Felix Hausdorff [2] introduced the term "**metric space**" as part of his pioneering work on abstract set theory, where he concentrated on the structure and properties of point sets. Hausdorff's terminology emphasized the role of distance in defining relationships within sets, influencing the development of topology and analysis.

Moreover, a **metric space** serves as an extension of the real number line $\mathbb{R}$ by establishing a "space" equipped with a concept of "distance" between any two points. This distance function, or metric, provides a framework for expressing foundational concepts from real analysis—like convergence and continuity—within a broader and often more versatile context. Interestingly, this generalization can simplify our understanding of these concepts, enabling us to explore them without being confined to the properties of real numbers alone.

However, many have attempted to broaden the notion of a metric space during the past century. The most frequently used generalizations are covered in this chapter, along with the recent spike in some generalizations and their significance for metric fixed point theory.

A **metric space** is a set equipped with a function that defines a notion of distance between any two elements of the set. Formally, a metric space is a pair (X, d), where X is a set and $d : X \times X \to \mathbb{R}$ is a **metric** (or distance function) that satisfies the following properties for all $x, y, z \in X$:

S. K. Panda et al., *Fixed Point Theory from Early Foundations to Contemporary Challenges*, Synthesis Lectures on Mathematics & Statistics,
https://doi.org/10.1007/978-3-032-03585-1_2

- (Non-negativity) $d(x, y) \geq 0$,
- (Identity) $d(x, y) = 0$ if and only if $x = y$,
- (Symmetry) $d(x, y) = d(y, x)$,
- (Triangle inequality) $d(x, z) \leq d(x, y) + d(y, z)$.

The pair (X, d) is known as **metric space**, often known as **standard metric space**.

These properties ensure that the concept of distance in a metric space behaves in an intuitive way, making it a fundamental structure in many areas of mathematics.

Example 2.1 In a general Euclidean space $\mathbb{R}^n$, the set $\mathbb{R}^n$ consists of all ordered n-tuples $(x_1, x_2, \ldots, x_n)$ where each x_i is a real number. The distance between any two points $(x_1, x_2, \ldots, x_n)$ and $(y_1, y_2, \ldots, y_n)$ in $\mathbb{R}^n$ can be measured by the **Euclidean metric**, defined by:

$$d((x_1, \ldots, x_n), (y_1, \ldots, y_n)) = \sqrt{(x_1 - y_1)^2 + (x_2 - y_2)^2 + \cdots + (x_n - y_n)^2}.$$

This formula calculates the straight-line (or Euclidean) distance between the two points by summing the squares of the differences between their corresponding coordinates, then taking the square root of that sum. The Euclidean metric is a generalization of the familiar distance formulas:

- For $n = 1$, this metric reduces to the absolute value distance on $\mathbb{R}$, $d(x, y) = |x - y|$.
- For $n = 2$, it gives the standard distance between points in the plane, $\mathbb{R}^2$.
- For $n = 3$, it calculates the usual distance in three-dimensional space, $\mathbb{R}^3$.

Thus, $\mathbb{R}^n$ with the Euclidean metric forms a **metric space** that enables us to extend many concepts from real analysis, such as limits and continuity, to n-dimensional spaces in a consistent and intuitive way.

Example 2.2 Consider $\mathscr{M}$, which represents the Euclidean plane consisting of all ordered pairs of real numbers. Suppose we have two typical points on this plane: $u = (x_1, y_1)$ and $v = (x_2, y_2)$. The traditional Euclidean distance between these points is expressed as:

$$\text{(Euclidean Distance)} \quad d(u, v) = \sqrt{(x_1 - x_2)^2 + (y_1 - y_2)^2}$$

Alternatively, we can define distance in $\mathscr{M}$ using two other functions:

$$\text{(Chebyshev/Max Distance)} \quad d(u, v) = \max(|x_1 - x_2|, |y_1 - y_2|)$$

$$\text{(Manhattan Distance)} \quad d(u, v) = |x_1 - x_2| + |y_1 - y_2|$$

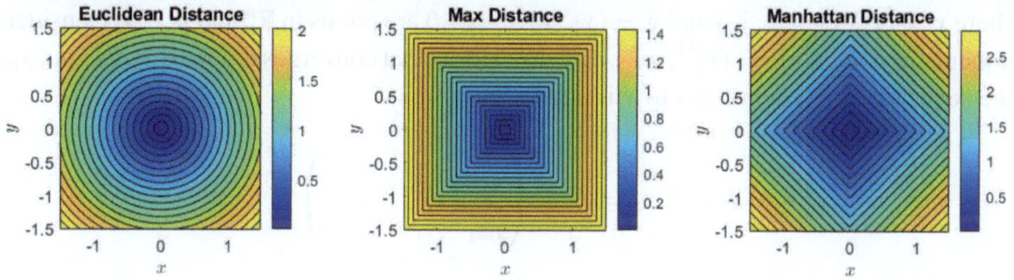

Fig. 2.1 Contours at distance 1.5

To visualize these different distance metrics, we can plot the points at distance 1.5 from the origin, as illustrated in Fig. 2.1.

Topological properties of metric spaces provide a framework to study the properties of spaces in a more abstract sense than just their geometric structure. Here are some key topological properties of metric spaces:

1. A subset U of a metric space (X, d) is called **open** if for every point $x \in U$, there exists a radius $r > 0$ such that the ball

$$B(x, r) = \{y \in X \mid d(x, y) < r\}$$

is entirely contained in U.
2. A subset C of X is **closed** if its complement $X \setminus C$ is open, or equivalently, if it contains all its limit points.
3. A point x in X is a **limit point** of a subset $A \subseteq X$ if every open ball centered at x contains at least one point from A different from x. The **closure** of A, denoted $\overline{A}$, is the set of all points in A together with its limit points.
4. A sequence (x_n) in X converges to a limit x if for every $\epsilon > 0$, there exists an $N \in \mathbb{N}$ such that for all $n \geq N$, $d(x_n, x) < \epsilon$.
5. A metric space (X, d) is **complete** if every Cauchy sequence (a sequence where the distances between successive terms eventually become arbitrarily small) converges to a limit in X.

Example 2.3 In this example, the authors of [3] examined the convergence of sequences in the n-dimensional real space $\mathbb{R}^n$ with a p-norm metric. Let the metric d_p on $\mathbb{R}^n$ be defined as:

$$d(x, y) = d_p(x, y) = \left(\sum_{j=1}^{n} |x_j - y_j|^p \right)^{1/p},$$

where $x = (x_1, x_2, \ldots, x_n)$ and $y = (y_1, y_2, \ldots, y_n)$ are points in $\mathbb{R}^n$ and $p \geq 1$. Consider a sequence of points $x^{(k)} = (x_1^{(k)}, x_2^{(k)}, \ldots, x_n^{(k)})$ in $\mathbb{R}^n$ that converges to $x = (x_1, x_2, \ldots, x_n)$. The sequence $\{x^{(k)}\}$ converges to x if and only if:

$$\lim_{k\to\infty} d(x^{(k)}, x) = \lim_{k\to\infty} \left(\sum_{j=1}^{n} |x_j^{(k)} - x_j|^p \right)^{1/p} = 0.$$

This implies that for each $j = 1, \ldots, n$, $x_j^{(k)}$ must converge to x_j as k approaches infinity, indicating that the sequence converges coordinate-wise. On the other hand, if $x_j^{(k)} \to x_j$ for each j, then for a given $\epsilon > 0$, there exists a sufficiently large integer k_0 such that for all $k \geq k_0$, the inequality $|x_j^{(k)} - x_j| < \epsilon/n^{1/p}$ holds for every j. Consequently,

$$d(x^{(k)}, x) = \left(\sum_{j=1}^{n} |x_j^{(k)} - x_j|^p \right)^{1/p} < \epsilon,$$

for all $k \geq k_0$. Thus, convergence of sequences in $(\mathbb{R}^n, d_p)$ is equivalent to coordinate-wise convergence.

Example 2.4 Consider a sequence $\{x^{(k)}\}$ in $\mathbb{R}^3$, where each point $x^{(k)}$ is given by:

$$x^{(k)} = \left(1 + \frac{1}{k}, 2 - \frac{1}{k^2}, 3 + \frac{1}{k^3}\right).$$

We want to determine if this sequence converges to the point $x = (1, 2, 3)$ using the p-norm metric. For two points $x = (x_1, x_2, x_3)$ and $y = (y_1, y_2, y_3)$ in $\mathbb{R}^3$, the p-norm distance $d(x, y)$ is defined as:

$$d(x, y) = d_p(x, y) = \left(|x_1 - y_1|^p + |x_2 - y_2|^p + |x_3 - y_3|^p\right)^{1/p}.$$

Using this metric, we calculate $d(x^{(k)}, x)$ to see if it approaches zero as k increases:

$$d(x^{(k)}, x) = \left(\left|1 + \frac{1}{k} - 1\right|^p + \left|2 - \frac{1}{k^2} - 2\right|^p + \left|3 + \frac{1}{k^3} - 3\right|^p \right)^{1/p}.$$

This simplifies to:

$$d(x^{(k)}, x) = \left(\left(\frac{1}{k}\right)^p + \left(\frac{1}{k^2}\right)^p + \left(\frac{1}{k^3}\right)^p \right)^{1/p}.$$

As $k \to \infty$, each term $\left(\frac{1}{k}\right)^p$, $\left(\frac{1}{k^2}\right)^p$, and $\left(\frac{1}{k^3}\right)^p$ tends to zero, implying:

$$\lim_{k\to\infty} d(x^{(k)}, x) = 0.$$

Thus, $\{x^{(k)}\} = \left(1 + \frac{1}{k}, 2 - \frac{1}{k^2}, 3 + \frac{1}{k^3}\right)$ converges to $x = (1, 2, 3)$ in the p-norm metric in $\mathbb{R}^3$. This shows that as each coordinate of $x^{(k)}$ independently approaches its target, the sequence converges to x, verifying coordinate-wise convergence in $\mathbb{R}^3$ with the p-norm metric.

As above example is associate with $\mathbb{R}^3$, we can visualize the convergence of the sequence $\{x^{(k)}\} = \left(1 + \frac{1}{k}, 2 - \frac{1}{k^2}, 3 + \frac{1}{k^3}\right)$ converges to $x = (1, 2, 3)$ by plotting each coordinate as a function of k along with the target coordinates. Each subplot shows the individual convergence of $x_1^{(k)}$, $x_2^{(k)}$ and $x_3^{(k)}$ to their respective targets as shown in Fig. 2.2.

Example 2.5 Consider the space $X = l^\infty$, the set of all bounded sequences of real numbers, with the metric defined by the sup norm:

$$d(x, y) = \sup_{n\in\mathbb{N}} |x_n - y_n| \quad \text{for } x = \{x_n\} \text{ and } y = \{y_n\} \text{ in } l^\infty.$$

Define the sequence $\{x^{(k)}\}_{k=1}^\infty$ in l^∞ by

$$x^{(k)} = \{x_n^{(k)}\}_{n=1}^\infty \quad \text{where} \quad x_n^{(k)} = \begin{cases} 1, & \text{if } n \le k, \\ 0, & \text{if } n > k. \end{cases}$$

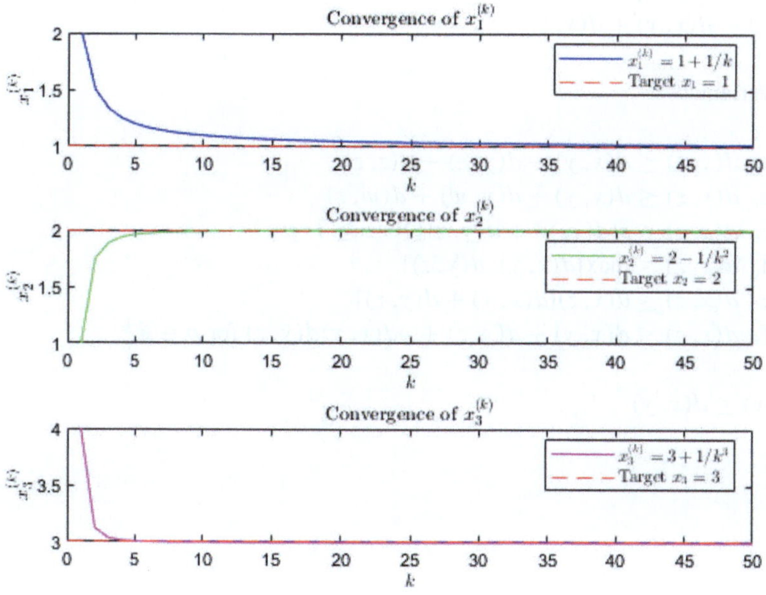

Fig. 2.2 Convergence of $x_1^{(k)}$, $x_2^{(k)}$ and $x_3^{(k)}$

This sequence $\{x^{(k)}\}$ is Cauchy with respect to the sup norm metric because for any $\epsilon > 0$, we can find an integer N such that $d(x^{(m)}, x^{(n)}) < \epsilon$ for all $m, n \geq N$. Specifically,

$$d(x^{(m)}, x^{(n)}) = \sup_{n \in \mathbb{N}} |x_n^{(m)} - x_n^{(n)}| \leq 1,$$

and becomes small as m and n increase.

However, $\{x^{(k)}\}$ does not converge to any point in l^∞. Intuitively, as $k \to \infty$, the sequence $x^{(k)}$ resembles the sequence that is 1 at every position, which is not a bounded sequence in l^∞. Hence, $\{x^{(k)}\}$ is a Cauchy sequence in l^∞ that does not converge within l^∞.

2.1 Beyond Traditional Metric Spaces: A Comprehensive Overview

This section examines various conditions for a distance function $d : X \times X \to [0, \infty)$, with X as a nonempty set and $\theta : X \times X \to [1, \infty)$. While some of these conditions were addressed earlier, they are revisited here for completeness.

Conditions for all d for $x, y, z, w \in X$:

1. $d(x, x) = 0$
2. $d(x, y) = d(y, x) = 0 \Rightarrow x = y$
3. $d(x, y) = d(y, x)$
4. $d(x, z) \leq d(x, y) + d(y, z)$

   - Variants:

     a. $d(x, z) \leq d(x, y) + d(y, z) - d(z, z)$
     b. $d(x, z) \leq d(x, y) + d(y, w) + d(w, z)$
     c. $d(x, z) \leq s[d(x, y) + d(y, z)]$ for $s \geq 1$
     d. $d(x, z) \leq \max\{d(x, y), d(y, z)\}$
     e. $d(x, z) \leq \theta(x, z)[d(x, y) + d(y, z)]$
     f. $d(x, z) \leq d(x, y) + d(y, z) + \rho d(x, y)d(y, z)$ for $\rho \in \mathbb{R}^+$

5. $d(x, x) \leq d(x, y)$

2.1.1 Classification of Generalized Metric Spaces

Below is an overview of well-known generalizations.

The pair (X, d) is called:

- **Metric Space** if d satisfies conditions 1 to 4.
- **Bianciari Distance Space** [4–13] if d satisfies conditions 1 to 3 along with variant 4b.
- **Dislocated Metric Space** [14–17] if d satisfies conditions 2 to 4.
- **Quasi-Metric Space** [18–22] if d satisfies conditions 1, 2, and 4.
- **Symmetric Space** (or **Semi-Metric space**) [23–27] if d satisfies only conditions 1 to 3.
- **Partial Metric Space** [28–31] if d satisfies conditions 2, 3, variant 4a, and condition 5.
- b-**Metric Space** [32–40] if d satisfies conditions 1 to 3 and the scalar-adjusted triangle inequality (variant 4c).
- **Ultra-Metric Space** [41, 42] if d satisfies conditions 1 to 3 with a max-distance variant in the triangle inequality (variant 4d).
- θ-**Metric Space** (or **an extended** b-**metric Space**) [43–47] if d satisfies conditions 1 to 3 with variant 4e.
- **Suprametric Space** [48–51] if d satisfies conditions 1 to 3 and variant 4f.

This overview highlights the variety of ways the concept of a metric space can be adapted and extended, each generalization offering unique properties and potential applications.

It is notable that numerous distance functions have emerged as generalizations and extensions of the metric space concept. Among these, we can cite, for example, fuzzy metric space [52], probabilistic metric space [53], Menger space [54], and various other metric space variations, each detailed with comprehensive definitions and significant observations in the literature [55–62]. This chapter, however, focuses exclusively on three key generalizations—symmetric space, b-metric space and θ-metric space—due to their relevance and frequent applications, which have attracted substantial academic interest and a wealth of research papers.

2.2 Symmetric Spaces: Convergence Axioms

W.A. Wilson [23] laid the groundwork for symmetric spaces by introducing two core axioms, (W3) and (W4), which became key to further developments in this field. Imdad et al. [24] capitalized on axiom (W3), using it to prove fundamental coincidence and common fixed-point theorems within symmetric spaces. Building on this foundation, Aliouche [25] introduced an additional axiom, (H.E), which, together with (W3) and (W4), allowed for the proof of a common fixed-point theorem under conditions for non-compatible self-mappings. Later, Cho et al. [26] contributed the axiom (C.C), which added a continuity aspect to symmetric metrics, providing insights into the interactions among (W3), (W4), and (C.C). Through examples, they illustrated the distinct roles of each axiom, highlighting their importance in advancing fixed-point theorems within symmetric spaces.

W.A. Wilson [23] and Imdad [24]'s work during 2005–15 saw considerable activity in fixed point theory through symmetric space.

Definition 2.1 [23] A **symmetric space** (X, d) consists of a nonempty set X together with a function $d : X \times X \to [0, \infty)$, called a **symmetric**, satisfying the following properties for all $x, y \in X$:

1. $d(x, y) = 0$ if and only if $x = y$,
2. $d(x, y) = d(y, x)$.

Given $x \in X$ and $\epsilon > 0$, define the open ball centered at x with radius ϵ as

$$B(x, \epsilon) = \{y \in X : d(x, y) < \epsilon\}.$$

These open balls generate a topology $\tau(d)$ on X, where a set $U \subseteq X$ belongs to $\tau(d)$ if, for every point $x \in U$, there exists $\epsilon > 0$ such that $B(x, \epsilon) \subseteq U$. A subset $S \subseteq X$ is said to be a **neighborhood** of a point $x \in X$ if there exists an open set $U \in \tau(d)$ such that $x \in U \subseteq S$.

d is called a **semimetric** if, for each $x \in X$ and $\epsilon > 0$, $B(x, \epsilon)$ is a neighborhood of x in the topology $\tau(d)$. Symmetric spaces differ from metric spaces primarily in that they do not require the **triangle inequality**, which yields symmetric spaces are not necessarily **Hausdorff**. To establish fixed-point theorems in symmetric spaces, certain additional axioms are needed:

- **(W3)** [23]: For a sequence $\{x_n\}$ in X and points $x, y \in X$, if $\lim_{n\to\infty} d(x_n, x) = 0$ and $\lim_{n\to\infty} d(x_n, y) = 0$, then $x = y$.

Example 2.6 Let $X = \mathbb{R}$ (the real line) with the usual distance $d(x, y) = |x - y|$. Take the sequence $x_n = \frac{1}{n}$, and suppose that $x = 0$ and $y = 0$. We have:

$$\lim_{n\to\infty} d(x_n, 0) = \lim_{n\to\infty} \left|\frac{1}{n} - 0\right| = 0,$$

and similarly,

$$\lim_{n\to\infty} d(x_n, y) = \lim_{n\to\infty} \left|\frac{1}{n} - 0\right| = 0.$$

Hence, $x = y = 0$, and the condition $\lim_{n\to\infty} d(x_n, x) = 0$ and $\lim_{n\to\infty} d(x_n, y) = 0$ leads to $x = y$.

- **(W4)** [23]: For sequences $\{x_n\}$ and $\{y_n\}$ in X and a point $x \in X$, if $\lim_{n\to\infty} d(x_n, x) = 0$ and $\lim_{n\to\infty} d(y_n, x_n) = 0$, then $\lim_{n\to\infty} d(y_n, x) = 0$.

Example 2.7 Let $X = \mathbb{R}$ again, with the usual distance $d(x, y) = |x - y|$. Take the sequences $x_n = \frac{1}{n}$ and $y_n = \frac{1}{n^2}$. We have:

$$\lim_{n\to\infty} d(x_n, 0) = \lim_{n\to\infty} \left|\frac{1}{n} - 0\right| = 0,$$

and

$$\lim_{n\to\infty} d(y_n, x_n) = \lim_{n\to\infty} \left|\frac{1}{n^2} - \frac{1}{n}\right| = 0.$$

Therefore, by (W4), we have:

$$\lim_{n\to\infty} d(y_n, 0) = \lim_{n\to\infty} \left|\frac{1}{n^2} - 0\right| = 0.$$

Hence, y_n converges to 0 as $n \to \infty$.

- **(H.E)** [24]: For sequences $\{x_n\}$ and $\{y_n\}$ in X and a point $x \in X$, if $\lim_{n\to\infty} d(x_n, x) = 0$ and $\lim_{n\to\infty} d(y_n, x) = 0$, then $\lim_{n\to\infty} d(x_n, y_n) = 0$.

Example 2.8 Consider $X = \mathbb{R}$ with the usual distance $d(x, y) = |x - y|$. Take $x_n = \frac{1}{n}$ and $y_n = \frac{1}{n} + \frac{1}{n^2}$. We have:

$$\lim_{n\to\infty} d(x_n, 0) = \lim_{n\to\infty} \left|\frac{1}{n} - 0\right| = 0,$$

and

$$\lim_{n\to\infty} d(y_n, 0) = \lim_{n\to\infty} \left|\frac{1}{n} + \frac{1}{n^2} - 0\right| = 0.$$

Then, by (H.E), we have:

$$\lim_{n\to\infty} d(x_n, y_n) = \lim_{n\to\infty} \left|\frac{1}{n} - \left(\frac{1}{n} + \frac{1}{n^2}\right)\right| = \lim_{n\to\infty} \frac{1}{n^2} = 0.$$

Thus, x_n and y_n converge to 0 in terms of the distance d.

- **(C.C)** [26]: For a sequence $\{x_n\}$ in X and points $x, y \in X$, if $\lim_{n\to\infty} d(x_n, x) = 0$, then $\lim_{n\to\infty} d(x_n, y) = d(x, y)$.

Example 2.9 Let $X = \mathbb{R}$, and again take $d(x, y) = |x - y|$. Take the sequence $x_n = \frac{1}{n}$ and the point $x = 0$, and $y = 1$. We have:

$$\lim_{n\to\infty} d(x_n, 0) = \lim_{n\to\infty} \left|\frac{1}{n} - 0\right| = 0.$$

Then, by (C.C), we must have:

$$\lim_{n\to\infty} d(x_n, 1) = d(0, 1) = |0 - 1| = 1.$$

Therefore, the distance from the sequence x_n to $y = 1$ converges to 1, as expected.

If d is a metric, then (W3), (W4), (H.E), and (C.C) are automatically satisfied. Furthermore, if $\tau(d)$ is Hausdorff, then (W3) is satisfied.

2.2.1 Implications and Non-implications Among the Axioms

Proposition 2.1 *[26] In a symmetric space (X, d), the following implications hold between the axioms:*

1. ***(W4) implies (W3)**: If d satisfies axiom (W4), it also satisfies axiom (W3).*
2. ***(C.C) implies (W3)**: If d satisfies axiom (C.C), it also satisfies axiom (W3).*

Proof Consider a sequence $\{x_n\}$ in X and points $x, y \in X$ such that $\lim_{n\to\infty} d(x_n, x) = 0$ and $\lim_{n\to\infty} d(x_n, y) = 0$.

1. **Proof of (W4) ⇒ (W3)**: Suppose that axiom (W4) holds. Set $y_n = y$ for each $n \in \mathbb{N}$, so that $\lim_{n\to\infty} d(x_n, y_n) = \lim_{n\to\infty} d(x_n, y) = 0$. According to (W4), since $\lim_{n\to\infty} d(x_n, x) = 0$ and $\lim_{n\to\infty} d(y_n, x_n) = 0$, it follows that $\lim_{n\to\infty} d(y_n, x) = 0$. This implies $d(y, x) = 0$, and therefore $x = y$, establishing (W3).
2. **Proof of (C.C) ⇒ (W3)**: Suppose that axiom (C.C) holds. By (C.C), we have that $\lim_{n\to\infty} d(x_n, x) = 0$ implies $d(x, y) = \lim_{n\to\infty} d(x_n, y) = 0$. This leads to $x = y$, proving (W3).

Example 2.10 Let $X = \{0\} \cup \{1/n : n \in \mathbb{N}\} \cup \{2\}$ and define a symmetric function $d : X \times X \to [0, \infty)$ by:

$$d(x, y) = \begin{cases} |x - y|, & \text{if } x, y \in \{1/n : n \in \mathbb{N}\} \cup \{0\}, \\ 1 + |x - y|, & \text{if } x = 2 \text{ or } y = 2, \text{ and } x, y \neq 0, \\ 2, & \text{if } x = 2 \text{ or } y = 2, \text{ and } x = 0 \text{ or } y = 0. \end{cases}$$

We will verify that this space satisfies (W4) and (H.E) axioms, but it does not satisfy (C.C).

1. **(W4) holds**: To satisfy (W4), we need to show that for sequences $\{x_n\}$ and $\{y_n\}$ in X with $\lim_{n\to\infty} d(x_n, 0) = 0$ and $\lim_{n\to\infty} d(y_n, x_n) = 0$, it follows that $\lim_{n\to\infty} d(y_n, 0) = 0$. Take $x_n = 1/n$ and $y_n = 0$ for all n. We have:

$$\lim_{n\to\infty} d(x_n, 0) = \lim_{n\to\infty} \left|\frac{1}{n} - 0\right| = 0,$$

and

$$\lim_{n\to\infty} d(y_n, x_n) = \lim_{n\to\infty} \left|0 - \frac{1}{n}\right| = 0.$$

And also, $\lim_{n\to\infty} d(y_n, 0) = 0$, thus (W4) is satisfied.

2. **(H.E) holds**: To check that (H.E) holds, take sequences $\{x_n = 1/n\}$ and $\{y_n = 1/(n+1)\}$. We have:

$$\lim_{n\to\infty} d(x_n, 0) = \lim_{n\to\infty} \left|\frac{1}{n} - 0\right| = 0,$$

and

$$\lim_{n\to\infty} d(y_n, 0) = \lim_{n\to\infty} \left|\frac{1}{n+1} - 0\right| = 0,$$

also

$$\lim_{n\to\infty} d(x_n, y_n) = \lim_{n\to\infty} \left|\frac{1}{n} - \frac{1}{n+1}\right| = 0.$$

This shows that (H.E) holds.

3. **(C.C) does not hold**: To verify that (C.C) does not hold, take the sequence $\{x_n = 1/n\}$ converging to 0 and select $y = 2$. Then:

$$\lim_{n\to\infty} d(x_n, 0) = 0,$$

but

$$\lim_{n\to\infty} d(x_n, 2) = \lim_{n\to\infty} \left(1 + \left|\frac{1}{n} - 2\right|\right) = 3 \neq d(0, 2) = 2.$$

Thus, (C.C) does not hold.

This example satisfies (W4) and (H.E) but fails to satisfy (C.C), as required.

Example 2.11 Consider the set $X = \{0\} \cup \{1, 1/2, 1/3, \dots\} \cup \{3\}$, and define a symmetric function $d : X \times X \to [0, \infty)$ by:

$$d(x, y) = \begin{cases} |x - y|, & \text{if } x, y \in \{1, 1/2, 1/3, \dots\} \cup \{0\}, \\ 2 + |x - y|, & \text{if } x = 3 \text{ or } y = 3, \text{ and } x, y \neq 0, \\ 4, & \text{if } x = 3 \text{ or } y = 3, \text{ and } x = 0 \text{ or } y = 0. \end{cases}$$

We will verify that this space satisfies the (H.E),(W3) and (W4) axioms, but it does not satisfy (C.C).

1. **(H.E) holds**: To satisfy (H.E), we need to show that if $\{x_n\}$ and $\{y_n\}$ are sequences in X with $\lim_{n\to\infty} d(x_n, 0) = 0$ and $\lim_{n\to\infty} d(y_n, 0) = 0$, then $\lim_{n\to\infty} d(x_n, y_n) = 0$. Take sequences $x_n = 1/n$ and $y_n = 1/(n+1)$. We have:

$$\lim_{n\to\infty} d(x_n, 0) = \lim_{n\to\infty} \left|\frac{1}{n} - 0\right| = 0,$$

and

$$\lim_{n\to\infty} d(y_n, 0) = \lim_{n\to\infty} \left|\frac{1}{n+1} - 0\right| = 0.$$

Now,

$$\lim_{n\to\infty} d(x_n, y_n) = \lim_{n\to\infty} \left|\frac{1}{n} - \frac{1}{n+1}\right| = 0.$$

Therefore, (H.E) is satisfied.

2. **(W3) holds**: To show that (W3) hold, consider the sequence $x_n = 1/n$ converging to $x = 0$ and the point $y = 3$. We have:

$$\lim_{n\to\infty} d(x_n, 0) = 0,$$

but

$$\lim_{n\to\infty} d(x_n, y) = \lim_{n\to\infty} \left(2 + \left|\frac{1}{n} - 3\right|\right) = 2 + |0 - 3| = 5 \neq 0.$$

This shows that (W3) is satisfied.

3. **(W4) holds**: To check that (W4) hold, take sequences $x_n = 1/n$ and $y_n = 1/(n+2)$. We have:

$$\lim_{n\to\infty} d(x_n, 0) = 0,$$

and

$$\lim_{n\to\infty} d(y_n, x_n) = \lim_{n\to\infty} \left|\frac{1}{n} - \frac{1}{n+2}\right| = 0,$$

but

$$\lim_{n\to\infty} d(y_n, 0) = \lim_{n\to\infty} \left|\frac{1}{n+2} - 0\right| = 0.$$

This shows that (W4) is satisfied.

4. **(C.C) does not hold**: To verify that (C.C) does not hold, take the sequence $x_n = 1/n$ converging to $x = 0$ and let $y = 3$. We need to check if $\lim_{n\to\infty} d(x_n, y) = d(0, 3)$.

- Calculate $\lim_{n\to\infty} d(x_n, y)$:

$$\lim_{n\to\infty} d(x_n, 3) = \lim_{n\to\infty}\left(2 + \left|\frac{1}{n} - 3\right|\right) = 2 + |0 - 3| = 5.$$

- Calculate $d(0, 3)$:

$$d(0, 3) = 4.$$

Since $\lim_{n\to\infty} d(x_n, 3) = 5 \neq d(0, 3) = 4$, (C.C) is not satisfied.

This example satisfies (H.E), (W3) and (W4), but fails to satisfy (C.C), as required.

Example 2.12 Let $X = \left\{\frac{1}{n} : n = 1, 2, 3, \ldots\right\} \cup \{0\}$, and define the symmetric distance function d on X as follows:

$$d(x, y) = \begin{cases} 0, & \text{if } x = y, \\ \left|\frac{1}{m} - \frac{1}{n}\right| + \frac{1}{m+n}, & \text{if } x = \frac{1}{m},\ y = \frac{1}{n},\ m + n \text{ is odd and } |m - n| = 1, \\ 2, & \text{if } x = \frac{1}{m},\ y = \frac{1}{n},\ m + n \text{ even}, \\ 1 + \frac{1}{m+n}, & \text{if } x = \frac{1}{m},\ y = \frac{1}{n},\ m + n \text{ is odd and } |m - n| > 1, \\ \frac{1}{n^2}, & \text{if } x = \frac{1}{n},\ y = 0, \text{ or } y = \frac{1}{n},\ x = 0. \end{cases}$$

In this symmetric space (X, d), the axioms $(C.C)$ and $(H.E)$ behave as follows:

- **Verification of** $(H.E)$: Now, consider two sequences $\{x_n\}$ and $\{y_n\}$ in X defined by:

$$x_n = \frac{1}{n},\quad y_n = \frac{1}{n+1} \quad \text{for } n = 1, 2, 3, \ldots.$$

Both sequences converge to 0 as $n \to \infty$:

$$\lim_{n\to\infty} d(x_n, 0) = \lim_{n\to\infty} \frac{1}{n^2} = 0, \quad \lim_{n\to\infty} d(y_n, 0) = \lim_{n\to\infty} \frac{1}{(n+1)^2} = 0.$$

However, evaluating the distance between x_n and y_n:

$$d(x_n, y_n) = \left|\frac{1}{n} - \frac{1}{n+1}\right| + \frac{1}{2n+1}, \text{ since } n + (n+1) = 2n + 1 \text{ is odd and } |n - (n+1)| = 1.$$

And,

$$\lim_{n\to\infty} d(x_n, y_n) = 0,$$

which holds the axiom $(H.E)$.

- **Verification of** $(C.C)$: Consider a sequence $\{x_n\}$ in X defined by $x_n = \dfrac{1}{n}$ for $n = 1, 2, 3, \ldots$. As $n \to \infty$, $x_n \to 0$, and we have:

$$\lim_{n\to\infty} d(x_n, 0) = \lim_{n\to\infty} \frac{1}{n^2} = 0.$$

For any fixed $y = \dfrac{1}{k} \in X$, where k is a positive integer, we observe:

$$\lim_{n\to\infty} d(x_n, y) \neq d(0, y).$$

This doesn't satisfies the axiom $(C.C)$ since the limit of $d(x_n, y)$ is not equals to $d(0, y)$.

Thus, this example demonstrates that the axiom $(H.E)$ does not imply $(C.C)$ in the symmetric space (X, d).

2.2.2 New Non-implications Among the Axioms

In a symmetric space, additional implications between the axioms (W3), (W4), (H.E), and (C.C) cannot generally be proven, as counterexamples show that these axioms are independent in certain contexts. Here is a summary of why other implications don't hold, supported by specific counterexamples.

1. **(W3) does not imply (W4)**:
   Let $X = \left\{\frac{1}{n} : n \in \mathbb{N}\right\} \cup \{0\}$, and define a symmetric function $d : X \times X \to [0, \infty)$ by:

$$d(x, y) = \begin{cases} |x - y|, & \text{if } x, y \in \left\{\frac{1}{n} : n \in \mathbb{N}\right\}, \\ 1, & \text{if } x = 0 \text{ and } y = \frac{1}{n} \text{ with } n \text{ even}, \\ \frac{1}{n}, & \text{if } x = 0 \text{ and } y = \frac{1}{n} \text{ with } n \text{ odd}, \\ d(y, x), & \text{if } d(x, y) \text{ is not explicitly defined.} \end{cases}$$

   Let $\{x_n\}$ be a sequence in X such that $\lim_{n\to\infty} d(x_n, x) = 0$ and $\lim_{n\to\infty} d(x_n, y) = 0$.

   - Since $d(x_n, x) \to 0$, the sequence $\{x_n\}$ converges to x.
   - Similarly, $d(x_n, y) \to 0$ implies $\{x_n\}$ converges to y.
   - In this space, the only sequences that can converge to 0 are those where $x_n = \frac{1}{n}$, and convergence to any other point $\frac{1}{k}$ is unique.
   - Therefore, if $x_n \to x$ and $x_n \to y$, we must have $x = y$.

   Thus, $(W3)$ is satisfied.

Now, consider:

- $x = 0$,
- $x_n = \frac{1}{2n+1}$ (denominator odd),
- $y_n = \frac{1}{2n}$ (denominator even).

$$d(x_n, 0) = \frac{1}{2n+1} \to 0 \text{ as } n \to \infty.$$

For $x_n = \frac{1}{2n+1}$ and $y_n = \frac{1}{2n}$,

$$d(y_n, x_n) = \left| \frac{1}{2n} - \frac{1}{2n+1} \right| = \frac{1}{(2n)(2n+1)} \to 0 \text{ as } n \to \infty.$$

For $y_n = \frac{1}{2n}$,

$$d(y_n, 0) = \begin{cases} 1, & \text{if } n \text{ is even,} \\ \frac{1}{n}, & \text{if } n \text{ is odd.} \end{cases}$$

Thus, $\lim_{n\to\infty} d(y_n, x) \neq 0$, and $(W4)$ fails.
The symmetric space (X, d) satisfies $(W3)$ but does not satisfy $(W4)$.

2. **(W4) does not imply (C.C)**:
Consider $X = \{0\} \cup \{1/n : n \in \mathbb{N}\}$ with a symmetric function d defined by:

$$d(x, y) = \begin{cases} 1/n, & \text{if } x = 1/n \text{ and } y = 0, \text{ (or) } y = 1/n \text{ and } x = 0, \\ 1, & \text{if } x \neq y \text{ and both } x, y \neq 0, \\ 0, & \text{if } x = y. \end{cases}$$

This space doesn't satisfies (C.C): since $x_n \to 0$, but $\lim_{n\to\infty} d(x_n, y) \neq d(0, y)$. However, it satisfies (W4). Take sequences $\{x_n = 1/n\}$ and $\{y_n = 0\}$; $d(x_n, 0) \to 0$ and $d(y_n, x_n) \to 0$, moreover, $d(y_n, 0) = 0$. Thus, (W4) holds, but (C.C) does not.

3. **(C.C) does not imply (H.E)**:
Let $X = \left\{\frac{1}{n} : n \in \mathbb{N}\right\} \cup \{0\}$, and define a symmetric function $d : X \times X \to [0, \infty)$ by:

$$d(x, y) = \begin{cases} |x - y|, & \text{if } x, y \in \left\{\frac{1}{n} : n \in \mathbb{N}\right\}, \\ \frac{1}{n}, & \text{if } x = 0 \text{ and } y = \frac{1}{n}, \\ 1, & \text{if } x = 0 \text{ and } y = \frac{1}{n}, \ n \text{ is even.} \end{cases}$$

One can easily check that (C.C) does not imply (H.E).

2.3 *b*-Metric Spaces: Topology and Some Inequalities

In 1993, Czerwik [32] introduced the concept of a *b-metric*, defined as follows:

Definition 2.2 [32] Let X be a non-empty set, and let $d : X \times X \to [0, +\infty)$ be a function that satisfies the following conditions for all $x, y, z \in X$:

1. $d(x, y) = 0$ if and only if $x = y$,
2. $d(x, y) = d(y, x)$,
3. $d(x, z) \leq 2[d(x, y) + d(y, z)]$.

If these conditions hold, d is called a *b-metric* on X, and the pair (X, d) is termed a *b-metric space*.
Later, in 1998, Czerwik [33] extended this concept by replacing the constant 2 with a general constant $s \geq 1$, while still referring to it as a *b-metric*, as described below:

Definition 2.3 [33] Let X be a non-empty set, and let $s \geq 1$ be a given real number. A function $d : X \times X \to [0, \infty)$ is called a **b-metric** if it satisfies the following properties for each $x, y, z \in X$:

1. $d(x, y) = 0$ if and only if $x = y$,
2. $d(x, y) = d(y, x)$ (symmetry),
3. $d(x, z) \leq s \cdot [d(x, y) + d(y, z)]$ (relaxed triangle inequality).

If these conditions hold, d is called a **b-metric** on X, and the pair (X, d) is termed a **b-metric space**.

The transformation from metric spaces to b-metric spaces represents a generalization to accommodate broader mathematical applications, particularly where traditional metric constraints are too restrictive. This adjustment preserves useful distance properties while enabling applications across various fields, including fixed-point theory and functional analysis, where a more flexible structure can yield significant analytical benefits.

Example 2.13 b-metric space that is not a metric space.
Let $Y = \{y_1, y_2, y_3, y_4\}$ be a set with a function $d : Y \times Y \to \mathbb{R}$ defined as follows:

- $d(y_1, y_2) = m$ with $m \geq 3$,
- $d(y_1, y_3) = d(y_1, y_4) = d(y_2, y_3) = d(y_2, y_4) = d(y_3, y_4) = 1$,
- $d(y_i, y_j) = d(y_j, y_i)$ for all $i, j = 1, 2, 3, 4$,
- $d(y_i, y_i) = 0$ for all $i = 1, 2, 3, 4$.

Verifying b-metric properties:

1. Non-negativity and Identity: Clearly, $d(y_i, y_j) \geq 0$ and $d(y_i, y_j) = 0$ if and only if $i = j$.
2. Symmetry: By definition, $d(y_i, y_j) = d(y_j, y_i)$, satisfying symmetry.
3. Relaxed Triangle Inequality: For a constant $s = \frac{m}{2}$, we check that:

$$d(y_i, y_j) \leq \frac{m}{2}\left(d(y_i, y_n) + d(y_n, y_j)\right), \quad \text{for any } i, j, n = 1, 2, 3, 4.$$

This inequality holds under the given conditions.

However, if $m > 3$, the standard triangle inequality:

$$d(y_i, y_j) \leq d(y_i, y_n) + d(y_n, y_j),$$

does not hold. For example, consider $d(y_1, y_2) = m$, while $d(y_1, y_3) + d(y_3, y_2) = 1 + 1 = 2 < m$, violating the triangle inequality when $m > 2$. This example illustrates a b-metric on Y that does not satisfy the traditional triangle inequality, making it a b-metric space but not a metric space.

Example 2.14 Let $X = C([a, b], \mathbb{R})$ be the space of all continuous real-valued functions defined on the closed interval $[a, b]$. Define the function $d : X \times X \to \mathbb{R}_+$ as

$$d(x, y) = \sup_{t \in [a,b]} |x(t) - y(t)|^2.$$

This d defines a b-metric with the coefficient $s = 2$. Here, the relaxed triangle inequality is satisfied because

$$d(x, z) \leq s \cdot [d(x, y) + d(y, z)] = 2 \cdot [d(x, y) + d(y, z)],$$

making (X, d) a b-metric space suitable for analyzing continuous functions with variations in their distances.

Example 2.15 Consider the set $X = \mathbb{N} \cup \{\infty\}$ (the set of natural numbers extended by the point at infinity) and define $d : X \times X \to \mathbb{R}_+$ by

$$d(x, y) = \begin{cases} 0, & \text{if } x = y, \\ \left|\frac{1}{x} - \frac{1}{y}\right|, & \text{if } x \text{ and } y \text{ are even or if } x = y = \infty, \\ 5, & \text{if } x \text{ and } y \text{ are odd and } x \neq y, \\ 2, & \text{otherwise.} \end{cases}$$

This function d is a b-metric with coefficient $s = 3$, though it does not satisfy continuity. This example showcases a more complex structure where distance relationships vary based on parity and the presence of infinity in the set X.

Example 2.16 Let $X = \ell_p(\mathbb{R})$ for $0 < p < 1$, where

$$\ell_p(\mathbb{R}) := \left\{ \{x_n\} \subset \mathbb{R} : \sum_{n=1}^{\infty} |x_n|^p < \infty \right\}.$$

Define a function $d : X \times X \to \mathbb{R}_+$ by

$$d(x, y) = \sum_{n=1}^{\infty} \frac{|x_n - y_n|^p}{1 + |x_n - y_n|^p},$$

where $x = \{x_n\}$ and $y = \{y_n\}$ are sequences in $\ell_p(\mathbb{R})$.
This function d satisfies the properties of a b-metric space with $s = 1 + 2^{1-p}$, specifically:

1. $d(x, y) = 0$ if and only if $x = y$,
2. $d(x, y) = d(y, x)$,
3. $d(x, z) \leq (1 + 2^{1-p}) \cdot [d(x, y) + d(y, z)]$.

In this case, the choice of $s = 1 + 2^{1-p}$ ensures that the relaxed triangle inequality holds. This b-metric is useful for controlling large distances, as the denominator $1 + |x_n - y_n|^p$ restricts the influence of outliers, especially in cases where individual terms $|x_n - y_n|$ can vary significantly.

Remark 2.1 In Definition 2.3,

- If we replace condition (3) with the following inequality:

$$d(x, y) \leq \alpha\, d(x, z) + \beta\, d(z, y),$$

where $\alpha, \beta \geq 1$, then the space is called a **generalized** b**-metric space** [63].
- If we instead replace condition (3) with:

$$d(x, y) \leq d(x, z) + K\, d(z, y),$$

where $K \geq 1$, then the space is known as a **strong** b**-metric space** [64].

These variations allow for further flexibility in defining distance functions, accommodating different analytical frameworks and applications in general topology and fixed-point theory.

2.3.1 Implications Among Metric, b-Metric, Generalized b-Metric, and Strong b-Metric Spaces

1. **Every Metric Space is a b-Metric Space**:

   - If $s = 1$ in the b-metric inequality, then (X, d) behaves like a standard metric space since the inequality $d(x, z) \leq d(x, y) + d(y, z)$ holds without any scaling.
   - Hence, every metric space is a b-metric space with $s = 1$.

2. **Generalized b-Metric Space Contains b-Metric Space**:

   - The generalized b-metric space with constants $\alpha, \beta \geq 1$ reduces to a b-metric space if we set $\alpha = s$ and $\beta = s$, indicating that generalized b-metric spaces can be seen as an extension of b-metric spaces.
   - Therefore, every b-metric space is a special case of a generalized b-metric space with $\alpha = \beta = s$.

3. **Strong b-Metric Space Includes b-Metric Space**:

   - The strong b-metric space has a relaxed triangle inequality of the form $d(x, y) \leq d(x, z) + Kd(z, y)$, where $K \geq 1$. If $K = 1$, the inequality simplifies to the standard triangle inequality, making every metric space a strong b-metric space with $K = 1$.
   - For b-metric spaces, setting $s = K$ in the strong b-metric inequality allows it to satisfy the b-metric conditions, so every b-metric space is also a strong b-metric space with $K = s$.

4. **Implications in Fixed-Point Theory and Convergence**:

   - Generalized and strong b-metric spaces allow for broader contraction conditions in fixed-point theory, enabling convergence proofs and applications in contexts where strict metric conditions do not hold.
   - Strong b-metric spaces, in particular, often support robust fixed-point results for mappings that may not be continuous in the conventional sense, widening their use in analysis and applied mathematics.

Thus, we can conclude that:

Metric spaces $\subset$ b-metric spaces $\subset$ strong b-metric spaces $\subset$ generalized b-metric spaces.

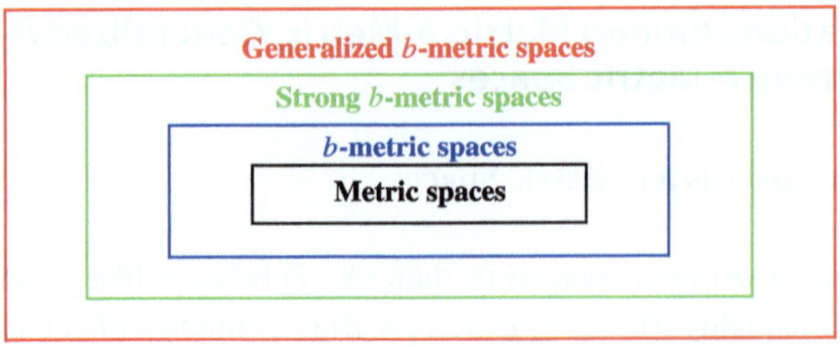

Each extension provides increased flexibility for different mathematical frameworks while retaining core distance properties that enable convergence and continuity-based analyses.

On the other hand, in 2010, Khamsi and Hussain [65] introduced a revised concept termed a *b-metric* within the broader context of *metric-type* spaces.

Definition 2.4 [65] Consider a non-empty set X, along with a constant $s > 0$ and a function $d : X \times X \to [0, +\infty)$ satisfying the following properties for any $x, y, z \in X$:

1. $d(x, y) = 0$ if and only if $x = y$,
2. $d(x, y) = d(y, x)$,
3. $d(x, z) \leq s\,[d(x, y) + d(y, z)]$.

When these conditions are met, d is referred to as a **metric-type** function on X, and the structure (X, d, s) is called a **metric-type space**. It follows from property (3) that $s \geq 1$. Also in 2010, Khamsi [66] proposed a different variant of the metric-type concept by altering condition (3) of Definition 2.4. The modified requirement stated that:

$$d(x, z) \leq s\,[d(x, y_1) + \cdots + d(y_n, z)]\,,$$

for any $x, y_1, \ldots, y_n, z \in X$ (see [66], Definition 2.7).

To visualize the generalized triangle inequality

$$d(x, z) \leq s\,[d(x, y_1) + \cdots + d(y_n, z)]\,,$$

for a function d in a metric-type space, we will create a plot that illustrates how the inequality holds over several points, $x, y_1, \ldots, y_n$, and z. To visualize this for a larger n, we will add more points and distances between them, forming a sequence from x to z with intermediate points $y_1, \ldots, y_n$ showing how the inequality holds as we increase the number of segments.

Let us illustrate this for $n = 5$ with the points connected in sequence.

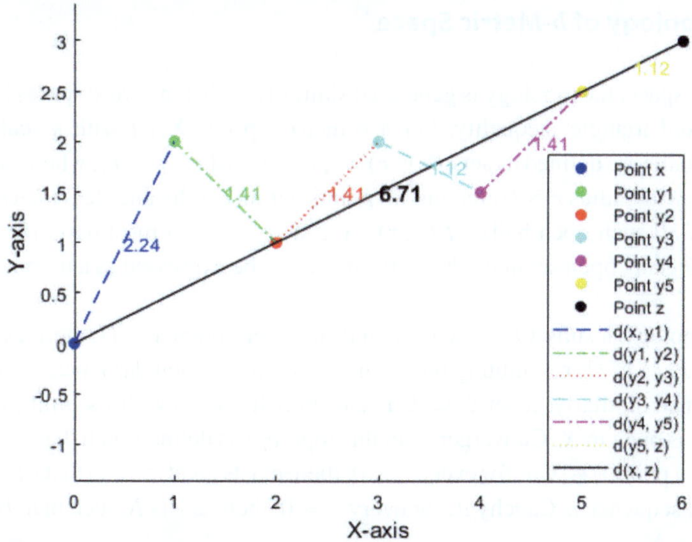

Fig. 2.3 Generalized triangle inequality for $n = 5$

In this setup:

- $d(x, y)$ represents the "distance" between points x and y
- The constant $s \geq 1$ scales the right-hand side of the inequality, adjusting how tightly it bounds $d(x, z)$.

We will illustrate the distances between points on a 2D plot, marking each segment and visualizing the inequality as shown in the Fig. 2.3.

We visualize the generalized triangle inequality with five intermediate points y_1, y_2, y_3, y_4, and y_5 between x and z. The inequality is given by:

$$d(x, z) \leq s\,[d(x, y_1) + d(y_1, y_2) + d(y_2, y_3) + d(y_3, y_4) + d(y_4, y_5) + d(y_5, z)],$$

where $s = 1.5$ is a scaling constant in the metric-type space.

This allowance enables the framework to accommodate a broader class of **distance-like functions**, particularly useful in settings where classical metrics might be too restrictive. Applications of "metric-type spaces" span various fields, from analysis and optimization to fixed point theory, as they provide a robust foundation for extending metric-based theorems in contexts where distances may exhibit non-standard behavior yet retain sufficient structure for meaningful analysis.

2.3.2 Topology of b-Metric Space

In a b-metric space, the topology is generated similarly to that in metric spaces but is adapted for the relaxed triangle inequality. For a b-metric space (X, d) with a scaling constant $s \geq 1$, open balls are defined as sets $B(x, r) = \{y \in X : d(x, y) < r\}$, where each open ball centered at x with radius $r > 0$ contains all points within a b-distance less than r from x. The collection of all such open balls, $\{B(x, r) : x \in X, r > 0\}$, forms a basis for the topology on X, meaning any open set in the b-metric space can be expressed as a union of these open balls.

In this topology, a subset $U \subseteq X$ is open if, for every point $x \in U$, there exists an $\epsilon > 0$ such that $B(x, \epsilon) \subseteq U$, resembling the open set structure in standard metric spaces. Closed sets are defined similarly: a set $C \subseteq X$ is closed if it contains all its limit points or if its complement is open in X. Convergence in this topology is defined such that a sequence $\{x_n\}$ converges to a point $x \in X$ if, for every $\epsilon > 0$, there exists an $N \in \mathbb{N}$ so that $d(x_n, x) < \epsilon$ for all $n \geq N$. A sequence is Cauchy if, for every $\epsilon > 0$, there exists N such that $d(x_n, x_m) < \epsilon$ for all $n, m \geq N$.

Completeness in a b-metric space is defined similarly to that in metric spaces: a b-metric space (X, d) is complete if every Cauchy sequence converges to a point in X. The **topology** induced by the "b-metric spaces" permits for a wide range of structures than "usual metric spaces", making them particularly beneficial in areas requiring adaptability, such as **operator theory** and **functional analysis**.

Example 2.17 Let us consider $X = C([a, b])$, which consists of all continuous real-valued functions defined on the closed interval $[a, b]$. Define a b-metric d on X by

$$d(f, g) = \left(\int_a^b |f(x) - g(x)|^p \, dx \right)^{1/p},$$

where $0 < p < 1$ and $f, g \in C([a, b])$. This d satisfies the properties of a b-metric with a constant $s = 2^{1/p}$, meaning that for all $f, g, h \in X$,

$$d(f, h) \leq s \left(d(f, g) + d(g, h) \right).$$

If a sequence $\{f_n\} \subset X$ converges to a function $f \in X$ in the b-metric d, this means that

$$\lim_{n \to \infty} d(f_n, f) = \lim_{n \to \infty} \left(\int_a^b |f_n(x) - f(x)|^p \, dx \right)^{1/p} = 0.$$

Thus, for every $\epsilon > 0$, there exists an integer N such that for all $n \geq N$,

$$\left(\int_a^b |f_n(x) - f(x)|^p \, dx \right)^{1/p} < \epsilon.$$

This implies that $f_n(x) \to f(x)$ pointwise for all $x \in [a, b]$ as $n \to \infty$.
Conversely, if $f_n(x) \to f(x)$ pointwise for each $x \in [a, b]$, then for any $\epsilon > 0$, there exists a sufficiently large integer N such that for $n \geq N$,

$$\left(\int_a^b |f_n(x) - f(x)|^p \, dx \right)^{1/p} < \epsilon.$$

Thus, $f_n \to f$ in the b-metric d, showing that pointwise convergence of $\{f_n\}$ to f implies convergence in the b-metric d.

- This example demonstrates how a b-metric can be applied in a function space with integral-based distance measures, allowing pointwise convergence to correspond to convergence in the b-metric space.

Example 2.18 Let X be the space of all sequences of real numbers, and define a b-metric d on X by

$$d(x, y) = \sum_{j=1}^{\infty} \frac{1}{2^j} \min\left(1, |x_j - y_j|^p\right),$$

where $x = \{x_j\}_{j=1}^{\infty}$ and $y = \{y_j\}_{j=1}^{\infty}$ are elements of X, and $0 < p < 1$. This function d satisfies the properties of a b-metric with a constant $s = 2^{1/p}$.

Suppose that a sequence $\{x^{(n)}\} = \{\{x_j^{(n)}\}_{j=1}^{\infty}\}_{n=1}^{\infty}$ in X converges to $x = \{x_j\}_{j=1}^{\infty}$ in the b-metric d. This means that

$$\lim_{n\to\infty} d(x^{(n)}, x) = \lim_{n\to\infty} \sum_{j=1}^{\infty} \frac{1}{2^j} \min\left(1, |x_j^{(n)} - x_j|^p\right) = 0.$$

Thus, for every $\epsilon > 0$, there exists an integer N such that $d(x^{(n)}, x) < \epsilon$ for all $n \geq N$. Since $d(x^{(n)}, x) \to 0$, it implies that $|x_j^{(n)} - x_j| \to 0$ as $n \to \infty$ for each coordinate j, meaning that $\{x_j^{(n)}\}$ converges to x_j coordinate-wise.

Conversely, suppose that $x_j^{(n)} \to x_j$ as $n \to \infty$ for each j. Then, for any $\epsilon > 0$, there exists an integer m such that

$$\sum_{j=m+1}^{\infty} \frac{1}{2^j} < \frac{\epsilon}{2}.$$

Since each $|x_j^{(n)} - x_j| \to 0$, for any fixed $j = 1, \ldots, m$, there exists an integer N such that $|x_j^{(n)} - x_j|^p < \frac{\epsilon}{2m}$ for all $n \geq N$. Therefore,

$$d(x^{(n)}, x) = \sum_{j=1}^{\infty} \frac{1}{2^j} \min\left(1, |x_j^{(n)} - x_j|^p\right) < \epsilon,$$

whenever n is sufficiently large. Thus, coordinate-wise convergence of $\{x^{(n)}\}$ to x implies convergence in the b-metric d.

- This example shows that "**convergence**" in this b-metric for sequence spaces corresponds to coordinate-wise convergence, demonstrating how a "b-**metric**" can be applied to analyze sequence convergence in a flexible way.

Example 2.19 Let $X = \ell_p$ for $0 < p < 1$, the space of all real-valued sequences $x = \{x_k\}_{k=1}^{\infty}$ such that

$$\sum_{k=1}^{\infty} |x_k|^p < \infty.$$

Define a b-metric d on X by

$$d(x, y) = \sum_{k=1}^{\infty} \frac{|x_k - y_k|^p}{1 + |x_k - y_k|^p},$$

where $x = \{x_k\}_{k=1}^{\infty}$ and $y = \{y_k\}_{k=1}^{\infty}$ are elements of X. This distance function satisfies the properties of a b-metric with a constant $s \geq 1$. Suppose a sequence $\{x^{(n)}\} = \{\{x_k^{(n)}\}_{k=1}^{\infty}\}_{n=1}^{\infty}$ in ℓ_p converges to $x = \{x_k\}_{k=1}^{\infty}$ in the b-metric d. This means that

$$\lim_{n\to\infty} d(x^{(n)}, x) = \lim_{n\to\infty} \sum_{k=1}^{\infty} \frac{|x_k^{(n)} - x_k|^p}{1 + |x_k^{(n)} - x_k|^p} = 0.$$

Thus, for any $\epsilon > 0$, there exists an integer N such that $d(x^{(n)}, x) < \epsilon$, for all $n \geq N$. This implies that $|x_k^{(n)} - x_k| \to 0$ for each coordinate k, meaning that coordinate-wise convergence of $x^{(n)} \to x$ is achieved.

Now let us establish a counterexample by defining

$$x_k^{(n)} = x_k + \delta_{kn},$$

where

$$\delta_{kn} = \frac{(-1)^n}{\sqrt{n+k}}.$$

This setup means $|x_k^{(n)} - x_k| = \left|\frac{(-1)^n}{\sqrt{n+k}}\right|$, which decreases as n and k grow. For each fixed k, it's easy to see that $x_k^{(n)} \to x_k$ as $n \to \infty$, so the sequence converges coordinate-wise.

However, for each n, we calculate

$$d(x^{(n)}, x) = \sum_{k=1}^{\infty} \frac{\left|\frac{(-1)^n}{\sqrt{n+k}}\right|^p}{1 + \left|\frac{(-1)^n}{\sqrt{n+k}}\right|^p}.$$

This expression does not necessarily tend to zero as $n \to \infty$, since the terms in the sum depend on both n and k in a way that may stabilize around a non-zero value. Thus, $d(x^{(n)}, x)$ may not approach 0 as $n \to \infty$, meaning that $x^{(n)}$ does not converge to x in the b-metric d, even though it converges coordinate-wise.

- This example highlights that with this choice of $x_k^{(n)} = x_k + \delta_{kn}$ and $\delta_{kn} = \frac{(-1)^n}{\sqrt{n+k}}$, we see that coordinate-wise convergence does not imply convergence in the b-metric d. This further illustrates the distinct behavior of convergence in b-metric spaces compared to traditional metric spaces.

Theorem 2.1 *Let (X, d) be a b-metric space with constant $s \geq 1$. Suppose $\{x^{(n)}\}_{n=1}^{\infty}$ is a sequence in ℓ_p such that $\lim_{n\to\infty} x_k^{(n)} = x_k$ for each k, where $x = \{x_k\}_{k=1}^{\infty}$ is an element of ℓ_p. Additionally, assume that for every $\epsilon > 0$, there exists an integer $m_0(\epsilon)$ such that*

$$\left(\sum_{k=m+1}^{\infty} |x_k^{(n)}|^p\right)^{1/p} < \frac{\epsilon}{4s^{\frac{1}{p}}} \quad \textit{for } m \geq m_0(\epsilon) \textit{ and for all } n.$$

Then $\lim_{n\to\infty} d(x^{(n)}, x) = 0$.

Proof *Let $\epsilon > 0$ be given. By assumption, for $m \geq m_0(\epsilon/(4s))$, we have*

$$\left(\sum_{k=m+1}^{\infty} |x_k^{(n)}|^p\right)^{1/p} < \frac{\epsilon}{4s^{\frac{1}{p}}} \quad \textit{for all } n.$$

Thus, for any $N \geq m+1$ and all n,

$$\left(\sum_{k=m+1}^{N} |x_k^{(n)}|^p\right)^{1/p} \leq \frac{\epsilon}{4s^{\frac{1}{p}}}.$$

Taking the limit as $n \to \infty$ in this finite sum, we obtain

$$\left(\sum_{k=m+1}^{N} |x_k|^p\right)^{1/p} \leq \frac{\epsilon}{4s^{\frac{1}{p}}}, \quad \textit{for all } N \geq m+1.$$

Letting $N \to \infty$, we conclude that

$$\left(\sum_{k=m+1}^{\infty} |x_k|^p\right)^{1/p} \leq \frac{\epsilon}{4s^{\frac{1}{p}}}.$$

By inequality properties, we know that,

$$\sum_{k=m+1}^{\infty} |x_k^{(n)} - x_k|^p \leq s\left(\sum_{k=m+1}^{\infty} |x_k^{(n)}|^p + \sum_{k=m+1}^{\infty} |x_k|^p\right),$$

where $s = 2^{p-1} > 1$.

Consider,

$$\begin{aligned}
d(x^{(n)}, x)^p &= \sum_{k=1}^{\infty} |x_k^{(n)} - x_k|^p \\
&= \sum_{k=1}^{m} |x_k^{(n)} - x_k|^p + \sum_{k=m+1}^{\infty} |x_k^{(n)} - x_k|^p \\
&\leq \sum_{k=1}^{m} |x_k^{(n)} - x_k|^p + s\left(\sum_{k=m+1}^{\infty} |x_k^{(n)}|^p + \sum_{k=m+1}^{\infty} |x_k|^p\right) \\
&\leq \sum_{k=1}^{m} |x_k^{(n)} - x_k|^p + s\left(\frac{\epsilon^p}{2s}\right) \\
&\leq \sum_{k=1}^{m} |x_k^{(n)} - x_k|^p + \left(\frac{\epsilon^p}{2}\right).
\end{aligned}$$

Since $x_k^{(n)} \to x_k$ *for each* k, *there exists an* $n_0(\epsilon)$ *such that*

$$\sum_{k=1}^{m} |x_k^{(n)} - x_k|^p < \frac{\epsilon^p}{2} \quad \text{for all } n \geq n_0(\epsilon).$$

Therefore,

$$d(x^{(n)}, x)^p \leq \frac{\epsilon^p}{2} + \frac{\epsilon^p}{2} = \epsilon^p \quad \text{for all } n \geq n_0(\epsilon),$$

which yields $d(x^{(n)}, x) < \epsilon$ *for all* $n \geq n_0(\epsilon)$.

Thus, we conclude that $\lim_{n\to\infty} d(x^{(n)}, x) = 0$, *completing the proof.*

Proposition 2.2 *Let* $X = \mathbb{R}^2$ *with elements represented as pairs* $x = (x_1, x_2)$ *and* $y = (y_1, y_2)$ *for* $x, y \in X$. *Define a b-metric* d *on* X *by*

$$d(x, y) = |x_1 - y_1| + |x_2 - y_2|,$$

which is also known as the ℓ_1 *or "taxicab" distance. The distance function* d *satisfies the relaxed triangle inequality with* $s = 1$, *making it both a metric and a b-metric. We aim to prove that* X *is complete under this b-metric, meaning that every Cauchy sequence in* X *converges to a point within* X.

Suppose $\{x^{(n)}\}_{n=1}^{\infty}$ *is a Cauchy sequence in X with respect to the b-metric d. This means that for every* $\epsilon > 0$*, there exists an integer N such that for all* $m, n \geq N$, $d(x^{(m)}, x^{(n)}) < \epsilon$. *Since*

$$d(x^{(m)}, x^{(n)}) = |x_1^{(m)} - x_1^{(n)}| + |x_2^{(m)} - x_2^{(n)}|,$$

it follows that each component sequence $\{x_1^{(n)}\}$ *and* $\{x_2^{(n)}\}$ *must individually be Cauchy in* $\mathbb{R}$*, since* $|x_1^{(m)} - x_1^{(n)}| < \epsilon$ *and* $|x_2^{(m)} - x_2^{(n)}| < \epsilon$ *for sufficiently large m and n.*

Since $\mathbb{R}$ *is complete, each component sequence converges to a real number. Let* $x_1 = \lim_{n\to\infty} x_1^{(n)}$ *and* $x_2 = \lim_{n\to\infty} x_2^{(n)}$*, and define* $x = (x_1, x_2) \in \mathbb{R}^2$*. We need to show that* $x^{(n)} \to x$ *in the b-metric d. For* $n \geq N$,

$$d(x^{(n)}, x) = |x_1^{(n)} - x_1| + |x_2^{(n)} - x_2|.$$

Since $x_1^{(n)} \to x_1$ *and* $x_2^{(n)} \to x_2$ *as* $n \to \infty$*, it follows that* $d(x^{(n)}, x) \to 0$ *as* $n \to \infty$.

Thus, $\{x^{(n)}\}$ *converges to x in X with respect to the b-metric d, proving that* $\mathbb{R}^2$ *is complete under this b-metric.*

Remark 2.2 Figure 2.4 showing the "circles" in the taxicab metric, represented as diamond-shaped contours centered at the origin. Each diamond represents a set of points at the same taxicab distance from the center, with distances increasing as we move outward. This illustrates how the taxicab metric differs from the Euclidean metric, where distance contours would be circular instead of diamond-shaped.

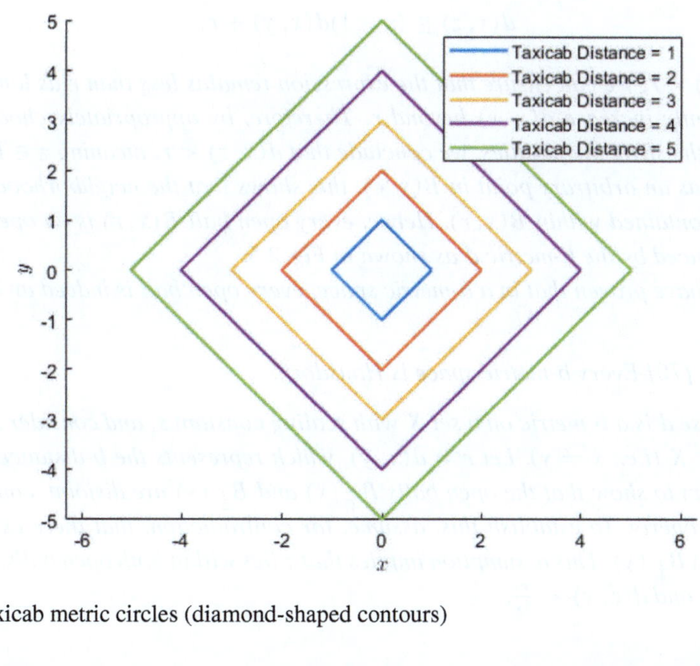

Fig. 2.4 Taxicab metric circles (diamond-shaped contours)

Proposition 2.3 *In a b-metric space (X, d), every open ball $B(x, r) = \{y \in X : d(x, y) < r\}$ is an open set.*

Proof *Let (X, d) be a b-metric space with a scaling constant s. Consider an open ball $B(x, r)$ centered at a point $x \in X$ with radius $r > 0$. To prove that this ball is an open set, we need to show that for any point y within the ball, there exists a neighborhood around y that is completely contained within $B(x, r)$.*

Assume $y \in B(x, r)$, which means that $d(x, y) < r$. We define $\delta = r - d(x, y)$, which is positive since $d(x, y) < r$. We can choose a smaller radius $\epsilon = \frac{\delta}{s}$, ensuring $\epsilon > 0$.

Now, consider any point z in the ball $B(y, \epsilon)$. This implies that $d(y, z) < \epsilon$. By applying the triangle inequality for b-metrics, we have:

$$d(x, z) \leq s(d(x, y) + d(y, z)).$$

Substituting $d(y, z) < \epsilon = \frac{r-d(x,y)}{s}$ into this inequality yields:

$$d(x, z) \leq s\left(d(x, y) + \frac{r - d(x, y)}{s}\right).$$

Simplifying this expression gives:

$$d(x, z) \leq s\left(d(x, y) + \frac{r - d(x, y)}{s}\right) = s \cdot d(x, y) + r - d(x, y).$$

Rearranging the terms results in:

$$d(x, z) \leq (s - 1)d(x, y) + r.$$

Since $d(x, y) < r$, we can ensure that the expression remains less than r as long as s does not significantly increase $d(x, y)$ beyond r. Therefore, by appropriately choosing s and considering the strict inequalities, we conclude that $d(x, z) < r$, meaning $z \in B(x, r)$.

Since z was an arbitrary point in $B(y, \epsilon)$, this shows that the neighborhood $B(y, \epsilon)$ is completely contained within $B(x, r)$. Hence, every open ball $B(x, r)$ is an open set in the topology induced by the b-metric d as shown in Fig. 2.5.

Thus, we have proven that in a b-metric space, every open ball is indeed an open set.

Lemma 2.1 *[70] Every b-metric space is Hausdorff.*

Proof *Suppose d is a b-metric on a set X with scaling constant s, and consider two distinct points $x, y \in X$ (i.e., $x \neq y$). Let $r = d(x, y)$, which represents the b-distance between x and y. We aim to show that the open balls $B_{\frac{r}{4s}}(x)$ and $B_{\frac{r}{4s}}(y)$ are disjoint, confirming the Hausdorff property. To establish this, assume, for contradiction, that there exists a point $z \in B_{\frac{r}{4s}}(x) \cap B_{\frac{r}{4s}}(y)$. This assumption implies that z lies within both open balls, so we have $d(x, z) < \frac{r}{4s}$ and $d(z, y) < \frac{r}{4s}$.*

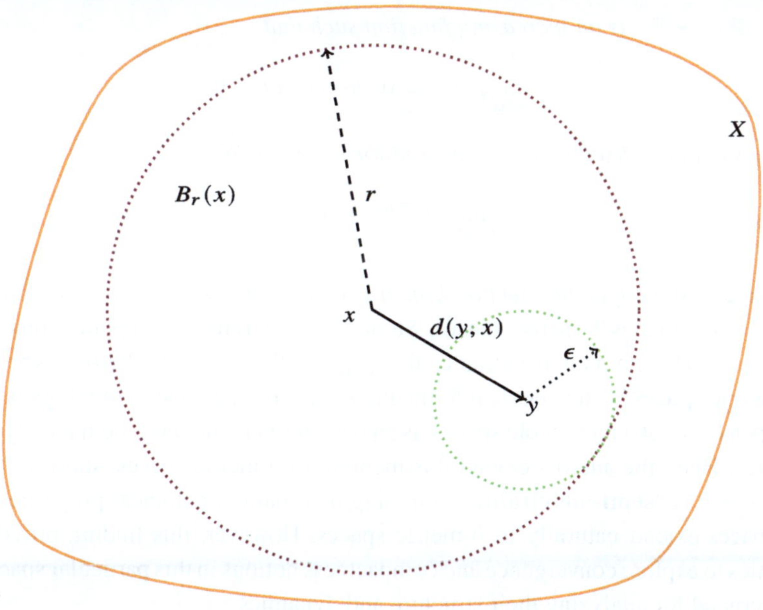

Fig. 2.5 An open ball is an open set

Using the properties of a b-metric, specifically the relaxed triangle inequality, we can write:

$$r = d(x, y) \le s\,(d(x, z) + d(z, y)).$$

Substituting the values for $d(x, z)$ and $d(z, y)$ from our assumption, we get:

$$d(x, y) < s\left(\frac{r}{4s} + \frac{r}{4s}\right) = s \cdot \frac{r}{2s} = \frac{r}{2}.$$

This results in the inequality $r < \frac{r}{2}$, which is clearly a contradiction, as r cannot be both less than and equal to $d(x, y)$. Therefore, no such point z exists, and we conclude that $B_{\frac{r}{4s}}(x) \cap B_{\frac{r}{4s}}(y) = \emptyset$. This proves that in any b-metric space, distinct points can be separated by disjoint open sets, fulfilling the Hausdorff property.

Remark 2.3 While a *b*-metric space can be complete (i.e., every Cauchy sequence converges to a limit within the space), it does not necessarily imply continuity of all functions. Each function needs to be checked individually for continuity in a *b*-metric space context.

Theorem 2.2 *[32] Let (X, d) be a complete b-metric space, and let $T : X \to X$ satisfy*

$$d(Tx, Ty) \le \varphi(d(x, y)) \quad \textit{for all } x, y \in X,$$

where $\varphi : \mathbb{R}_+ \to \mathbb{R}_+$ is an increasing function such that

$$\lim_{n\to\infty} \varphi^n(t) = 0 \quad \textit{for each } t > 0.$$

Then T has a unique fixed point $u \in X$, and for every $x \in X$,

$$\lim_{n\to\infty} d(T^n(x), u) = 0.$$

The paper *"Stone-type theorem on b-metric spaces and applications"* by Tran Van An et al., [67] investigates b-metric spaces, Stone-type theorems, and applications to related metric spaces. The main contribution of this paper is the proof of a "Stone-type theorem" for "**b-metric spaces**". The classical Stone theorem, a cornerstone in topology, asserts that "every open cover of a metrizable space has an open refinement that is both locally finite and σ-discrete". Here, the authors extend this theorem to b-metric spaces, showing that every b-metric space is "**semi-metrizable**", implying that many topological properties found in metric spaces extend naturally to b-metric spaces. However, this finding provides novel possibilities to explore convergence and compactness notions in this particular spaces, which are very crucial for analyzing their structure and dynamics.

Moreover, the above stated article discusses applications to "**2-metric** [68]" and "**S-metric spaces** [69]". The authors of [67] demonstrate that a bounded 2-metric space can induce a b-metric space with **an equivalent topological structure**, further suggesting that metric-type spaces can be viewed as a subset within the broader framework of b-metric spaces. Similarly, they show that "**S-metric spaces are metrizable**", which simplifies the analysis of "**convergence and continuity**" in these spaces.

The paper *"b-metrics"* by H. Kunzi and H. Pajoohesh [70] presents an analysis of b-metric spaces, exploring their properties, applications, and connections to general topological constructs. The above stated paper's primary contribution is the rigorous demonstration of "**metrizability**" in topologies induced by b-metrics. Metrizability is a desirable trait in topology, allowing spaces to inherit certain characteristics of metric spaces, such as the existence of a compatible metric. The authors successfully prove that every topology induced by a b-metric is "**metrizable**" by establishing that it is a T_1, "**regular space**" with a "**σ-discrete base**". This finding is important as it connects b-metric spaces to "**metrizable topologies**", thereby enabling the use of well-known metric tools for analysis.

A noteworthy aspect of the paper is its discussion on the extension of b-metrics to "**induced topologies**" and their "**regularity**". The authors use lemmas to confirm that b-metric spaces exhibit both "**Hausdorff**" and "**regular properties**", foundational traits for metrizable spaces. Additionally, they demonstrate that each **open cover** in a b-metric space has an "**open σ-discrete refinement**", a property essential for proving metrizability in such spaces.

The paper further explores the uniformity in b-metric spaces, establishing a parallel between "**uniform structures**" in metric spaces and those induced by b-metrics. This exploration into the "**uniformity**" and completeness of b-metric spaces highlights the robustness of these spaces in satisfying topological requirements.

2.3.3 Discussion on Certain Inequality-Related Results

2.3.3.1 Basic Inequality on a b-Metric Space

In 2017, the study of Suzuki [71] introduced a foundational inequality in b-metric spaces, building upon Czerwik's [32] concept of a relaxed triangle inequality. Suzuki [71] explored fixed-point theorems by examining sequences in b-metric spaces that satisfied specific conditions, improving existing lemmas that addressed Cauchy sequences and the convergence properties in these spaces. One of the key contributions was the demonstration of a generalized fixed-point theorem, which extended applications in multivalued mappings. This paper not only solidified the understanding of b-metric spaces but also paved the way for further inquiries into metric-type spaces and their unique characteristics.

In [71], the basic inequality considered in a b-metric space provides an essential framework for analyzing distances within sequences. The inequality taken into account is:

$$d(x_0, x_n) \leq \omega^{\mathbb{T}(n)} \sum_{j=0}^{n-1} d(x_j, x_{j+1}), \quad \text{where } \omega \geq 1 \text{and } \mathbb{T}(n) = -[-\log_2 n].$$

Above inequality grabs how the total distance between two points in a sequence can be bounded by the sum of intermediate distances, scaled by a factor of $\omega^{\mathbb{T}(n)}$. However, this inequality is essential in examining convergence within b-metric spaces, particularly when the distances $d(x_{j+1}, x_{j+2})$ are relatively small as compared to $d(x_j, x_{j+1})$. The structure of $\mathbb{T}(n)$, with its reliance on logarithmic growth, adds depth to the analysis, enabling applications in fixed-point theorems and further generalizations. By employing this inequality, Suzuki's [71] work advances the understanding of b-metric spaces, extending their utility in fixed-point theory and beyond.

Remark 2.4 [71] The function $\mathbb{T}(n)$ is defined as follows:
$\mathbb{T}(1) = 0, \mathbb{T}(2) = 1, \mathbb{T}(3) = \mathbb{T}(4) = 2, \mathbb{T}(5) = \cdots = \mathbb{T}(8) = 3,$
$\mathbb{T}(9) = \cdots = \mathbb{T}(16) = 4$, and so on.

To prove the inequality in a b-metric space, induction on n is used.

Starting with the base case where $n = 1$, the inequality holds trivially because $\mathbb{T}(1) = 0$, which means $\omega^{\mathbb{T}(1)} = 1$. Therefore, the inequality $d(x_0, x_1) \leq \omega^{\mathbb{T}(1)} d(x_0, x_1)$ reduces to an equality, confirming that the base case is true.

Assuming that the inequality holds for all values of n up to 2^k, consider a value n such that $2^k < n \leq 2^{k+1}$. For this range of n, $\mathbb{T}(n) = k + 1$ by the definition of f. The goal is to show that the inequality also holds for this n with $\mathbb{T}(n) = k + 1$.

By applying the triangle inequality in the b-metric space and splitting the distance $d(x_0, x_n)$ at the point x_{2^k},

$$d(x_0, x_n) \leq \omega \left(d(x_0, x_{2^k}) + d(x_{2^k}, x_n) \right).$$

Using the inductive hypothesis, $d(x_0, x_{2^k})$ can be bounded by $\omega^{\mathbb{T}(2^k)} \sum_{j=0}^{2^k-1} d(x_j, x_{j+1})$ and $d(x_{2^k}, x_n)$ by $\omega^{\mathbb{T}(n-2^k)} \sum_{j=2^k}^{n-1} d(x_j, x_{j+1})$. Substituting these bounds gives

$$d(x_0, x_n) \leq \omega \left(\omega^{\mathbb{T}(2^k)} \sum_{j=0}^{2^k-1} d(x_j, x_{j+1}) + \omega^{\mathbb{T}(n-2^k)} \sum_{j=2^k}^{n-1} d(x_j, x_{j+1}) \right).$$

Since $\mathbb{T}(n) = k + 1,\ \mathbb{T}(2^k) = k$ and $\mathbb{T}(n - 2^k) \leq k$; $\omega^{\mathbb{T}(n)}$ can be factored out, yielding

$$d(x_0, x_n) \leq \omega^{\mathbb{T}(n)} \sum_{j=0}^{n-1} d(x_j, x_{j+1}).$$

This completes the inductive step, demonstrating that the inequality holds for n in the range $2^k < n \leq 2^{k+1}$. By the principle of mathematical induction, the inequality holds for all n in the b-metric space.

2.3.3.2 Dragomir and Gosa-Type Inequalities in b-Metric Space

In [72], the authors investigated "**Dragomir and Gosa-type inequalities** [73]" within the framework of b-metric spaces. The study explored the adaptation of these inequalities in b-normed spaces and presented proofs that provided a geometrical interpretation. In their research, they mainly aimed to generalize and expand upon existing metric inequalities by introducing a weaker space notion, the b-**metric**, which relaxed the conditions of a standard metric space. However, this innovative method had implications for both fixed-point theory and nonlinear analysis, showcasing the adaptability and application of inequalities in varied metric structures.

The main outcome of [72] is as shown below:

Theorem 2.3 *[72] Let (X, d) be a b-metric space with $s \geq 1$. Let $x_i \in X$ and $p_i \geq 0$ for all $i \in \{1, 2, \ldots, n\}$, satisfying*

$$\sum_{i=1}^{n} p_i = \frac{1}{s}.$$

Then, the following inequality holds:

$$\sum_{1\leq i<j\leq n} p_i p_j d(x_i, x_j) \leq \inf_{x\in X}\left[\sum_{i=1}^{n} p_i d(x_i, x)\right].$$

This inequality is sharp, which means that the constant $c = 1$ *before the infimum cannot be replaced by a smaller value.*

Sketch of the Proof. In order to prove the theorem, take, for any $x \in X$ and $i, j \in \{1, 2, \ldots, n\}$, we have:

$$d(x_i, x_j) \leq s\left[d(x_i, x) + d(x, x_j)\right].$$

Multiplying this inequality by p_i and p_j, and summing over all i and j from 1 to n, we obtain:

$$\sum_{i,j=1}^{n} p_i p_j d(x_i, x_j) \leq s\left[\sum_{i,j=1}^{n} p_i p_j \left(d(x_i, x) + d(x, x_j)\right)\right].$$

By symmetry, we know that:

$$\sum_{i,j=1}^{n} p_i p_j d(x_i, x_j) = 2\sum_{1\leq i<j\leq n} p_i p_j d(x_i, x_j).$$

Using the given condition $\sum_{i=1}^{n} p_i = \frac{1}{s}$, we can rewrite the sum as follows:

$$\sum_{i,j=1}^{n} p_i p_j \left(d(x_i, x) + d(x, x_j)\right) = \frac{2}{s}\sum_{i=1}^{n} p_i d(x_i, x).$$

So, from above hypothesis, we get,

$$\begin{aligned}\sum_{1\leq i<j\leq n} p_i p_j d(x_i, x_j) &= \frac{1}{2}\sum_{i,j=1}^{n} p_i p_j d(x_i, x_j)\\ &\leq \frac{s}{2}\left[\sum_{i,j=1}^{n} p_i p_j\left(d(x_i, x) + d(x, x_j)\right)\right]\\ &= \sum_{i=1}^{n} p_i d(x_i, x).\end{aligned}$$

Thus, we derive that:

$$\sum_{1\le i<j\le n} p_i p_j d(x_i, x_j) \le \sum_{i=1}^{n} p_i d(x_i, x).$$

Since the infimum represents the greatest lower bound, we conclude the inequality in the theorem statement:

$$\sum_{1\le i<j\le n} p_i p_j d(x_i, x_j) \le \inf_{x\in X}\left[\sum_{i=1}^{n} p_i d(x_i, x)\right].$$

Finally, to verify the sharpness of this inequality, assume that there exists a constant $c > 0$ such that

$$\sum_{1\le i<j\le n} p_i p_j d(x_i, x_j) \le c \inf_{x\in X}\left[\sum_{i=1}^{n} p_i d(x_i, x)\right].$$

By choosing $n = 2$, $p_1 = p$, and $p_2 = 1 - p$ where $p \in (0, 1)$, and setting $x = x_1$, we can deduce that $c \ge 1$, proving that the inequality is indeed sharp.

Corollary 2.1 *Let (X, d) be a b-metric space with constant $s \ge 1$, and let $x_i \in X$ for $i \in \{1, 2, \dots, n\}$. Suppose there exists a point $z \in X$ and a radius $r > 0$ such that a closed ball $B(z, r) = \{y \in X : d(z, y) \le r\}$ contains all points x_i. For any non-negative weights p_i satisfying $\sum_{i=1}^{n} p_i = \frac{1}{s}$, we have:*

$$\sum_{1\le i<j\le n} p_i p_j d(x_i, x_j) \le \frac{r}{s}.$$

Proof By the definition of $B(z, r)$, we know that $d(z, x_i) \le r$ for each $i \in \{1, 2, \dots, n\}$. From Theorem 2.3, we have:

$$\sum_{1\le i<j\le n} p_i p_j d(x_i, x_j) \le \inf_{x\in X}\left[\sum_{i=1}^{n} p_i d(x_i, x)\right].$$

Choosing $x = z$ in the infimum term, we get:

$$\sum_{1\le i<j\le n} p_i p_j d(x_i, x_j) \le \sum_{i=1}^{n} p_i d(x_i, z).$$

Since $d(x_i, z) \le r$ for each x_i, we can further bound this by:

$$\sum_{1\le i<j\le n} p_i p_j d(x_i, x_j) \le \sum_{i=1}^{n} p_i r = \frac{r}{s}.$$

2.4 θ-Metric Spaces: Understanding Their Structure and Properties

Due to more adaptability and flexibility, the notion of a "θ-**metric space**" [43] provides a huge benefit over **standard metric** and/or b-**metric spaces**. The regular triangle inequality in metric space is:

$$d(x, z) \leq d(x, y) + d(y, z), \quad \forall\, x, y, z \in X,$$

requiring exact proportionality among distances. Considering a constant $s \geq 1$ in the triangle inequality weakens this limitations in b-**metric space**, this phenomena leads to:

$$d(x, z) \leq s[d(x, y) + d(y, z)], \quad \forall\, x, y, z \in X.$$

Moreover, this fixed constant "s" allows some adaptability in applications, but it remains uniform across all points in the space.

The modified version of b-metric space, the θ-metric space additionally generalizes this by introducing a novel scaling function "$\theta(x, z)$", which adapts dynamically based on the specific points. This means that rather than a fixed constant factor, each point pair "(x, z)" can have its own scaling influence via "$\theta(x, z)$", accommodating frameworks with varying point relationships. Above stated dynamic scaling is more advantageous in fields such as nonlinear integral equations, chaos theory, and mathematical modelling, where the links among points may exhibit complex, nonlinear variations and cannot be appropriately described by a metric and/or b-metric structure.

Definition 2.5 [43] Given a function $\theta : X \times X \to [1, \infty)$, where X be a non-empty set. The function $d : X \times X \to [0, \infty)$ is said to be θ-**metric** if for all $x, y, z \in X$,

- $d(x, y) = 0 \Leftrightarrow x = y$,
- $d(x, y) = d(y, x)$,
- $d(x, z) \leq \theta(x, z)[d(x, y) + d(y, z)]$ (generalized triangle inequality).

Then, the pair (X, d) is called an extended b-metric space. For our convenience, we call it, θ-**metric space**.

This generalization introduces a control function θ, allowing for a more flexible distance measurement beyond traditional metric spaces. The generalized triangle inequality is an extension of the traditional concept of the triangle inequality, allowing greater flexibility in how distances between points are measured.

Example 2.20 Consider $X = \{1, 2, 3\}$ with the control function $\theta(x, y) = |x - y| + 2$ and the distance function d defined as follows:

$$d(1, 1) = d(2, 2) = d(3, 3) = 0, \quad d(1, 2) = d(2, 1) = 50,$$

$$d(1, 3) = d(3, 1) = 200, \quad d(2, 3) = d(3, 2) = 120.$$

To verify the θ-metric space conditions, observe that $d(x, y) = 0$ only if $x = y$, and symmetry $d(x, y) = d(y, x)$ holds. For the triangle inequality, consider $x = 1$, $y = 2$, and $z = 3$: here, $d(1, 3) = 200$ and

$$\theta(1, 3)[d(1, 2) + d(2, 3)] = 680,$$

satisfying $200 \le 680$. Thus, (X, d) be a θ-metric space.

Definition 2.6 [43] Let (X, d) be a θ-metric space. Then:

1. A sequence $\{x_n\} \subset X$ **converges** to a point $x \in X$ if, for every $\epsilon > 0$, there exists an integer N such that $d(x_n, x) < \epsilon$, for all $n \ge N$. This is denoted as $\lim_{n\to\infty} x_n = x$.
2. A sequence $\{x_n\} \subset X$ is called a **Cauchy sequence** if, for every $\epsilon > 0$, there exists an integer N such that $d(x_m, x_n) < \epsilon$, for all $m, n \ge N$. This implies that the terms of the sequence become arbitrarily close to each other as $m, n \to \infty$.

Definition 2.7 [43] The θ-metric space (X, d) is **complete** if every Cauchy sequence in X converges to a point within X.

2.4.1 Some Key Generalizations of θ-Metric Space

1. **Controlled Metric Type Space** [74]:

    - Let X be a nonempty set with $d : X \times X \to [0, \infty)$ and a control function $\alpha : X \times X \to [1, \infty)$. The function d is called a **controlled metric type** if it satisfies:

        a. $d(x, y) = 0$ if and only if $x = y$,
        b. $d(x, y) = d(y, x)$,
        c. $d(x, y) \le \alpha(x, z)\, d(x, z) + \alpha(z, y)\, d(z, y)$ for all $x, y, z \in X$.

    The pair (X, d) is called **controlled metric type space.**
2. **Double Controlled Metric Type Space** [75]:

    - If $d : X \times X \to [0, \infty)$ is associated with two non-comparable control functions $\alpha, \mu : X \times X \to [1, \infty)$, then d is called a **double controlled metric type** if it satisfies:

        a. $d(x, y) = 0$ if and only if $x = y$,
        b. $d(x, y) = d(y, x)$,
        c. $d(x, y) \le \alpha(x, z)\, d(x, z) + \mu(z, y)\, d(z, y)$ for all $x, y, z \in X$.

    The pair (X, d) is called **double controlled metric type space.**

These generalizations provide flexibility in modifying the triangle inequality using single or multiple control functions, enabling applications in fixed-point theory and other analytical contexts. It is notable that numerous distance functions have emerged as generalizations and extensions of the θ-metric space concept. Among these, we can cite, for example, an extended suprametric space [49], an extended partial b-metric space [76], an extended rectangular b-metric space [77] and various other θ-metric space variations, each detailed with comprehensive definitions and significant observations in the literature. These generalizations offer broader applicability in various mathematical and applied contexts:

- The flexibility provided by the variable control function α in controlled metric type spaces supports analyses in systems where the contraction factor may differ across the space.
- The double control functions α and μ in double controlled metric type spaces allow for scaling adjustments between pairs of points, enabling fixed-point results in systems with non-uniform interactions or asymmetric distance dynamics.

Overall, each generalization brings further adaptability, which is beneficial in settings requiring diverse contraction conditions and specialized convergence behaviors, particularly in fields like dynamical systems and iterative processes.

Example 2.21 Let $X = [1, \infty)$, and define $d : X \times X \to [0, \infty)$ by:

$$d(x, y) = \begin{cases} 0, & x = y, \\ |x - y|, & x \neq y. \end{cases}$$

Now, define control functions $\alpha, \mu : X \times X \to [1, \infty)$ as follows:

$$\alpha(x, y) = \begin{cases} x, & x, y \in [1, 2), \\ 2, & \text{otherwise,} \end{cases} \quad \text{and} \quad \mu(x, y) = \begin{cases} y, & x, y \in [2, 3), \\ 3, & \text{otherwise.} \end{cases}$$

This example demonstrates that by carefully defining α and μ, we can create a nontrivial double controlled metric type space with different behaviors depending on intervals in X.

Example 2.22 Let $X = \{a, b, c, d\}$ and define $d : X \times X \to [0, \infty)$ by:

$d(x, y) =$

	a	b	c	d
a	0	2	3	1
b	2	0	1	4
c	3	1	0	2
d	1	4	2	0

Now, define control functions $\alpha, \mu : X \times X \to [1, \infty)$ as follows:

$\alpha(x, y) =$

	a	b	c	d
a	1	2	1.5	3
b	2	1	1.2	2.5
c	1.5	1.2	1	2
d	3	2.5	2	1

$\mu(x, y) =$

	a	b	c	d
a	1	1.5	2	2.5
b	1.5	1	1.8	2
c	2	1.8	1	3
d	2.5	2	3	1

By assigning different control values for α and μ, we illustrate a nontrivial double controlled metric type space on a finite set, showing flexibility in controlling the distance calculations.

The adaptability of the θ-metric space allows it to extend the applicability of fixed-point theorems, fundamental in optimization, economic modeling, and theoretical computer science, to cases where traditional metrics and fixed constants do not apply. Moreover, θ-metric spaces can effectively model asymmetric relationships, as $\theta(x, z)$ can vary according to each pair of points, enabling a flexible structure where distance relationships are not uniformly symmetric.

Thus, this resourcefulness makes θ-metric spaces powerful tools for analyzing real-world scenarios with complex, asymmetric, or varying point interactions, advancing beyond the capabilities of both metric and b-metric spaces.

Example 2.23 Let us consider a lattice $L = \{0, 1, 2, 3, 4, 5, 6\}$ with the partial order $\leq$, where elements are ordered according to divisibility. In this pursuit, L is structured such that each pair of elements $x, y \in L$ has both a "**meet**" and a "**join**". In order to define a distance function and a scaling function, we can use these lattice operations.
Now, define $d : L \times L \to [0, \infty)$ which is based on the difference between the join and meet of two elements x and y. Specifically, we set

$$d(x, y) = |(x \vee y) - (x \wedge y)|,$$

where $x \vee y$ denotes the join (LCM) of x and y, and $x \wedge y$ denotes their meet (GCD) and with the scaling function θ,

$$\theta(x, z) = 1 + |x \wedge z|.$$

Consider elements 2, 3, and 4 in L: The distance between 2 and 3 is calculated as follows:

$$2 \vee 3 = 6 \quad \text{and} \quad 2 \wedge 3 = 1,$$

so,

$$d(2, 3) = |6 - 1| = 5.$$

The distance between 3 and 4 is determined similarly:

$$3 \vee 4 = 12 \quad \text{and} \quad 3 \wedge 4 = 1,$$

which gives

$$d(3, 4) = |12 - 1| = 11.$$

The distance between 2 and 4 is calculated as:

$$2 \vee 4 = 4 \quad \text{and} \quad 2 \wedge 4 = 2,$$

leading to

$$d(2, 4) = |4 - 2| = 2.$$

Next, we compute the scaling factor $\theta(2, 4)$:

$$\theta(2, 4) = 1 + |2 \wedge 4| = 1 + 2 = 3.$$

To confirm that these values satisfy the triangle inequality, we check if:

$$d(2, 4) \leq \theta(2, 4)[d(2, 3) + d(3, 4)].$$

Substituting the values, we have: $d(2, 4) = 2$ and $\theta(2, 4)(d(2, 3) + d(3, 4)) = 3 \times (5 + 11) = 3 \times 16 = 48$. As a result, it holds the generalized triangle inequality.

Figure 2.6 represents the Hasse diagram of the lattice $L = \{0, 1, 2, 3, 4, 5, 6\}$ with the partial order relationships arranged in a hierarchical structure.

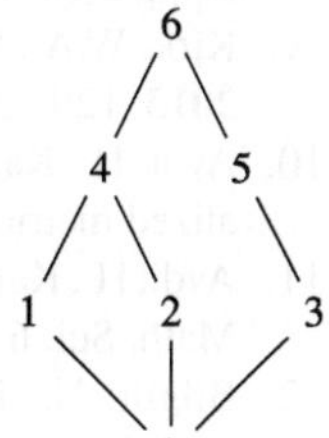

Fig. 2.6 Hasse diagram of the lattice $L = \{0, 1, 2, 3, 4, 5, 6\}$

Open Questions

2.1 How does the "Stone type theorem" adapt to generalized metric spaces according to [4], and what are the additional/necessary conditions needed for their validity?

2.2 In what ways do the "Dragomir and Gosa-type Inequalities [73]" apply to symmetric spaces, and what other requirements are required for their applicability?

2.3 **Prove or disprove**: Open sets in a "Double Controlled Metric Type Space" X satisfy the below stated properties:

- $\emptyset$ and X are open sets.
- The union of any collection of open sets is open.
- If A and B are open, then $A \bigcap B$ is open.
- An open ball is an open set.

References

1. Fréchet, M.R.: Sur quelques points du calcul fonctionnel. Rend. Circ. Mat. Palermo **22**, 1–74 (1906). https://doi.org/10.1007/BF03018603
2. Hausdorff, F.: Mengenle hre. W. de Gruyter & Co (1927)
3. Shirali, S., Vasudeva, H.L.: Metric Spaces. Springer (2006)
4. Branciari, A.: A fixed point theorem of banach-caccioppoli type on a class of generalized metric spaces. Pub. Math. Dubrecen **57**, 31–37 (2000)
5. Das, P.: A fixed point theorem on a class of generalized metric spaces. Korean J. Math. Sci **9**, 29–33 (2002)
6. Kadelburg, Z., Radenovic, S.: Fixed point results in generalized metric spaces without Hausdorff property. TWMS J. Pure Appl. Math. **5**(1), 3–13 (2014)
7. Samet, B.: Discussion on "a fixed point theorem of Banach-Caccioppoli type on a class of generalized metric spaces". Pub. Math. Dubrecen **57**, 93–494 (2010)
8. Sarma, I.R., Rao, J.M., Rao, S.S.: Contractions over generalized metric spaces. J. Nonlinear Sci. Appl. **2**(3), 180–182 (2009)
9. Kirk, W.A., Shahzad, N.: Generalized metrics and Caristi's theorem. Fixed Point Theory Appl. **2013**, 129 (2013). https://doi.org/10.1186/1687-1812-2013-129
10. Aydi, H., Karapınar, E., Samet, B.: Fixed points for generalized $(\alpha - \psi)$-contractions on generalized metric spaces. J. Inequal. Appl. **2014**, 229 (2014)
11. Aydi, H., Karapınar, E., Lakzian, H.: Fixed point results on the class of generalized metric spaces. Math. Sci. **6** (2012)
12. Bilgili, N., Karapınar, E.: A note on common fixed points for (ψ, α, β)-weakly contractive mappings in generalized metric spaces. Fixed Point Theory Appl. **2013**, 287 (2013)
13. Fulga A., E. Karapınar E.: Revisiting Meir-Keeler type fixed operators on Branciari distance space. Tbilisi Math. J. **12**(4), 97–110 (2019)

14. Hitzler, P: Generalized metrics and topology in logic programming semantics. PhD thesis, School of Mathematics, Applied Mathematics and Statistics, National University Ireland, University College Cork (2001)
15. Karapinar, E., Salimi, P.: Dislocated metric space to metric spaces with some fixed point theorems. Fixed Point Theory Appl. **2013**, 222 (2013)
16. Sarma, I.R., Kumari, P.S.: On dislocated metric spaces. Int. J. Math. Arch. **3**(1), 72–77 (2012)
17. Kumari, P.S., Zoto, K., Panthi, D.: D-neighborhood system and generalized f-contraction in dislocated metric space. Springerplus **4**, 368 (2015). https://doi.org/10.1186/s40064-015-1095-3
18. Hicks, T.L.: Fixed point theorems for quasi-metric spaces. Math. Japonica. **33**, 231–236 (1988)
19. Alsulami, H.H., Karapınar, E., Khojasteh, F., Roldán-López-de-Hierro, A.F.: A proposal to the study of contractions in quasi-metric spaces. Discret. Dyn. Nat. Soc. **2014**, Article ID 269286 (2014)
20. Aydi, H., Bilgili, N., Karapınar, E.: Common fixed point results from quasi-metric spaces to g-metric spaces. J. Egyptian Math. Soc. **23**(2), 356–361 (2015)
21. Bilgili, N., Karapınar E., Samet, B.: Generalized $(\alpha - \psi)$-contractive mappings in quasi-metric spaces and related fixed-point theorems. J. Inequal. Appl. **2014**, 36 (2014)
22. Chen, C.M., Karapınar, E., Lin, I.J.: Periodic points of weaker Meir-Keeler contractive mappings on generalized quasi-metric spaces. Abstr. Appl. Anal. **2014**, Article ID: 490450 (2014)
23. Wilson, W.A.: On semi-metric spaces. Am. J. Math. **53**(2), 361–373 (1931)
24. Imdad, M., Ali, J., Khan, L.: Coincidence and fixed points in symmetric spaces under strict contractions. J. Math. Anal. Appl. **320**(1), 352–360 (2006)
25. Aliouche, A.: A common fixed point theorem for weakly compatible mappings in symmetric spaces satisfying contractive condition of integral type. J. Math. Anal. Appl. **332**(2), 796–802 (2006)
26. Cho, S., Lee, G., Bae, J.: On coincidence and fixed-point theorems in symmetric spaces. Fixed Point Theory and Applications (2008)
27. Sarma, I.R., Rao, J.M., Kumari, P.S., Panthi, D.: Convergence axioms on dislocated symmetric spaces. Abstr. Appl. Anal. **2014**, Article ID 745031, 7 pages (2014). https://doi.org/10.1155/2014/745031
28. Matthews, S.G.: Partial metric topology, Research Report 212. University of Warwick, Department of Computer Science (1992)
29. Matthews, S.G.: Partial metric topology. In: Proceedings of the 8th Summer Conference on General Topology and Applications, Annals of the New York Academy of Sciences, vol. 728, pp. 183–197 (1994)
30. Abedelljawad, T., Karapınar, E., Taş, K.: Existence and uniqueness of common fixed point on partial metric spaces. Appl. Math. Lett. **24**, 1894–1899 (2011)
31. Karapınar, E.: A note on common fixed point theorems in partial metric spaces. Miskolc Math. Notes. **12**, 185–191 (2011)
32. Czerwik, S.: Contraction mappings in b-metric spaces. Acta Math. Univ. Ostrav. **1**(1), 5–11 (1993)
33. Czerwik, S.: Nonlinear set-valued contraction mappings in b-metric spaces. Atti Semin. Mat. Fis. Univ. Modena **46**, 263–276 (1998)
34. Akkouchi, M.: A common fixed point theorem for expansive mappings under strict implicit conditions on b-metric spaces. Acta Univ. Palacki. Olomuc., Fac. Rerum Nat., Math. **50**(1), 5–15 (2011)
35. Aydi, H., Bota, M.F., Karapinar, E., Mitrović, S.: A fixed point theorem for set-valued quasi-contractions in b-metric spaces. Fixed Point Theory Appl. **2012**(88), 1–8 (2012)
36. Aydi, H., Bota, M.F., Karapinar, E., Mitrović, S.: A common fixed point for weak ϕ-contractions on b-metric spaces. Fixed Point Theory **13**(2), 1–10 (2012)

37. Karapınar, E., Du, W.S.: A note on b-cone metric and its related results: generalizations or equivalence? Fixed Point Theory Appl. **2013**, 210 (2013)
38. Alsulami, H.H., Karapınar, E., Rakocevic, V.: Ciric type nonunique fixed point theorems on b-metric spaces. Filomat. **31**, 3147–3156 (2017)
39. Aksoy, U., Karapınar, E., Erhan, I.M.: Fixed points of generalized α-admissible contractions on b-metric spaces with an application to boundary value problems. J. Nonlinear Convex **A17**, 1095–1108 (2016)
40. Afshari, H., Aydi, H., Karapınar, E.: Existence of fixed points of set-valued mappings in b-metric spaces. East Asian Math. J. **32**, 319–332 (2016)
41. Van Rooij, A.C.M.: Non-Archimedean Functional Analysis. Marcel Dekker, New York (1978)
42. Gajic, L.: On ultrametric space. Novi Sad J. Math. **31**(2), 69–71 (2001)
43. Kamran, T., Samreen, M., UL Ain, Q.: A generalization of b-metric space and some fixed point theorems. Mathematics **5**, 19 (2017). https://doi.org/10.3390/math5020019
44. Abdeljawad, T., et al.: Solutions of the nonlinear integral equation and fractional differential equation using the technique of a fixed point with a numerical experiment in extended b-metric space. Symmetry **11**(5), 686 (2019)
45. Abdeljawad, T., et al.: Solutions of boundary value problems on extended-Branciari b-distance. J. Inequalities Appl. **2020**, 1–16 (2020)
46. Alqahtani, B., Fulga, A., Karapınar, E.: Common fixed point results on an extended b-metric space. J. Inequalities Appl. **2018**, 1–15 (2018)
47. Karapınar, E., Kumari, P.S., Lateef, D.: A new approach to the solution of the Fredholm integral equation via a fixed point on extended b-metric spaces. Symmetry **10**(10), 512 (2018)
48. Berzig, M.: First results in suprametric spaces with applications. Mediterr. J. Math. **19**, 226 (2022). https://doi.org/10.1007/s00009-022-02148-6
49. Panda, S.K., Agarwal, R.P., Karapınar, E.: Extended suprametric spaces and stone-type theorem. AIMS Math. **8**(10), 23183–23199 (2023). https://doi.org/10.3934/math.20231179
50. Berzig, M.: Nonlinear contraction in b-suprametric spaces. J. Anal. **32**, 2401–2414 (2024). https://doi.org/10.1007/s41478-024-00732-5
51. Berzig, M.: Strong b-suprametric spaces and fixed point principles. Complex Anal. Oper. Theory **18**, 148 (2024). https://doi.org/10.1007/s11785-024-01594-2
52. Kramosil, O., Michalek, J.: Fuzzy metric and statistical metric spaces. Kybernetika **11**, 326–334 (1975)
53. Hicks, T.L.: Fixed point theory in probabilistic metric spaces. Univ. u Novom Sadu Zb. Rad. Prirod.-Mat. Fak. Ser. Mat. **13**, 63–72 (1983)
54. Menger, K.: Statistical metrics. Proc. Nat. Acad. Sci. USA **28**, 535–537 (1942)
55. Hadžic, O., Pap, E.: A fixed point theorem for multivalued mappings in probabilistic metric spaces and an application in fuzzy metric spaces. Fuzzy Sets Syst. **127**, 333–344 (2002)
56. Mustafa, Z., Sims, B.: A new approach to generalized metric spaces. J. Nonlinear Convex Anal. **7**(2), 289–297 (2006)
57. Huang, L.G., Zhang, X.: Cone metric spaces and fixed point theorems of contractive mappings. J. Math. Anal. Appl. **332**(2), 1468–1476 (2007)
58. Berinde, V.: Common fixed points of noncommuting almost contractions in cone metric spaces. Math. Commun. **15**(1), 229–241 (2010)
59. Jleli, M., Samet, B.: A generalized metric space and related fixed point theorems. Fixed Point Theory Appl. **2015**, 61 (2015). https://doi.org/10.1186/s13663-015-0312-7
60. Jankovic, S., Kadelburg, Z., Radenovic, S.: On cone metric spaces: a survey. Nonlinear Anal. **74**, 2591–2601 (2011)
61. Agarwal, R.P., Alsulami, H., Karapınar, E., Khojasteh, F.: Remarks on some recent fixed point results in quaternion-valued metric spaces. Abstr. Appl. Anal. **2014**, Article ID: 171624 (2014)

62. Dosenovic, T., Postolache, M., Radenovic, S.: On multiplicative metric spaces: survey. Fixed Point Theory Appl. **2016**, 92 (2016)
63. Singh, P., Singh, V., Jele, C.M.T.: A new relaxed b-metric type and fixed point results. Aust. J. Math. Anal. Appl. **18**(2), Art. 8, 8 Pages (2021)
64. Kirk, W., Shahzad, N.: Fixed Point Theory in Distance Spaces. Springer (2014)
65. Khamsi, M.A.: Hussain, n, kkm mappings in metric type spaces. Nonlinear Anal. **7**(9), 3123–3129 (2010)
66. Khamsi, M.A.: Remarks on cone metric spaces and fixed point theorems of contractive mappings. Fixed Point Theory Appl. **2010**, 1–7 (2010)
67. An, T.V., Tuyen, L.Q., Dung, N.V.: Stone-type theorem on b-metric spaces and applications. Topol. Appl. **185**, 50–64 (2015)
68. Gähler, V.S.: 2-metrische Räume und ihre topologische struktur. Math. Nachr. **26**, 115–118 (1963/1964)
69. Sedghi, S., Shobe, N., Aliouche, A.: A generalization of fixed point theorem in S-metric spaces. Mat. Vesn. **64**(3), 258–266 (2012)
70. Kunzi, H., Pajoohesh, H.: *b*-metrics. Topol. Appl. **309**, 107905 (2022)
71. Suzuki, T.: Basic inequality on a *b*-metric space and its applications. J. Inequal. Appl. **2017**, 256 (2017). https://doi.org/10.1186/s13660-017-1528-3
72. Karapınar, E., Noorwali, M.: Dragomir and gosa type inequalities on *b*-metric spaces. J. Inequal. Appl. **2019**, 29 (2019). https://doi.org/10.1186/s13660-019-1979-9
73. Dragomir, S.S., Gos, A.C.: An inequality in metric spaces (2004)
74. Mlaiki, N., Aydi, H., Souayah, N., Abdeljawad, T.: Controlled metric type spaces and the related contraction principle. Mathematics **6**, 194 (2018)
75. Abdeljawad, T., Mlaiki, N., Aydi, H., Souayah, N.: Double controlled metric type spaces and some fixed point results. Mathematics **6**, 320 (2018). https://doi.org/10.3390/math6120320
76. Parvaneha, V., Kadelburg, Z.: Extended partial *b*-metric spaces and some fixed point results. Filomat **32**(8), 2837–2850 (2018). https://doi.org/10.2298/FIL1808837P
77. Asim, M., Imdad, M., Radenovic, S.: Fixed point results in extended rectangular *b*-metric spaces with an application. UPB Sci. Bull. Ser. A **81**(2), 43–50 (2019)

Renowned Theorems in Fixed Point Theory: Topological and Discrete Perspectives

3

The theory of fixed points is one of the most powerful tools of modern mathematics.

Felix Browder

As we all know that, fixed point theory is generally divided into three key categories:

1. ***Topological Fixed Point Theory***-originated from ***Brouwer's*** fixed point theorem [1], which was established in 1912;
2. ***Metric Fixed Point Theory***-originated from ***Banach's*** fixed point theorem [2], which was established in 1922;
3. ***Discrete Fixed Point Theory***-originated from ***Tarski's*** fixed point theorem [3], which was established in 1955.

We have already discussed the developments in metric fixed point theory in previous chapter. In this chapter, we aim to present several renowned fixed point theorems, focusing on their statements and significance, without delving into their proofs. These theorems form the backbone of *topological* and *discrete fixed point theory*, highlighting key principles and their wide-ranging applications.

3.1 Some Standard Fixed Point Theorems

The list that follows is simply ordered by year, not by category.

Theorem 3.1 [1] (Brouwer Fixed Point Theorem-1911) *Every compact and convex nonempty subset of* $\mathbb{R}^n$ *has a fixed point.*

S. K. Panda et al., *Fixed Point Theory from Early Foundations to Contemporary Challenges*, Synthesis Lectures on Mathematics & Statistics,
https://doi.org/10.1007/978-3-032-03585-1_3

- The Brouwer Fixed Point Theorem is fundamental in "**topology**", "**economics**", and "**game theory**", particularly in proving the existence of "**equilibria**". It is widely used in the analysis of continuous systems where one needs to demonstrate the existence of steady states or solutions.

However, it is *non-constructive*, its applications in proving Nash equilibria and in optimization make it indispensable in both theoretical and applied mathematics.

A general criterion provided by the "***Banach fixed-point theorem-1922***" ensures that the method of iterating a function yields a fixed point [4].

Here is a visual example that demonstrate the iterative constructive nature of the "***Banach fixed-point theorem***".

Example 3.1 We want to demonstrate the Banach fixed-point theorem using the contraction mapping:

$$f(x) = \frac{x}{2}.$$

This function satisfies the contraction condition:

$$|f(x) - f(y)| \leq k|x - y| \quad \text{where } k = 0.5 \ (|f'(x)| = 0.5 < 1).$$

The fixed point is the value x^* where:

$$f(x^*) = x^*.$$

Solving for $f(x) = \frac{x}{2} = x$, we get:

$$x = 0.$$

Thus, the fixed point is $x^* = 0$.

Starting with an initial value $x_0 = 1$, we apply the iteration rule:

$$x_{n+1} = f(x_n) = \frac{x_n}{2}.$$

The sequence generated is:

$$x_1 = f(x_0) = \frac{1}{2}, \quad x_2 = f(x_1) = \frac{1}{4}, \quad x_3 = f(x_2) = \frac{1}{8}, \ldots$$

This sequence converges to $x^* = 0$.

The Fig. 3.1 below shows the iterative process graphically. Each orange step represents an application of the function $f(x)$ or the mapping back to the diagonal $y = x$, converging to the fixed point $x = 0$.

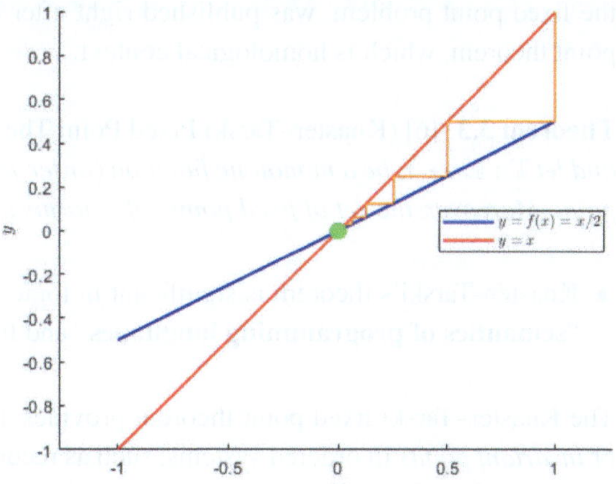

Fig. 3.1 Banach fixed-point theorem: convergence to fixed point

Theorem 3.2 [5] (Nielsen Fixed-Point Theorem-1927) *Let $\mathbb{T} : X \to X$ be a continuous map on a compact, connected, triangulable space X. The number of fixed points of $\mathbb{T}$ that cannot be removed by a homotopy of $\mathbb{T}$, known as the Nielsen number $N(\mathbb{T})$, is a homotopy-invariant lower bound for the minimum number of fixed points of $\mathbb{T}$. Specifically:*

$$\mathrm{MF}(\mathbb{T}) \geq N(\mathbb{T}),$$

where $\mathrm{MF}(\mathbb{T})$ *is the minimum number of fixed points of all maps homotopic to* $\mathbb{T}$.

- The Nielsen Fixed-Point Theorem extends the Lefschetz fixed-point theorem by refining the count of fixed points to account for their homotopy classes. It is a fundamental result in algebraic topology, particularly in the study of fixed points of maps on surfaces and higher-dimensional spaces. Applications include analyzing periodic orbits in dynamical systems, computing invariants in geometric group theory, and solving problems in topological data analysis. The concept of the Nielsen number has practical significance in minimizing the fixed points in deformable systems.

Nielsen fixed point theory is a topic of research in algebraic and geometric topology. The maps are permitted to undergo continuous deformation, unlike the metric fixed point theory that was derived from Banach's contraction principle. A major focus of Nielsen theory is how the fundamental group affects how fixed points behave on maps. Jakob Nielsen is the person who developed the Nielsen fixed point theory, hence the name. He identified the smallest set of fixed points for every isotopy category of torus self-homeomorphisms in early of 19th century. Brouwer quickly expanded this to include self-maps of the torus, and both papers were published next to one another in Math. Annalen. Nielsen's significant study was published in 1927 [5] on self-homeomorphisms of hyperbolic surfaces, with a focus on the fixed point problem, was published right after Lefschetz proved his well-known fixed point theorem, which is homological context.

Theorem 3.3 [6] (Knaster–Tarski Fixed Point Theorem-1928) *Let $\mathbb{L}$ be a complete lattice, and let $\mathbb{T} : \mathbb{L} \to \mathbb{L}$ be a monotone function (order-preserving). Then $\mathbb{T}$ has at least one fixed point. Moreover, the set of fixed points of $\mathbb{T}$ forms a complete lattice.*

- Knaster–Tarski's theorem is significant in logic and computer science, especially in the "**semantics of programming languages**" and fixed-point induction.

The Knaster–Tarski fixed point theorem provides a structured way to find *stable solutions* or *invariant points* in ordered systems, such as recursive definitions and fixed-point computations in "**functional programming**".

Theorem 3.4 [7] (Schauder Fixed Point Theorem-1930) *Let $\mathscr{M}$ be a closed, convex, and bounded subset of a Banach space, and let $\mathbb{T} : \mathscr{M} \to \mathscr{M}$ be a continuous and compact mapping. Which implies that, $\mathbb{T}$ has at most one fixed point.*

- Schauder's theorem generalizes Brouwer's result to infinite-dimensional spaces, making it essential in the study of "**nonlinear analysis**" and "**differential equations**".

Its compactness condition is very important in many scenarions, such as proving the existence of solutions for "**boundary value problems**" and "**integral equations**". The Schauder's fixed point theorem extends *fixed point theory* into *functional analysis*, bridging *finite and infinite-dimensional spaces*.

Theorem 3.5 [8, 9] *Let $\mathscr{M}$ be a compact convex subset of a Banach space X. Assume that $\mathbb{T}$ is a continuous operator. Then $\mathbb{T}$ has at least one fixed point in $\mathscr{M}$.*

Sketch of the Proof. Let $\mathscr{M}$ be a compact convex subset of a Banach space X, and let $\mathbb{T} : \mathscr{M} \to \mathscr{M}$ be a continuous operator. Since $\mathscr{M}$ is compact, the image $\mathbb{T}(\mathscr{M})$ is also relatively compact in $\mathscr{M}$. Define $\mathscr{C}(\mathbb{T}(\mathscr{M}))$ as the closure of the convex hull of $\mathbb{T}(\mathscr{M})$. By the Mazur Theorem, $\mathscr{C}(\mathbb{T}(\mathscr{M}))$ is a compact, convex subset of $\mathscr{M}$.

Because $\mathbb{T}$ is continuous and $\mathscr{C}(\mathbb{T}(\mathscr{M}))$ is compact and convex, $\mathbb{T}(\mathscr{C}(\mathbb{T}(\mathscr{M}))) \subseteq \mathscr{C}(\mathbb{T}(\mathscr{M}))$. By the Theorem 3.4, $\mathbb{T}$ has at least one fixed point in $\mathscr{C}(\mathbb{T}(\mathscr{M}))$. Since $\mathscr{C}(\mathbb{T}(\mathscr{M})) \subseteq \mathscr{M}$, the fixed point lies in $\mathscr{M}$. Thus, $\mathbb{T}$ has a fixed point in $\mathscr{M}$.
To apply Theorem 3.4, it is essential to define the pair $(\mathbb{T}, \mathscr{M})$, where $\mathscr{M} \subset X$ is a closed, bounded, and convex subset of the Banach space X, and $\mathbb{T} : \mathscr{M} \to \mathscr{M}$ is a compact operator. The Leray–Schauder Fixed Point Theorem provides a precise and illustrative example of this framework.

Theorem 3.6 [8, 9] (Leray-Schauder Fixed Point Theorem-1934) *Let X be a Banach space and $\mathbb{T} : X \to X$ a compact operator. Assume there exists a positive number $r > 0$ such that:*

$$\|x\| \leq r \quad \textit{for all } x \textit{ satisfying } x = t\mathbb{T}(x) \textit{ with } t \in [0, 1].$$

Then $\mathbb{T}$ has at least one fixed point $x \in X$.

Sketch of the Proof. Let $\mathscr{B}_r = \{x \in X : \|x\| \leq r\}$, the closed ball of radius r in X. By the assumption, any x satisfying $x = t\mathbb{T}(x)$ for $t \in [0, 1]$ lies within $\mathscr{B}_r$.

Define a modified operator $\psi : \mathscr{B}_r \to \mathscr{B}_r$ by:

$$\psi(x) = \begin{cases} \mathbb{T}(x), & \text{if } \|\mathbb{T}(x)\| \leq r, \\ r\frac{\mathbb{T}(x)}{\|\mathbb{T}(x)\|}, & \text{if } \|\mathbb{T}(x)\| > r. \end{cases}$$

The operator ψ is well-defined, continuous, and compact because:

1. $\mathbb{T}$ is compact, so it maps bounded subsets of X to relatively compact subsets.
2. ψ is continuous since $\mathbb{T}$ is continuous and the truncation at r is smooth.

Next, consider the homotopy $H : \mathscr{B}_r \times [0, 1] \to \mathscr{B}_r$ defined by:

$$H(x, t) = t\psi(x).$$

For $t = 0$, $H(x, 0) = 0$, and for $t = 1$, $H(x, 1) = \psi(x)$. The homotopy ensures that $H(x, t)$ is continuous in $\mathscr{B}_r$. Since ψ maps $\mathscr{B}_r$ into itself and is compact, the Leray–Schauder degree theory applies.

By assumption, there are no fixed points of $H(x, t)$ on the boundary of $\mathscr{B}_r$, $\|x\| = r$, because any x on the boundary satisfying $x = t\psi(x)$ would contradict $\|x\| \leq r$. Therefore, the degree of ψ in $\mathscr{B}_r$ is nonzero.

By the fixed point index theory, ψ has a fixed point $x \in \mathscr{B}_r$. Since $\psi(x) = \mathbb{T}(x)$ for $\|x\| \leq r$, x is also a fixed point of $\mathbb{T}$. Thus, $\mathbb{T}$ has a fixed point in X.

Remark 3.1 Note that Leray-Schauder is usually proven by using the hypotheses to construct a mapping that satisfies the conditions of the Schauder fixed point theorem, and then appealing to the Schauder fixed point theorem.

Theorem 3.7 [10] (Lefschetz Fixed Point Theorem-1937) *Let X be a compact topological space and $\mathbb{T} : X \to X$ a continuous map. If the Lefschetz number $\mathscr{L}(\mathbb{T}) \neq 0$, then $\mathbb{T}$ has at least one fixed point.*

- The Lefschetz theorem links "**algebraic topology**" and "**fixed point theory**", allowing the existence of fixed points to be deduced from *algebraic invariants*.

The Lefschetz fixed point theorem has broad applications in "**dynamical systems**", particularly in the study of periodic orbits, and serves as a foundation for more advanced results like the "**Nielsen fixed point theory**". Its algebraic approach makes it a versatile tool in both theoretical and applied mathematics.

Theorem 3.8 [11, 12] (Markov–Kakutani Fixed-Point Theorem-1936–38) *Let X be a nonempty compact convex subset of a locally convex topological vector space, and let $\{\mathbb{T}_i\}_{i\in I}$ be a commuting family of continuous affine self-maps of X. Then there exists a common fixed point $x^* \in X$, i.e., $\mathbb{T}_i(x^*) = x^*$ for all $i \in I$.*

- The Markov–Kakutani Fixed-Point Theorem generalizes classical fixed-point theorems by considering families of commuting affine maps rather than a single map. This theorem is essential in functional analysis, particularly in the study of invariant measures and ergodic theory. Applications include demonstrating the existence of invariant states in operator algebras, fixed points in convex sets under group actions, and equilibrium problems in game theory and economics.

Theorem 3.9 [13] (Kakutani Fixed Point Theorem-1941) *Let $\mathscr{M}$ be a nonempty, compact, and convex subset of a topological vector space, and let $\mathbb{T} : \mathscr{M} \to 2^{\mathscr{M}}$ be an upper hemicontinuous, convex-valued correspondence. Then $\mathbb{T}$ has at least one fixed point.*

- Kakutani's result is very important in "**game theory**" and "**economics**", particularly in proving the existence of "**mixed-strategy Nash equilibria**".

However, by extending Brouwer's theorem to set-valued mappings, it enables analysis in scenarios involving *multifunctions*, such as "**equilibria in markets**" or "**dynamic games**". It provides a powerful tool for working with more complex systems in applied mathematics.

Theorem 3.10 [14] (Kleene Fixed-Point Theorem-1952) *Let $(\boldsymbol{L}, \leq)$ be a complete lattice and $\mathbb{T} : \boldsymbol{L} \to \boldsymbol{L}$ a monotone function. Then:*

- $\mathbb{T}$ has a least fixed point, which is equal to the supremum of the ascending chain $\{\mathbb{T}^n(\bot)\}_{n\geq 0}$, where $\bot$ is the least element of $\mathbf{L}$ and $\mathbb{T}^n$ denotes the n-fold composition of $\mathbb{T}$.
- If $\mathbb{T}$ is also continuous (preserves suprema of directed sets), then the least fixed point is achieved as the limit of this ascending chain.

The Kleene fixed-point theorem is a pivotal result in "**lattice theory**" and "**theoretical computer science**". The Kleene fixed-point theorem provides a constructive method for finding the "**least fixed point of a monotone function**", which is especially important in the semantics of programming languages, such as defining recursive functions and solving domain equations. The iterative approach constructs fixed-point computations in regular techniques, abstract perception, and logic programming semantics.

Example 3.2 Compute the least fixed point of the function $\mathbb{T}(x) = \frac{1}{10}x^2 + \arctan(x) + 1$ using the Kleene Fixed-Point Theorem.

Solution: The function $\mathbb{T}(x) = \frac{1}{10}x^2 + \arctan(x) + 1$ is defined for all $x \in \mathbb{R}$, and its derivative is:

$$\mathbb{T}'(x) = \frac{1}{5}x + \frac{1}{1+x^2}.$$

Since $\mathbb{T}'(x) > 0$ for all x, the function is monotone increasing. This satisfies the monotonicity condition of the Kleene Fixed-Point Theorem, ensuring that the sequence generated by iterating $\mathbb{T}(x)$ will converge to the least fixed point if we start from the least element of the domain.

We start with $x_0 = 0$ as the initial value and iteratively compute $x_{n+1} = \mathbb{T}(x_n)$ until convergence.

$$\mathbb{T}(x) = \frac{1}{10}x^2 + \arctan(x) + 1.$$

Iteration Steps:

- **Step 1**: Starting with $x_0 = 0$,

$$x_1 = \mathbb{T}(0) = \frac{1}{10}(0)^2 + \arctan(0) + 1 = 1.$$

- **Step 2**: Using $x_1 = 1$,

$$x_2 = \mathbb{T}(1) = \frac{1}{10}(1)^2 + \arctan(1) + 1 = 0.1 + \frac{\pi}{4} + 1 \approx 1.8854.$$

- **Step 3**: Using $x_2 \approx 1.8854$,

$$x_3 = \mathbb{T}(1.8854) = \frac{1}{10}(1.8854)^2 + \arctan(1.8854) + 1 \approx 2.4237.$$

This process continues until the difference $|x_{n+1} - x_n|$ becomes smaller than a predefined tolerance, such as 10^{-6}.

The sequence $\{x_n\}$ converges to the least fixed point x^* of $\mathbb{T}(x)$. For this function, numerical computations show that the least fixed point is approximately at

$x^* \approx 3.579$ as shown in Fig.3.2.

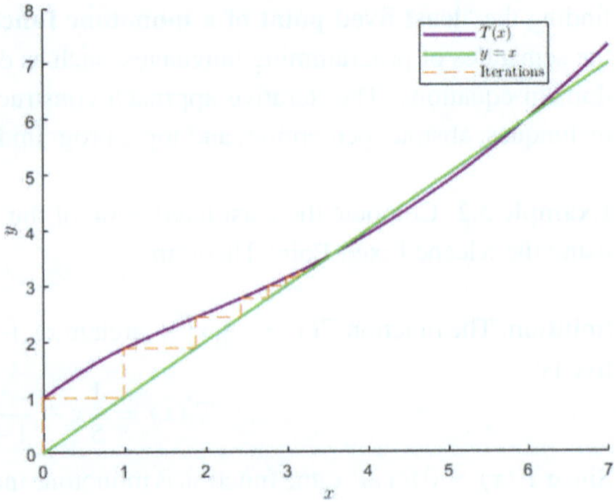

Fig. 3.2 Graphical representation of $\mathbb{T}(x)$ and iterative convergence

The Kleene Fixed-Point Theorem guarantees this convergence due to the monotonicity and continuity of $\mathbb{T}(x)$.

Example 3.3 Compute the least fixed point of the function $\mathbb{T}(x) = \frac{1}{5}x^3 - sin(x) + 1$ on [0,3] using the Kleene Fixed-Point Theorem.

We iteratively compute:

$$x_{n+1} = \mathbb{T}(x_n) = \frac{1}{5}x_n^3 - \sin(x_n) + 1,$$

starting with $x_0 = 0$, until $|x_{n+1} - x_n| < 10^{-6}$. Fig. 3.3

Iteration (n)	x_n	$\mathbb{T}(x_n)$	$\lvert x_{n+1} - x_n \rvert$
0	0.000000	1.000000	1.000000
1	1.000000	0.799037	0.200963
2	0.799037	0.581875	0.217162
3	0.581875	0.540444	0.041431
4	0.540444	0.518313	0.022131
5	0.518313	0.510760	0.007553
6	0.510760	0.507941	0.002819
7	0.507941	0.506891	0.001050
8	0.506891	0.506497	0.000394
9	0.506497	0.506345	0.000152
10	0.506345	0.506283	0.000062

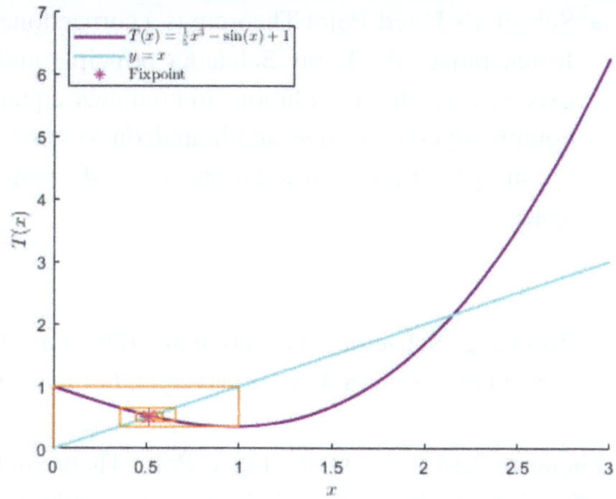

Fig. 3.3 Illustration of Kleene fixed-point iteration for $\mathbb{T}(x)$

The sequence converges to the least fixed point

$$x^* \approx 0.506 \text{ as shown in Fig. 3.3}$$

where $\mathbb{T}(x^*) = x^*$.

Theorem 3.11 [3] (Tarski Fixed Point Theorem-1955) *Let* $(\mathbb{P}, \leq)$ *be a partially ordered set where every chain has a supremum, and let* $f : \mathbb{P} \to \mathbb{P}$ *be a monotone function. Then* f *has at least one fixed point.*

- Tarski's theorem is pivotal in "**lattice theory**" and widely applied in "**computer science**", "**formal semantics**", and "**economics**".

The Tarski's theorem especially helpful in proving the existence of "***equilibria***" in ordered structures, such as fixed points in iteration algorithms or stable outcomes in social choice theory. Its lattice-theoretic framework makes it flexible for both theoretical exploration and practical implementation.

Theorem 3.12 [15] (Schaefer's Fixed-Point Theorem-1955) *Let* X *be a Banach space, and let* $\mathbb{T} : X \to X$ *be a continuous and compact operator (i.e.,* $\mathbb{T}$ *maps bounded subsets of* X *to relatively compact subsets). Suppose the set:*

$$\bigcup_{0 \leq \lambda \leq 1} \{x \in X : x = \lambda \mathbb{T}(x) \text{ for some } \lambda \in [0, 1]\}$$

is bounded. Then $\mathbb{T}$ *has at least one fixed point, i.e., there exists* $x^* \in X$ *such that* $\mathbb{T}(x^*) = x^*$.

- Schaefer's Fixed-Point Theorem is a cornerstone result in nonlinear functional analysis. It generalizes the Leray-Schauder principle and provides a powerful tool for proving existence results for solutions to nonlinear equations and boundary value problems. By combining compactness and boundedness conditions, the theorem is widely applied in the study of differential equations, integral equations, and variational problems in Banach spaces.

Remark 3.2 Schaefer's fixed point theorem, is a version of Schauder's theorem. Sometimes it is called the Leray-Schauder principle, For more info, refer, [16].

Theorem 3.13 [17] (Darbo Fixed-Point Theorem-1955) *Let X be a Banach space, and let $\mathscr{M} \subseteq X$ be a closed, bounded, and convex subset. Suppose $\mathbb{T} : \mathscr{M} \to \mathscr{M}$ is a continuous operator, and there exists a measure of noncompactness α on $\mathscr{M}$ such that:*

$$\alpha(\mathbb{T}(\mathscr{A})) \leq k \cdot \alpha(\mathscr{A}) \quad \text{for all bounded subsets } \mathscr{A} \subseteq \mathscr{M},$$

where $k \in [0, 1)$ is a constant. Then $\mathbb{T}$ has at least one fixed point in $\mathscr{M}$.

- The Darbo Fixed-Point Theorem generalizes the Theorem 3.4 by incorporating the concept of a "**measure of noncompactness**" instead of "**compactness**".

The Darbo fixed-point theorem is a crucial tool in the study of nonlinear functional equations, integral equations, and operator equations where *compactness is not present* but can be replaced by a "**measure of noncompactness**". The theorem finds applications in solving differential and integral equations and in studying *"iterative methods"* in Banach spaces.

Theorem 3.14 [18] (Borel Fixed-Point Theorem-1956) *Let G be a connected solvable algebraic group acting on a nonempty complete algebraic variety X over an algebraically closed field. Then G has a fixed point in X.*

- The Borel Fixed-Point Theorem is a cornerstone result in "***algebraic geometry***" and "***the theory of algebraic groups***".

The Borel fixed-point theorem guarantees the existence of fixed points under solvable groups on complete varieties, this practice is very predominant in the notion of invariant theory and geometric representation theory. However, this renowned result is highly utilized in understanding the notion of algebraic groups, especially in the interpretation of parabolic subgroups.

Theorem 3.15 [19] (Krasnoselskii's Fixed-Point Theorem-1958) *Let $\mathscr{M}$ be a closed, convex, and nonempty subset of a Banach space X. Suppose $\mathscr{A} : \mathscr{M} \to \mathscr{M}$ and $\mathscr{B} : \mathscr{M} \to \mathscr{M}$ are two mappings such that:*

- *$\mathscr{A}$ is a contraction mapping.*
- *$\mathscr{B}$ is completely continuous (i.e., continuous and compact).*
- *For all $x \in \mathscr{M}$, $\mathscr{A}(x) + \mathscr{B}(x) \in \mathscr{M}$.*

Then there exists at least one fixed point $x^* \in \mathscr{M}$ such that:

$$x^* = \mathscr{A}(x^*) + \mathscr{B}(x^*).$$

- Krasnoselskii's Fixed-Point Theorem is a vital result in nonlinear functional analysis.

The Krasnoselskii's fixed-point theorem extends the "***Banach contraction principle***" by combining contraction and compactness conditions, making it particularly useful in solving operator equations of the form "$x = \mathscr{A}(x) + \mathscr{B}(x)$". More clearly, the Banach contraction theorem and Schauder's theorem are two of fixed-point theory's primary findings. Krasnoselskii combined them into the above stated result. Applications include differential equations, integral equations, and iterative methods in Banach spaces. This theorem acts as a cornerstone for complex fixed-point problems in "***science and technology***" associated with "***applied mathematics***".

Theorem 3.16 [20] (Ryll-Nardzewski Fixed-Point Theorem-1962) *Let X be a nonempty compact convex subset of a locally convex topological vector space, and let $\mathscr{G}$ be a group of affine transformations acting continuously on X. If $\mathscr{G}$ is amenable (i.e., there exists a left-invariant mean on $\mathscr{G}$), then there exists a fixed point in X for the action of $\mathscr{G}$.*

- The Ryll-Nardzewski Fixed-Point Theorem is a fundamental result in the "**theory of group actions**" on convex sets.

The Ryll-Nardzewski fixed-point theorem is a key tool in functional analysis, particularly in the study of "**invariant measures**" and "**ergodic theory**". The theorem has significant applications in proving the existence of equilibria in *"game theory"*, analyzing fixed-point properties of amenable groups, and studying representations of groups in *"convex compact"* settings.

Bekić's theorem [21], often known as Bekić's lemma, is a fixed-point theorem in computability theory that permits dividing an equitable recursion through recursions on a single variable at a time.

Theorem 3.17 [21] (Bekić Fixed Point Theorem-1964) *Let X and Y be complete lattices, and let* $\mathscr{A} : X \times Y \to X$ *and* $\mathscr{B} : X \times Y \to Y$ *be monotone functions. Then the pair of equations*

$$x = \mathscr{A}(x, y), \quad y = \mathscr{B}(x, y).$$

has a unique solution (x^*, y^*) *in* $X \times Y$ *that can be computed iteratively. Furthermore, the solution can be represented as:*

$$x^* = \boldsymbol{sup}_{y \in Y} \mathscr{A}(x^*, y), \quad y^* = \boldsymbol{sup}_{x \in X} \mathscr{B}(x, y^*).$$

- Bekić's theorem is fundamental in computer science and formal methods, particularly in fixed-point computations for systems of equations involving lattices.

Theorem 3.18 [22, 23] (Browder Fixed Point Theorem-1965) *Let* $\mathscr{M}$ *be a nonempty, compact, convex subset of a Banach space X, and let* $\mathbb{T} : \mathscr{M} \to \mathscr{M}$ *be a continuous mapping that is non-expansive, i.e.,* $\|\mathbb{T}(x) - \mathbb{T}(y)\| \leq \|x - y\|$ *for all* $x, y \in \mathscr{M}$*. Then* $\mathbb{T}$ *has at least one fixed point in* $\mathscr{M}$.

- The Browder fixed Point Theorem is a generalization of Schauder's fixed point theorem and extends fixed point results to non-expansive mappings in Banach spaces.

Numerous generalisations or variations of Browder's fixed point theorem have been seen in the literature due to its numerous applicability mainly in ***nonlinear analysis*** and ***variational inequalities***. However, this theorem also utilized in ***game-theoretic results*** and ***economic equilibria***.

Theorem 3.19 [24, 25] (Atiyah–Bott Fixed Point Theorem-1966) *Let X be a smooth manifold and let* $\mathbb{T} : X \to X$ *be a smooth mapping such that the fixed points of* $\mathbb{T}$ *are isolated. Then the Lefschetz number of* $\mathbb{T}$*, defined in terms of traces of induced maps on cohomology, is equal to the sum of contributions from each fixed point. Specifically, if* $\mathbb{T}$ *has isolated fixed points* $p_1, p_2, \ldots, p_k$*, then:*

$$\mathscr{L}(\mathbb{T}) = \sum_{i=1}^{k} \frac{\boldsymbol{sign}(\det(I - d\mathbb{T}_{p_i}))}{|\det(I - d\mathbb{T}_{p_i})|},$$

where $d\mathbb{T}_{p_i}$ *is the derivative of* $\mathbb{T}$ *at* p_i.

- The Atiyah–Bott fixed point theorem is a powerful result in differential geometry and algebraic topology, generalizing the Theorem 3.7 to smooth manifolds.

The Atiyah–Bott Fixed Point Theorem provides a way to compute fixed points in terms of "**cohomological invariants**" and is widely used in mathematical physics, particularly in "**gauge theory**" and the study of "**moduli spaces**". The theorem has deep implications in enumerative "**geometry**" and "**topological quantum field theory**".

Theorem 3.20 [26] (Sadovskii's Fixed-Point Theorem-1967) *Let $\mathscr{M}$ be a closed, bounded, and convex subset of a Banach space X, and let $\mathbb{T} : \mathscr{M} \to \mathscr{M}$ be a continuous and condensing map with respect to a measure of noncompactness α, which means,*

$$\alpha(\mathbb{T}(\mathcal{A})) < \alpha(\mathcal{A}) \quad \textit{for every bounded subset } \mathcal{A} \subseteq \mathscr{M} \textit{ with } \alpha(\mathcal{A}) > 0.$$

Then $\mathbb{T}$ has at least one fixed point in $\mathscr{M}$.

- Sadovskii's Fixed-Point Theorem generalizes the Theorem 3.4 by relaxing the "**compactness**" requirement, replacing it with the condition of being "**condensing with respect to a measure of noncompactness**".

The Sadovskii's fixed-point theorem is instrumental in "**nonlinear functional analysis**" and has significant applications in the study of differential equations, integral equations, and operator equations where the compactness of the operator cannot be **assumed**.

Theorem 3.21 [27] (Lawvere's Fixed-Point Theorem-1969) *Let $\mathscr{C}$ be a Cartesian closed category, and let $\mathbb{T} : \mathcal{A} \to \mathcal{A}$ be a morphism in $\mathscr{C}$. Suppose there exists an object $\mathcal{B}$ in $\mathscr{C}$ and a morphism $\mathbb{S} : \mathcal{B} \to \mathcal{A}^{\mathcal{B}}$ such that $\mathbb{S}$ is a diagonal universal (exhibiting $\mathcal{B}$ as an exponent). Then $\mathbb{S}$ has a fixed point.*

- Lawvere's Fixed-Point Theorem is foundational in category theory and theoretical computer science, particularly in understanding the nature of self-referential and recursive definitions. It generalizes fixed-point results by embedding them in the framework of Cartesian closed categories. This theorem has profound implications in logic, type theory, and the semantics of programming languages, especially in characterizing recursive functions and the concept of self-replication in formal systems.

Theorem 3.22 [28] (Mönch's Fixed-Point Theorem-1980) *Let X be a Banach space, and let $\mathscr{M} \subseteq X$ be a closed, convex, and bounded subset. Suppose $\mathbb{T} : \mathscr{M} \to 2^{\mathscr{M}}$ is a set-valued map such that:*

- *$\mathbb{T}$ is upper semicontinuous with nonempty, compact, and convex values.*
- *Every sequence $\{x_n\}$ in $\mathscr{M}$ with $x_n \in \mathbb{T}(x_{n-1})$ (for $n \geq 1$) has a subsequence converging to a point in $\mathscr{M}$.*

Then $\mathbb{T}$ *has at least one fixed point, i.e., there exists* $x^* \in \mathcal{M}$ *such that* $x^* \in \mathbb{T}(x^*)$.

- Mönch's Fixed-Point Theorem is a powerful generalization of Theorem 3.4, particularly applicable to "**set-valued mappings**".

The Mönch's fixed-point theorem is frequently used in nonlinear analysis and differential inclusions, providing existence results where *"continuity and single-valued assumptions are relaxed"*. The theorem has applications in control theory, optimization, and dynamical systems, especially for problems involving ***"multifunctions"*** or ***"discontinuous mappings"***.

Open Questions

3.1 Let $f : J \times X^2 \to X$ satisfy the growth condition

$$\|f(t, x, y)\| \leq C\big(1 + \|x\|^{p(t)} + \|y\|^{q(t)}\big), \quad t \in [a, b],$$

where $p, q \in C([a, b], [1, \infty))$.
Can the method of measures of noncompactness (MNC) used by Mönch [28] be extended to such nonlinearities with nonuniform variable exponent growth?

3.2 The article [28] obtains findings in weakly compactly generated (wcg) spaces. Is it possible to obtain similar existence results in non-reflexive, non-wcg Banach spaces by means of newer generalized topological measures of noncompactness, e.g. Kuratowski–Ryll-Nardzewski measures?

3.3 Suppose $\mathbb{T}$ is not compact, but weakly sequentially continuous and condensing in regard to a weak MNC. Is it possible to get multiplicity of solutions or nontrivial weak solutions to the BVP via extensions of Schauder-type fixed point theory?

References

1. Brouwer, L.: Uber abbildungen von mannigfaltigkeiten. Math. Ann. **71**, 97–115 (1912)
2. Banach, S.: Sur les opérations dans les ensembles abstraits et leur application aux équations intégrales. Fund. Math. **3**, 133–181 (1922)
3. Tarski, A.: A lattice theoretical fixpoint theorem and its applications. Pacific J. Math. **5**, 285–309 (1955)
4. Giles, J.R.: Introduction to the Analysis of Metric Spaces. Cambridge University Press (1987)
5. Nielsen, J.: Untersuchungen zur topologie der geschlossenen zweiseitigen flachen. Acta Math. **50**, 189–358 (1927)

6. Knaster, B., Tarski, A.: Un théorème sur les fonctions d'ensembles. Ann. Soc. Polon. Math. **6**, 133–134 (1928)
7. Schauder, J.: Der fixpunktsatz in funktionalrumen. Studia Math. **2**, 171–180 (1930)
8. Leray, J., Schauder, J.: Topologie et équations fonctionnelles. Ann. Sci. École Norm. Sup. **51**, 45–78 (1934)
9. Zou, H.: Handbook of Differential Equations: Stationary Partial Differential Equations, vol. 6, pp. 503–583. North-Holland (2008)
10. Lefschetz, S.: On the fixed point formula. Ann. Math. **38**(4), 819–822 (1937). https://doi.org/10.2307/1968838
11. Markov, A.: Quelques théorèmes sur les ensembles abéliens. Dokl. Akad. Nauk SSSR **10**, 311–314 (1936)
12. Kakutani, S.: Two fixed point theorems concerning bicompact convex sets. Proc. Imp. Akad. Tokyo **14**, 242–245 (1938)
13. Kakutani, S.: A generalization of Brouwer's fixed point theorem. Duke Math. J. **8**(3), 457–459 (1941). https://doi.org/10.1215/S0012-7094-41-00838-4
14. Kleene, S.C.: Introduction to Metamathematics. North-Holland Pub. Co., Amsterdam (1952). zbMATH: 0047.00703 (1955)
15. Schaefer, H.: Uber die method der a priori-schranken. Math. Ann. **126**, 415–416 (1955)
16. Smart, D.R.: Fixed Point Theorems. Cambridge University Press, London (1974)
17. Darbo, G.: Punti uniti in transformazioni a condomino non compatto. Rend. Sem. Mat. Univ. Padova **24**, 84–92 (1955)
18. Borel, A.: Groupes linéaires algébriques. Ann. Math. 2 **64**(1) (1956). Ann. Math. 20–82. https://doi.org/10.2307/1969949
19. Krasnoselskii, M.A.: Some problems of nonlinear analysis. Am. Math. Soc. Transl. **10**(2), 345–409 (1958)
20. Ryll-Nardzewski, C.: Generalized random ergodic theorems and weakly almost periodic functions. Bull. Acad. Polon. Sci. Sér. Sci. Math. Astron. Phys. **10**, 271–275 (1962)
21. Bekić, H.: Definable operations in general algebras, and the theory of automata and flowcharts. Programming languages and their definition. Lecture Notes in Computer Science, vol. 177, pp. 30–55 (1984). https://doi.org/10.1007/BFb0048939
22. Felix, E.: Browder, nonexpansive nonlinear operators in a Banach space. Proc. Natl. Acad. Sci. U.S.A. **54**, 1041–1044 (1965)
23. Browder, F.E.: Problèmes Non-Linéaires, Séminaire de Mathématiques Supérieures, 15. Les Presses de l'Université de Montréal, Montréal (1966)
24. Atiyah, M.F., Bott, R.: A Lefschetz fixed point formula for elliptic differential operators. Bull. Am. Math. Soc. **72**(2), 245–50 (1966). https://doi.org/10.1090/S0002-9904-1966-11483-0
25. Atiyah, M.F., Bott, R.: A Lefschetz fixed point formula for elliptic complexes: i. Ann. Math. (2) **86**(2), 374–407 (1967)
26. Sadovskii, B.N.: A fixed point principle. Funktsional. Anal. Prilozhen. **1**, 74–76 (1967)
27. Lawvere, F.W.: Diagonal arguments and Cartesian closed categories. Category Theory, Homology Theory and their Applications II (Lecture Notes in Mathematics, vol. 92). Springer, Berlin (1969)
28. Mönch, H.: Boundary value problems for nonlinear ordinary differential equations of second order in Banach spaces. Nonlinear Anal. Theory Methods Appl. **4**, 985–999 (1980). https://doi.org/10.1016/0362-546X(80)90010-3

Foundational Extensions of the Contraction Mapping

4

Mathematics is the most beautiful and most powerful creation of the human spirit.

Stefan Banach

Stefan Banach published his doctoral dissertation, "On operations on abstract sets and their applications to integral equations" in Fundamenta Mathematicae in 1922 [1]. The Banach Fixed Point Theorem or Banach Contraction Principle was one of them. Of the many Banach theorems, it is considered one of the most famous. It defines conditions for the existence and uniqueness of a fixed point of certain mappings (called contractions) of a complete metric space into itself. Banach's fundamental fixed point theorem on a complete metric space is the origin of the metric fixed point theory. This chapter will present its main generalizations, contractive conditions, simpler proof and converse theorems. It is impossible to give all the information about the Banach Fixed Point Theorem that has been written in the past years since its publication. So, we are only highlighting the key milestones in the "**metric fixed point theory**".

4.1 The Banach Fixed Point Theorem

The *Banach Fixed Point Theorem* was established by Banach in the second section of the article [1]. It was stated as follows, assuming that a space E satisfies all of the axioms I, II, and III:

S. K. Panda et al., *Fixed Point Theory from Early Foundations to Contemporary Challenges*, Synthesis Lectures on Mathematics & Statistics,
https://doi.org/10.1007/978-3-032-03585-1_4

Theorem 4.1 [1] (Banach Fixed Point Theorem (Original Statement)) *Let E be a Banach space satisfying the following axioms:*

(I) ***Linearity**: E is a linear space with vector addition and scalar multiplication.*
(II) ***Norm**: A norm $\|\cdot\|$ is defined on E, satisfying $\|x\| \geq 0$, $\|x\| = 0 \iff x = 0$, and the triangle inequality $\|x + y\| \leq \|x\| + \|y\|$.*
(III) ***Completeness**: Every Cauchy sequence in E converges to an element in E.*

Suppose $\mathbb{T} : E \to E$ is a continuous operator satisfying:

(a) There exists $M \in (0, 1)$ such that $\|\mathbb{T}(x_1) - \mathbb{T}(x_2)\| \leq M\|x_1 - x_2\|$ for all $x_1, x_2 \in E$.

Then there exists a unique $x^* \in E$ such that $\mathbb{T}(x^*) = x^*$.

Proof of Theorem 4.1 The fixed point x^* is obtained constructively using an iterative process. For an arbitrary initial point $x_0 \in E$, define the sequence $x_{n+1} = \mathbb{T}(x_n)$. Observe:

$$\|x_{n+1} - x_n\| = \|\mathbb{T}(x_n) - \mathbb{T}(x_{n-1})\| \leq M\|x_n - x_{n-1}\|.$$

Inductively, this implies $\|x_{n+1} - x_n\| \leq M^n\|x_1 - x_0\|$, and since $M < 1$, the series $\sum_{n=1}^{\infty} \|x_{n+1} - x_n\|$ converges. Thus, $\{x_n\}$ is a Cauchy sequence, and by completeness, it converges to some $x^* \in E$.

Continuity of $\mathbb{T}$ ensures $\mathbb{T}(x^*) = \lim_{n\to\infty} \mathbb{T}(x_n) = x^*$. Uniqueness follows from the contraction property: if x_1^*, x_2^* are fixed points, then $\|x_1^* - x_2^*\| \leq M\|x_1^* - x_2^*\|$, forcing $x_1^* = x_2^*$.

In modern times, the aforementioned theorem is expressed in a somewhat different manner.

Theorem 4.2 [2] (Modern Formulation) *Let (X, d) be a* ***complete metric space****, and let $\mathbb{T} : X \to X$ be a* ***contraction mapping****, i.e., there exists $\alpha \in [0, 1)$ such that:*

$$d(\mathbb{T}(x), \mathbb{T}(y)) \leq \alpha d(x, y) \quad \textit{for all } x, y \in X.$$

Then:

1. *$\mathbb{T}$ has a unique fixed point $x^* \in X$, i.e., $\mathbb{T}(x^*) = x^*$.*
2. *For any initial point $x_0 \in X$, the sequence $\{x_n\}$ defined by $x_{n+1} = \mathbb{T}(x_n)$ converges to x^*.*
3. *The convergence rate is geometric and satisfies:*

$$d(x_n, x^*) \leq \frac{\alpha^n}{1-\alpha} d(x_0, x_1).$$

Proof of Theorem 4.2 Let (X, d) be a complete metric space, and let $\mathbb{T} : X \to X$ be a contraction mapping with contraction constant α, where $0 < \alpha < 1$. To prove the theorem, we construct a sequence starting from an arbitrary point $x_0 \in X$, defining $x_{n+1} = \mathbb{T}(x_n)$. Since $\mathbb{T}$ maps X into itself, the sequence $\{x_n\}$ is well-defined.
By the contraction property of $\mathbb{T}$, for any $n \geq 0$, we have:

$$d(x_{n+1}, x_n) = d(\mathbb{T}(x_n), \mathbb{T}(x_{n-1})) \leq \alpha \cdot d(x_n, x_{n-1}).$$

By applying this inequality recursively, we observe that:

$$d(x_{n+1}, x_n) \leq \alpha^n \cdot d(x_1, x_0).$$

This shows that the distances between consecutive terms in the sequence decrease geometrically. Next, we estimate the distance between any two terms x_m and x_n, where $m > n$, using the triangle inequality:

$$d(x_m, x_n) \leq d(x_m, x_{m-1}) + d(x_{m-1}, x_{m-2}) + \cdots + d(x_{n+1}, x_n).$$

Substituting the contraction property, we get:

$$d(x_m, x_n) \leq \alpha^n \cdot d(x_1, x_0) \cdot (1 + \alpha + \alpha^2 + \cdots + \alpha^{m-n-1}).$$

The sum $1 + \alpha + \alpha^2 + \cdots + \alpha^{m-n-1}$ is a finite geometric series, which evaluates to:

$$\frac{1 - \alpha^{m-n}}{1 - \alpha}.$$

Thus:

$$d(x_m, x_n) \leq \frac{\alpha^n}{1 - \alpha} \cdot d(x_1, x_0). \tag{4.1}$$

As $m, n \to \infty$, the term $\alpha^n \to 0$, so $d(x_m, x_n) \to 0$. Therefore, $\{x_n\}$ is a Cauchy sequence in X.

Since X is a complete metric space, the Cauchy sequence $\{x_n\}$ converges to a point $x^* \in X$. To verify that x^* is a fixed point, consider:

$$\mathbb{T}(x^*) = \mathbb{T}\left(\lim_{n\to\infty} x_n\right).$$

Using the continuity of $\mathbb{T}$, we exchange the limit and $\mathbb{T}$, obtaining:

$$\mathbb{T}(x^*) = \lim_{n\to\infty} \mathbb{T}(x_n) = \lim_{n\to\infty} x_{n+1} = x^*.$$

Thus, $\mathbb{T}(x^*) = x^*$, confirming that x^* is a fixed point.

To prove uniqueness, suppose x^* and y^* are two fixed points of $\mathbb{T}$. Then:

$$d(x^*, y^*) = d(\mathbb{T}(x^*), \mathbb{T}(y^*)) \leq \alpha \cdot d(x^*, y^*).$$

Since $0 < \alpha < 1$, the only solution to this inequality is $d(x^*, y^*) = 0$, which implies $x^* = y^*$.

Finally, the convergence rate of the sequence $\{x_n\}$ to x^* is geometric. For any $n \geq 0$, letting $m \to \infty$ in the inequality (4.1), we derive that

$$\begin{aligned} d(x_n, x^*) &= \lim_{m\to\infty} d(x_n, x_m) \\ &\leq \frac{\alpha^n}{1-\alpha} d(x_1, x_0), \end{aligned}$$

which yields that

$$d(x_n, x^*) \leq \frac{\alpha^n}{1-\alpha} d(x_1, x_0).$$

This completes the proof of the theorem.

On the other hand, when a topologist encounters the phrase "**fixed point theory**," the first thought that likely comes to mind is Brouwer's result.

The **Brouwer Fixed Point Theorem** [3] is a fundamental result in topology. It states:

Theorem 4.3 *For any continuous function* $\mathbb{T} : B \to B$*, where* B *is a closed ball in a Euclidean space* $\mathbb{R}^n$*, there exists at least one fixed point* $x \in B$ *such that:*

$$\mathbb{T}(x) = x.$$

Above result demonstrates the existence of a fixed point; however, it does not provide guidance on how to find this fixed point or how to establish its uniqueness.

Table 4.1 provides a straightforward comparison of the Banach fixed-point theorem and the Brouwer fixed-point theorem.

4.1.1 A Simple Proof of Banach Fixed Point Theorem According to *Richard S. Palais* [4]

Theorem 4.4 [4] *Let* (X, d) *be a complete metric space, and suppose* $\mathbb{T} : X \to X$ *is a contraction mapping with contraction constant* K *such that* $0 < K < 1$*. That is, for all* $x_1, x_2 \in X$*, the inequality*

$$d(\mathbb{T}(x_1), \mathbb{T}(x_2)) \leq K d(x_1, x_2)$$

Table 4.1 Simple comparison

Aspect	Banach fixed point theorem (1922)	Brouwer fixed point theorem (1912)
Domain	Applies to **complete metric spaces**	Applies to **compact, convex subsets of Euclidean spaces**
Function conditions	Requires the function to be a **contraction mapping**	Works for **continuous functions**, no contraction needed
Existence and uniqueness	Guarantees a **unique fixed point**	Ensures existence of at least one fixed point, but not uniqueness
Constructiveness	Provides an **iterative method** to approximate the fixed point	Non-constructive; does not provide a method to find the fixed point
Applicability	Suited for **numerical and computational methods**	Theoretical, useful in topology and proofs of existence
Scope of spaces	More general; applies to any **complete metric space**	Restricted to **finite-dimensional Euclidean spaces**

holds. Then, $\mathbb{T}$ *has a unique fixed point* $p \in X$*, and for any* $x \in X$*, the sequence of iterates* $\{\mathbb{T}^n(x)\}_{n=0}^{\infty}$ *converges to* p*. Furthermore, the convergence is quantifiable by the inequality:*

$$d(\mathbb{T}^n(x), p) \leq \frac{K^n}{1-K} d(x, \mathbb{T}(x)).$$

The proof proceeds as follows: Assume $\mathbb{T}$ has two fixed points x_1 and x_2, so $\mathbb{T}(x_1) = x_1$ and $\mathbb{T}(x_2) = x_2$. Using the contraction property, we have:

$$d(x_1, x_2) = d(\mathbb{T}(x_1), \mathbb{T}(x_2)) \leq K d(x_1, x_2).$$

Since $0 < K < 1$, this implies $d(x_1, x_2) = 0$, and hence $x_1 = x_2$. Thus, $\mathbb{T}$ has at most one fixed point.

Next, consider any $x \in X$ and the sequence $\{\mathbb{T}^n(x)\}$. Using the contraction property repeatedly, we find:

$$d(\mathbb{T}^n(x), \mathbb{T}^m(x)) \leq \frac{K^n + K^m}{1-K} d(x, \mathbb{T}(x)).$$

As $n, m \to \infty$, the terms K^n and K^m approach 0 since $0 < K < 1$. Thus, $\{\mathbb{T}^n(x)\}$ is a Cauchy sequence. Since X is complete, $\{\mathbb{T}^n(x)\}$ converges to some $p \in X$.

To verify that p is a fixed point, note that $\mathbb{T}$ is continuous, and hence:

$$\mathbb{T}(p) = \mathbb{T}\left(\lim_{n\to\infty} \mathbb{T}^n(x)\right) = \lim_{n\to\infty} \mathbb{T}^{n+1}(x) = p.$$

Thus, p is the unique fixed point of $\mathbb{T}$.

Finally, the error bound is derived from the contraction property as:

$$d(\mathbb{T}^n(x), p) \leq \frac{K^n}{1-K} d(x, \mathbb{T}(x)).$$

This establishes the rate of convergence of the sequence $\{\mathbb{T}^n(x)\}$ to the fixed point p.

4.1.2 Short Notes on Converse of the Banach Fixed Point Theorem

The converse of the Banach Fixed Point Principle addresses scenarios where a unique fixed point exists, and the iterations of a function converge to that fixed point. The question it seeks to answer is whether there always exists a metric under which the iterative function becomes a contraction mapping, making Banach's theorem applicable in reverse.

- **Some Well-Known Converses**:
  - ***Bessaga's Converse Theorem*** [5]: If $\mathbb{T} : X \to X$ has unique fixed points for $\mathbb{T}^n$ for all $n \in \mathbb{N}$, then for every $c \in (0, 1)$, there exists a metric d_c such that $\mathbb{T}$ becomes a contraction under d_c.
  - ***Meyers's Converse Theorem*** [6]: Extends Bessaga's result to ensure d_c is topologically equivalent to the original metric d, assuming $\mathbb{T}$ is continuous, X is compact, and $\mathbb{T}^n(U) \to x^*$ for an open neighborhood U around x^*.

The classical converse theorems, while providing existence results for contraction metrics, have limitations. They do not ensure that the constructed metric d_c directly relates to the original metric d when approximating fixed points. Moreover, these results lack practical tools to determine the number of iterations required for convergence under d, leaving a gap in their applicability to iterative methods.

In 2018, the authors of [7] presented a stronger converse to the Banach Fixed Point Principle. They demonstrated that if $\mathbb{T} : X \to X$ globally converged to a unique fixed point x^* under a metric d, then it was always possible to construct a modified metric $d_{c,\epsilon}$ such that $\mathbb{T}$ became a contraction under $d_{c,\epsilon}$. This metric also allowed analysts to bound the number of iterations needed to approximate x^* to a given accuracy ϵ under the original metric d. Their work illustrated the universality of Banach's theorem for analyzing iterative methods.

They applied their results to the power iteration method for computing principal eigenvectors of matrices. They showed that, although this method was not contracting under standard norms, a specialized metric could be constructed, under which the power iteration converged with an optimal contraction rate determined by the eigenvalue ratio.

4.2 Some Definitions of Contractive Mappings

In the context of complete metric spaces, a variety of definitions have been proposed to generalize the well-known Banach contraction principle. These definitions describe mappings that guarantee the existence of a unique fixed point, often determined via iterative procedures such as Picard iteration.

In [8], B.E. Rhoades compiled 25 contraction conditions, many of which were well-established in the literature, with the goal of comparing them and deriving fixed point theorems. Beyond the Banach contraction principle, the study included conditions introduced by researchers such as Rakotch [9], Edelstein [10], Kannan [11], Bianchini [12], Reich [13, 14], Sehgal [15], Chatterjea [16], Hardy and Rogers [17], Zamfirescu [18], and Ciric [19, 20]. These comparisons are significant because if one condition (m) implies another (n), any fixed point theorem proven for condition (n) can automatically extend to condition (m).

Although the comparisons in [8] were not exhaustive, they also explored generalizations of these 25 conditions to one or two iterates of the function, applied uniformly across one, both, or none of the variables. However, the primary focus was on the comparison of the 25 fundamental conditions, as this serves as the foundation for understanding other groups of conditions applied to iterates of the function $\mathbb{T}$.

To maintain consistency, we will refer to the contractive conditions using the same numbering system as presented by B.E. Rhoades in [8]. However, below is a list of notable definitions:

1. **Banach Contraction** [1]: There exists a constant $a \in [0, 1)$ such that for all $x, y \in X$:

$$d(\mathbb{T}(x), \mathbb{T}(y)) \leq a \cdot d(x, y).$$

2. **Rakotch's Condition** [9]: There exists a monotone decreasing function $a : (0, \infty) \to [0, 1)$ such that for all $x, y \in X$:

$$d(\mathbb{T}(x), \mathbb{T}(y)) \leq a \cdot d(x, y).$$

3. **Edelstein's Weak Contraction** [10]: For all $x, y \in X$ with $x \neq y$:

$$d(\mathbb{T}(x), \mathbb{T}(y)) < d(x, y).$$

4. **Kannan's Contraction** [11]: There exists a constant $a \in (0, \frac{1}{2})$ such that for all $x, y \in X$:

$$d(\mathbb{T}(x), \mathbb{T}(y)) \leq a \cdot [d(x, \mathbb{T}(x)) + d(y, \mathbb{T}(y))].$$

5. **Bianchini's Metric Dominance** [12]: There exists a constant $h \in [0, 1)$ such that for all $x, y \in X$:

$$d(\mathbb{T}(x), \mathbb{T}(y)) \leq h \cdot \max\{d(x, \mathbb{T}(x)), d(y, \mathbb{T}(y))\}.$$

6. **Simple Maximum Condition** [8]: For all $x, y \in X$ and x is distinct to y:

$$d(\mathbb{T}(x), \mathbb{T}(y)) < \max\{d(x, \mathbb{T}(x)), d(y, \mathbb{T}(y))\}.$$

7. **Reich's Averaging** [13]: There exist nonnegative constants a, b, c with $a + b + c \leq 1$ such that for all $x, y \in X$:

$$d(\mathbb{T}(x), \mathbb{T}(y)) \leq a \cdot d(x, \mathbb{T}(x)) + b \cdot d(y, \mathbb{T}(y)) + c \cdot d(x, y).$$

8. **Reich's Generalized Averaging** [14]: There exist monotone decreasing functions $a, b, c : (0, \infty) \to [0, 1)$ with $a(t) + b(t) + c(t) < 1$, such that for all $x, y \in X$:

$$d(\mathbb{T}(x), \mathbb{T}(y)) \leq a(d(x, y)) \cdot d(x, \mathbb{T}(x)) + b(d(x, y)) \cdot d(y, \mathbb{T}(y)) + c(d(x, y)) \cdot d(x, y).$$

9. **Variable Coefficients (Supremum)** [8]: There exist nonnegative functions: $a(x, y), b(x, y), c(x, y)$ such that

$$\sup_{x,y \in X} (a(x, y) + b(x, y) + c(x, y)) < 1,$$

and for all $x, y \in X$:

$$d(\mathbb{T}(x), \mathbb{T}(y)) \leq a(x, y) \cdot d(x, \mathbb{T}(x)) + b(x, y) \cdot d(y, \mathbb{T}(y)) + c(x, y) \cdot d(x, y).$$

10. **Sehgal's Condition** [15]: For all $x, y \in X$ and x is distinct to y:

$$d(\mathbb{T}(x), \mathbb{T}(y)) < \max\{d(x, \mathbb{T}(x)), d(y, \mathbb{T}(y)), d(x, y)\}.$$

11. **Chatterjea's Symmetry Constraint** [16]: There exists a constant $a \in (0, \frac{1}{2})$ such that for all $x, y \in X$:

$$d(\mathbb{T}(x), \mathbb{T}(y)) \leq a \cdot [d(x, \mathbb{T}(y)) + d(y, \mathbb{T}(x))].$$

12. **Symmetric Maximum Condition** [8]: There exists a constant $h \in [0, 1)$ such that for all $x, y \in X$:

$$d(\mathbb{T}(x), \mathbb{T}(y)) \leq h \cdot \max\{d(x, \mathbb{T}(y)), d(y, \mathbb{T}(x))\}.$$

13. **Maximum Symmetric Values** [8]: For all $x, y \in X$ and x is distinct to y:

$$d(\mathbb{T}(x), \mathbb{T}(y)) < \max\{d(x, \mathbb{T}(y)), d(y, \mathbb{T}(x))\}.$$

14. **Reich's Symmetric Coefficients** [8]: There exist nonnegative constants a, b, c with $a + b + c < 1$, such that for all $x, y \in X$:

$$d(\mathbb{T}(x), \mathbb{T}(y)) \leq a \cdot d(x, \mathbb{T}(y)) + b \cdot d(y, \mathbb{T}(x)) + c \cdot d(x, y).$$

15. **Monotone Symmetric Coefficients** [8]: There exist monotone decreasing functions $a, b, c : (0, \infty) \to [0, 1)$ with $a(t) + b(t) + c(t) < 1$, such that for all $x, y \in X$ and x is distinct to y:

$$d(\mathbb{T}(x), \mathbb{T}(y)) \le a(d(x, y)) \cdot d(x, \mathbb{T}(y)) + b(d(x, y)) \cdot d(y, \mathbb{T}(x)) + c(d(x, y)) \cdot d(x, y).$$

16. [8] There exist nonnegative functions $a(x, y)$, $b(x, y)$, and $c(x, y)$ defined for all $x, y \in X$, such that:

$$\sup_{x,y\in X} \{a(x, y) + b(x, y) + c(x, y)\} < 1,$$

and for all $x, y \in X$, the following inequality holds:

$$d\big(\mathbb{T}(x), \mathbb{T}(y)\big) \le a(x, y)d\big(x, \mathbb{T}(y)\big) + b(x, y)d\big(y, \mathbb{T}(x)\big) + c(x, y)d(x, y).$$

17. **Symmetric Maximum Bound** [8]: For all $x, y \in X$:

$$d(\mathbb{T}(x), \mathbb{T}(y)) \le \max\{d(x, \mathbb{T}(y)), d(y, \mathbb{T}(x)), d(x, y)\}.$$

18. **Hardy and Rogers (Extended)** [17]: There exist nonnegative constants a_1, a_2, a_3, a_4, a_5 with $a_1 + a_2 + a_3 + a_4 + a_5 < 1$, such that for all $x, y \in X$:

$$\begin{aligned} d(\mathbb{T}(x), \mathbb{T}(y)) \le\, & a_1 \cdot d(x, y) + a_2 \cdot d(x, \mathbb{T}(x)) + a_3 \cdot d(y, \mathbb{T}(y)) \\ & + a_4 \cdot d(x, \mathbb{T}(y)) + a_5 \cdot d(y, \mathbb{T}(x)). \end{aligned}$$

19. **Zamfirescu's Hybrid Contractive Conditions** [18]: For each $x, y \in X$, there exist constants α, β, γ, where $0 \le \alpha < 1$ and $0 \le \beta, \gamma < \frac{1}{2}$, such that at least one of the following holds:

$$\begin{aligned} &\text{(i)}\ \ d(\mathbb{T}(x), \mathbb{T}(y)) < \alpha \cdot d(x, y), \\ &\text{(ii)}\ \ d(\mathbb{T}(x), \mathbb{T}(y)) < \beta \cdot \big[d(x, \mathbb{T}(x)) + d(y, \mathbb{T}(y))\big], \\ &\text{(iii)}\ \ d(\mathbb{T}(x), \mathbb{T}(y)) < \gamma \cdot \big[d(x, \mathbb{T}(y)) + d(y, \mathbb{T}(x))\big]. \end{aligned}$$

20. [8] For all $x, y \in X$:

$$d(\mathbb{T}(x), \mathbb{T}(y)) < \max\left\{d(x, y), \frac{d(x, \mathbb{T}(x)) + d(y, \mathbb{T}(y))}{2}, \frac{d(x, \mathbb{T}(y)) + d(y, \mathbb{T}(x))}{2}\right\}.$$

21. **Ciric's Advanced Averaging** [20]: There exist nonnegative functions q, r, s, t such that

$$\sup_{x,y\in X} (q(x, y) + r(x, y) + s(x, y) + 2t(x, y)) < 1,$$

and for all $x, y \in X$:

$$d(\mathbb{T}(x), \mathbb{T}(y)) \leq q(x, y) \cdot d(x, y) + r(x, y) \cdot d(x, \mathbb{T}(x)) + s(x, y) \cdot d(y, \mathbb{T}(y)) \\ + t(x, y) \cdot [d(x, \mathbb{T}(y)) + d(y, \mathbb{T}(x))].$$

22. [8] For all $x, y \in X$:

$$d(\mathbb{T}(x), \mathbb{T}(y)) < \max \left\{ d(x, y), d(x, \mathbb{T}(x)), d(y, \mathbb{T}(y)), \frac{d(x, \mathbb{T}(y)) + d(y, \mathbb{T}(x))}{2} \right\}.$$

23. [8] There exist monotonically decreasing functions $a_i : (0, \infty) \to [0, 1)$ satisfying $\sum_{i=1}^{5} a_i(t) < 1$, such that, for each $x, y \in X, x \neq y$:

$$d(\mathbb{T}(x), \mathbb{T}(y)) \leq a_1(d(x, y))d(x, \mathbb{T}(x)) + a_2(d(x, y))d(y, \mathbb{T}(y)) + a_3(d(x, y))d(x, \mathbb{T}(y)) \\ + a_4(d(x, y))d(y, \mathbb{T}(x)) + a_5(d(x, y))d(x, y).$$

24. **Ciric** [20]: There exists a constant h, $0 \leq h < 1$, such that, for each $x, y \in X$:

$$d(\mathbb{T}(x), \mathbb{T}(y)) \leq h \max\{d(x, y), d(x, \mathbb{T}(x)), d(y, \mathbb{T}(y)), d(x, \mathbb{T}(y)), d(y, \mathbb{T}(x))\}.$$

25. [8] For each $x, y \in X, x \neq y$:

$$d(\mathbb{T}(x), \mathbb{T}(y)) < \max\{d(x, y), d(x, \mathbb{T}(x)), d(y, \mathbb{T}(y)), d(x, \mathbb{T}(y)), d(y, \mathbb{T}(x))\}.$$

26. **Bailey's Condition** [21]: For all $x, y \in X$, if $0 < d(x, y)$, there exists an integer $p(x, y)$ such that:

$$d(\mathbb{T}^{p(x,y)}(x), \mathbb{T}^{p(x,y)}(y)) < d(x, y).$$

For each $x, y \in X$ with $x \neq y$, there exists an integer $p(x, y)$ (dependent on both x and y) such that the distance between the iterated points is strictly less than the original distance $d(x, y)$.

27. **Guseman's Condition** [22]: There exists $a \in (0, 1)$ such that for each $x \in X$, there exists an integer $p(x)$ satisfying:

$$d(\mathbb{T}^{p(x)}(x), \mathbb{T}^{p(x)}(y)) < a \cdot d(x, y).$$

For each $x \in X$, there exists an iterate $p(x)$ (dependent on x) such that the contractive condition holds for all $y \in X$, with a constant a, $0 < a < 1$.

28. **Yen's Condition** [23]: There exist integers p, q and $a \in (0, 1)$ such that:

$$d(\mathbb{T}^p(x), \mathbb{T}^q(y)) < a \cdot d(x, y).$$

Two different iterates of $\mathbb{T}$, namely $\mathbb{T}^p$ and $\mathbb{T}^q$, satisfy a contractive condition for fixed positive integers p, q, and a constant a with $0 < a < 1$. This holds for all $x, y \in X$.

29. **Singh's Iterative Condition** [24]: For a positive integer p and $a \in (0, \frac{1}{2})$:

$$d(\mathbb{T}^p(x), \mathbb{T}^p(y)) \leq a \cdot [d(x, \mathbb{T}^p(x)) + d(y, \mathbb{T}^p(y))].$$

This equation generalizes contractive mappings to iterates of the function $\mathbb{T}$. For some integer p and constant a with $0 < a < 1$, the p-th iterate of $\mathbb{T}$ satisfies the contractive condition. This applies to all $x, y \in X$, where X is the metric space.

While Rhoades' findings [8] have been generalized by subsequent works, including those by Park [25] and Rhoades [26] himself, none of these extensions comprehensively compared all 25 fundamental conditions. Similarly, while Kineses and Totik [27] examined the fixed point theorems associated with these conditions, they did not explore their interrelationships. Thereafter, the authors of [28] presented a detailed comparison of various contraction conditions, organizing them into a table to clearly illustrate the relationships between these conditions. They also derived all implications and non-implications using a minimal set of results, as outlined in below theorems.

Theorem 4.5 *[28] Let $\{(i, j)\}$ represent pairs of indices corresponding to the 25 contraction conditions. The theorem identifies the key implications necessary to establish all relationships among these conditions. Specifically:*

1. (1) $\Longrightarrow$ (2) $\Longrightarrow$ (3) $\Longrightarrow$ (10) $\Longrightarrow$ (22) $\Longrightarrow$ (25),
2. (1) $\Longrightarrow$ (7) $\Longrightarrow$ (8) $\Longrightarrow$ (23) $\Longrightarrow$ (20) $\Longrightarrow$ (22),
3. (2) $\Longrightarrow$ (8) $\Longrightarrow$ (10),
4. (2) $\Longrightarrow$ (15) $\Longrightarrow$ (17) $\Longrightarrow$ (25),
5. (4) $\Longrightarrow$ (5) $\Longrightarrow$ (6) $\Longrightarrow$ (10),
6. (4) $\Longrightarrow$ (7) $\Longrightarrow$ (18) $\Longrightarrow$ (19) $\Longrightarrow$ (21) $\Longrightarrow$ (22),
7. (5) $\Longrightarrow$ (9) $\Longrightarrow$ (10),
8. (7) $\Longrightarrow$ (9) $\Longrightarrow$ (21) $\Longrightarrow$ (24),
9. (11) $\Longrightarrow$ (12) $\Longrightarrow$ (13) $\Longrightarrow$ (17),
10. (11) $\Longrightarrow$ (14) $\Longrightarrow$ (15) $\Longrightarrow$ (23),
11. (1) $\Longrightarrow$ (14) $\Longrightarrow$ (16) $\Longrightarrow$ (24) $\Longrightarrow$ (25),
12. (14) $\Longrightarrow$ (18) $\Longrightarrow$ (23),
13. (12) $\Longrightarrow$ (16) $\Longrightarrow$ (17),
14. (3) $\Longrightarrow$ (20),
15. (3) $\Longrightarrow$ (17),
16. (19) $\Longrightarrow$ (20).

Above theorem in [28] ensures that these implications provide a sufficient foundation to derive all existing relationships among the 25 contraction conditions.

Theorem 4.6 *[28] Let* (m) *and* (n) *denote two contraction conditions from the set of 25 conditions. The theorem identifies specific cases where* (m) *does not imply* (n)*, thereby completing the comparison using counterexamples. The key non-implications are as follows:*
$(3) \nRightarrow (23)$, $(13) \nRightarrow (24)$, $(6) \nRightarrow (24)$, $(1) \nRightarrow (6)$, $(1) \nRightarrow (13)$, $(5) \nRightarrow (20)$,
$(2) \nRightarrow (24)$, $(12) \nRightarrow (22)$, $(4) \nRightarrow (17)$, $(11) \nRightarrow (10)$, $(19) \nRightarrow (23)$.

The authors of [28] explicitly state that while they list the functions used as examples, they omit the detailed proofs verifying that these functions meet the required criteria satisfying (m) but not (n).

4.2.1 Some Special Cases Among Contractive Mappings

There are some special conditions or relations among these definitions by observing their structural similarities, dependencies, or logical hierarchies. Some definitions are generalizations or extensions of others, while some may imply certain properties under specific assumptions.

- **Banach Contraction (1)** is the most fundamental definition. Many other definitions, such as **Edelstein's Weak Contraction (3)**, **Kannan's Condition (4)**, and **Reich's Averaging (7)**, are generalizations or extensions.
- **Reich's Generalized Averaging (8)** reduces to **Reich's Averaging (7)** if the functions $a(t), b(t), c(t)$ are constant.
- **Ciric's Maximum Form (21)** and **Zamfirescu's Hybrid Conditions (18)** generalize multiple definitions by combining contractive constraints, including Banach, Kannan, and Reich.
- **Reich's Generalized Averaging (8)** is equivalent to **Reich's Averaging (7)** when the monotone decreasing functions a, b, c are constants.
- **Symmetric Maximum Bound (26)** reduces to **Simple Maximum Condition (6)** if the symmetry condition is removed.
- **Hardy and Rogers (17)** reduces to **Banach Contraction (1)** when $a_2 = a_3 = a_4 = a_5 = 0$ and $a_1 < 1$.
- Definitions like **Bailey's Definition (22)** and **Guseman's Definition (27)** depend on specific integers $p(x)$ and are specialized forms of iterative mappings.
- **Singh's Iterative Condition (29)** applies **Kannan's Condition (4)** iteratively.
- **Yen's Definition (28)** generalizes **Bailey's Definition (22)** by introducing separate powers p and q.
- **Ciric's Maximum Form (21)** and **Zamfirescu's Maximum Combination (19)** encompass several definitions, including **Banach (1)**, **Edelstein (3)**, and **Reich (7)**, depending on the dominating maximum term.

- **Chatterjea's Symmetry Constraint (11)** introduces symmetry and serves as a counterpart to **Kannan's Condition (4)**.
- **Reich's Symmetric Coefficients (14)** generalizes **Reich's Averaging (7)** to include symmetry.
- **Continuity**: If $\mathbb{T}$ is continuous, definitions like **Banach (1)**, **Reich (7)**, and **Ciric (21)** ensure unique fixed points.
- **Monotonicity**: Definitions such as **Reich's Generalized Averaging (8)** benefit from monotone decreasing functions for a, b, c.
- **Additivity or Weighting**: Combine weighted conditions from definitions like **Reich's Averaging (7)** and **Hardy and Rogers (17)** for customized mappings.
- **Iterative Constraints**: Use conditions like **Singh (29)** and **Yen (28)** for analyzing dynamical systems.
- **Maximal Symmetry**: Explore symmetric conditions such as **Chatterjea (11)** and **Reich's Symmetric Coefficients (14)** for mappings with undirected distance relationships.

4.2.2 Pair of Mappings

In fixed point theory, the study of a pair of mappings involves two functions $\mathbb{T}$ and $\mathbb{S}$ defined on a metric space (X, d). These mappings are analyzed jointly to identify conditions under which they have a shared fixed point, i.e., a point $x^* \in X$ such that $\mathbb{T}(x^*) = \mathbb{S}(x^*) = x^*$. The interaction between the mappings introduces additional flexibility, allowing for broader applications than single mapping scenarios. Commonly, contractive conditions for pairs of mappings extend those for single mappings, often incorporating distances like $d(\mathbb{T}(x), \mathbb{S}(y))$, $d(\mathbb{T}(x), \mathbb{T}(y))$, and $d(\mathbb{S}(x), \mathbb{S}(y))$.

For instance, a pair of mappings may satisfy a condition such as:

For all $x, y \in X$,

- [29] There exists a constant a, where $0 < a < 1$, such that:

$$d(\mathbb{T}(x), \mathbb{S}(y)) \leq a \cdot \left[d(x, \mathbb{T}(x)) + d(y, \mathbb{S}(y))\right].$$

- [30] There exists a constants $a, b, c \geq 0$ and $a + b + c < 1$ such that:

$$d(\mathbb{T}(x), \mathbb{S}(y)) \leq a \cdot d(x, y) + b \cdot d(x, \mathbb{T}(x)) + c \cdot d(y, \mathbb{S}(y)).$$

For more details on the topic "pair of mappings", the reader can refer [31–34].

4.2.3 Some Well-Known Fixed Point Theorems for Various Contractive Mappings

The nonlinear extensions of Banach's contraction principle have expanded its utility in fixed-point theory by relaxing its conditions. The first major generalization was introduced by Rakotch, who replaced strict contractions with contractive mappings dependent on a monotonic function. Subsequent extensions, developed by various mathematicians, further broadened the scope of fixed-point theorems. Here are some key results extracted from the work of Erdal and Ravi [35]:

Theorem 4.7 [36] (Rakotch) *This theorem extends Banach's principle by incorporating a monotonic decreasing function,* $\alpha : (0, \infty) \to [0, 1)$. *For a complete metric space* (X, d), *if a self-mapping* $\mathbb{T}$ *satisfies:*

$$d(\mathbb{T}x, \mathbb{T}y) \leq \alpha(d(x, y))d(x, y), \quad \forall x, y \in X,$$

then $\mathbb{T}$ *has a unique fixed point* x^*. *The sequence* $\{\mathbb{T}^n x\}$ *converges to* x^* *for any* $x \in X$.

Theorem 4.8 [37] (Browder) *Browder introduced a fixed-point theorem for bounded subsets* Y *of a complete metric space* (X, d). *If* $\mathbb{T} : Y \to Y$ *satisfies:*

$$d(\mathbb{T}x, \mathbb{T}y) \leq \phi(d(x, y)), \quad \forall x, y \in Y,$$

where $\phi : [0, \infty) \to [0, \infty)$ *is a monotonic, right-continuous function satisfying* $\phi(t) < t$ *for* $t > 0$, *then* $\mathbb{T}$ *has a unique fixed point* x^*, *and* $\{\mathbb{T}^n x\}$ *converges to* x^*.

Theorem 4.9 [38] (Boyd and Wong) *Boyd and Wong extended fixed-point results to upper semi-continuous functions. For a self-mapping* $\mathbb{T}$ *on a complete metric space* (X, d), *if there exists an upper semi-continuous function* $\phi : [0, \infty) \to [0, \infty)$ *with* $\phi(t) < t$ *for all* $t > 0$, *and* $\mathbb{T}$ *satisfies:*

$$d(\mathbb{T}x, \mathbb{T}y) \leq \phi(d(x, y)), \quad \forall x, y \in X,$$

then $\mathbb{T}$ *has a unique fixed point* x^*. *The sequence* $\{\mathbb{T}^n x\}$ *converges to* x^*.

Theorem 4.10 [39] (Matkowski) *Matkowski's theorem is a significant step toward broader applicability. For a complete metric space* (X, d), *if there exists a monotonic nondecreasing function* $\phi : (0, \infty) \to (0, \infty)$ *with* $\lim_{n\to\infty} \phi^n(t) = 0$ *for all* $t > 0$, *and* $\mathbb{T}$ *satisfies:*

$$d(\mathbb{T}x, \mathbb{T}y) \leq \phi(d(x, y)), \quad \forall x, y \in X.$$

Then, $\mathbb{T}$ *has a unique fixed point* x^*. *Thus, the sequence* $\{\mathbb{T}^n x\}$ *converges to* x^*.

Theorem 4.11 [40] (Meir and Keeler) *If $\mathbb{T}$ is a self-mapping on a complete metric space (X, d), and for any $\epsilon > 0$ there exists a $\delta > 0$ such that:*

$$\epsilon \leq d(x, y) < \epsilon + \delta \implies d(\mathbb{T}x, \mathbb{T}y) < \epsilon, \quad \forall x, y \in X.$$

Then $\mathbb{T}$ has a unique fixed point x^*. Moreover, the sequence $\{\mathbb{T}^n x\}$, starting from any $x \in X$, converges to x^*.

Theorem 4.12 [41] (Krasnosel'skii and Zabrejko) *If $\mathbb{T}$ is a self-mapping on a complete metric space (X, d), and for each $0 < \alpha < \beta$, there exists a constant $k(\alpha, \beta)$ such that:*

$$\alpha \leq d(x, y) \implies d(\mathbb{T}x, \mathbb{T}y) \leq k(\alpha, \beta)d(x, y), \quad \forall x, y \in X.$$

Then $\mathbb{T}$ has a unique fixed point x^*. For any starting point $x \in X$, the sequence $\{\mathbb{T}^n x\}$ converges to x^*.

Theorem 4.13 [42] (Burton) *If $\mathbb{T}$ is a self-mapping on a complete metric space (X, d), and for every $\lambda > 0$, there exists a constant $k(\lambda)$ such that:*

$$d(x, y) \geq \lambda \implies d(\mathbb{T}x, \mathbb{T}y) \leq k(\lambda)d(x, y), \quad \forall x, y \in X.$$

Then $\mathbb{T}$ has a unique fixed point x^*. For any $x \in X$, the sequence $\{\mathbb{T}^n x\}$ converges to x^*.

Theorem 4.14 [43] (Geraghty) *If $\mathbb{T}$ is a self-mapping on a complete metric space (X, d), and there exists a function $\beta : (0, \infty) \to [0, 1)$ such that:*

$$\beta(t_n) \to 1 \implies t_n \to 0,$$

and:

$$d(\mathbb{T}x, \mathbb{T}y) \leq \beta(d(x, y))d(x, y), \quad \forall x, y \in X.$$

Then $\mathbb{T}$ has a unique fixed point x^*. For any $x \in X$, the sequence $\{\mathbb{T}^n x\}$ converges to x^*.

Geraghty's work represents a significant refinement, offering fixed-point results in settings where traditional contractivity is relaxed.

Theorem 4.15 *This theorem aggregates several important extensions to Banach's principle, each addressing different variations of contractivity conditions. It states that $\mathbb{T}$ has a unique fixed point x^*, and $\{\mathbb{T}^n x\}$ converges to x^* under the following scenarios:*

1. *[44]* ***Das and Gupta****: If there exist constants $k_1, k_2 \geq 0$, $0 < k_1 + k_2 < 1$, such that:*

$$d(\mathbb{T}x, \mathbb{T}y) \leq k_1 \frac{[1 + d(x, \mathbb{T}x)]d(y, \mathbb{T}y)}{1 + d(x, y)} + k_2 d(x, y), \quad \forall x, y \in X.$$

2. *[45]* ***Khan and Fisher****: If there exists* $k \in [0, 1)$ *such that:*

$$d(\mathbb{T}x, \mathbb{T}y) < k\frac{d(x, \mathbb{T}x)d(y, \mathbb{T}y) + d(x, \mathbb{T}y)d(y, \mathbb{T}x)}{d(x, \mathbb{T}y) + d(y, \mathbb{T}x)},$$

when $d(x, \mathbb{T}y) + d(y, \mathbb{T}x) \neq 0$*; otherwise,* $d(\mathbb{T}x, \mathbb{T}y) = 0$.

3. *[46]* ***Alber and Guerre-Delabriere****: If there exists a lower semi-continuous function* $\phi : [0, \infty) \to [0, \infty)$ *such that:*

$$d(\mathbb{T}x, \mathbb{T}y) \leq d(x, y) - \phi(d(x, y)), \quad \forall x, y \in X.$$

4. *[47]* ***Dutta and Choudhury****: If there exist alternating distance functions* $\vartheta, \phi : [0, \infty) \to [0, \infty)$ *such that:*

$$\vartheta(d(\mathbb{T}x, \mathbb{T}y)) \leq \vartheta(d(x, y)) - \phi(d(x, y)), \quad \forall x, y \in X.$$

Each condition accommodates fixed-point existence in contexts with specific contractivity behaviors.

Theorem 4.16 [48] (Caristi) *Caristi introduced an energy-based perspective on fixed points. For a continuous self-mapping* $\mathbb{T}$ *on a complete metric space* (X, d)*, if there exists a function* $\phi : X \to [0, \infty)$ *such that:*

$$d(x, \mathbb{T}x) \leq \phi(x) - \phi(\mathbb{T}x), \quad \forall x \in X,$$

then $\mathbb{T}$ *has a unique fixed point* x^*. *The sequence* $\{\mathbb{T}^n x\}$ *converges to* x^*. *Additionally, the convergence rate can be estimated using:*

$$\sum_{i=n}^{m-1} d(\mathbb{T}^i x, \mathbb{T}^{i+1} x) \leq \phi(\mathbb{T}^n x) - \phi(\mathbb{T}^m x).$$

Caristi's theorem broadens the fixed-point framework by incorporating a control function ϕ*, highlighting a monotonic decrease in the "behaviour" of iterations until convergence.*

While Banach contraction mapping principle assumes continuity of the mapping, many real-world problems involve discontinuous functions. This discrepancy raises the question: **"can discontinuous mappings possess fixed points under some contractive conditions?"** This question was addressed affirmatively with several theoretical advancements, as illustrated below.

Example 4.1 [49] Consider a mapping $\mathbb{T}$ on [0, 3] defined as:

$$\mathbb{T}(x) = \begin{cases} 0, & \text{if } x \in [0, 2], \\ 2, & \text{if } x \in (2, 3]. \end{cases}$$

Although $\mathbb{T}$ is discontinuous at $x = 2$, iterating $\mathbb{T}$ twice yields:

$$\mathbb{T}^2(x) = 0, \quad \forall x \in [0, 3].$$

Thus, $\mathbb{T}^2$ is a contraction with the fixed point 0, demonstrating that a mapping need not be continuous to yield contractive iterations.

Example 4.2 Let $X = [0, \infty)$, $d(x, y) = |x - y|$, and define $\mathbb{T} : X \to X$ as:

$$\mathbb{T}(x) = \begin{cases} 3, & \text{if } x \in [0, 3], \\ \frac{2x}{x+1}, & \text{if } x > 3. \end{cases}$$

While $\mathbb{T}$ is not contractive, iterating $\mathbb{T}$ twice produces:

$$\mathbb{T}^2(x) = 3, \quad \forall x \in [0, \infty).$$

Thus, $\mathbb{T}^2$ has a fixed point at 3, fulfilling contractive conditions over its iterations.

Theorem 4.17 [49] (Bryant's Theorem) *Let $\mathbb{T}$ be a self-mapping on a complete metric space (X, d). If there exists $k \in [0, 1)$ such that:*

$$d(\mathbb{T}^m x, \mathbb{T}^m y) \leq k \cdot d(x, y),$$

for all $x, y \in X$ and some positive integer m, then $\mathbb{T}$ has a unique fixed point. The condition ensures that $\mathbb{T}^m$, rather than $\mathbb{T}$ itself, is contractive, allowing for fixed points without requiring $\mathbb{T}$ to be continuous.

Example 4.3 Consider the metric space $X = [0, \infty)$ with the standard metric $d(x, y) = |x - y|$. Define the mapping $\mathbb{T} : X \to X$ as follows:

$$\mathbb{T}(x) = \begin{cases} \frac{x}{2}, & \text{if } 0 < x < 2, \\ 2 + \frac{x-2}{3}, & \text{if } 2 \leq x \leq 5, \\ 5 + \frac{x-5}{4}, & \text{if } x > 5. \end{cases}$$

The first iteration is given by:

$$\mathbb{T}(x) = \begin{cases} \frac{x}{2}, & \text{if } 0 < x < 2, \\ 2 + \frac{x-2}{3}, & \text{if } 2 \leq x \leq 5, \\ 5 + \frac{x-5}{4}, & \text{if } x > 5. \end{cases}$$

The second iteration $\mathbb{T}^2(x) = \mathbb{T}(\mathbb{T}(x))$ is:

$$\mathbb{T}^2(x) = \begin{cases} \frac{x}{4}, & \text{if } 0 < x < 2, \\ 2 + \frac{x-2}{9}, & \text{if } 2 \leq x \leq 5, \\ 5 + \frac{x-5}{16}, & \text{if } x > 5. \end{cases}$$

The third iteration $\mathbb{T}^3(x) = \mathbb{T}(\mathbb{T}^2(x))$ is:

$$\mathbb{T}^3(x) = \begin{cases} \frac{x}{8}, & \text{if } 0 < x < 2, \\ 2 + \frac{x-2}{27}, & \text{if } 2 \leq x \leq 5, \\ 5 + \frac{x-5}{64}, & \text{if } x > 5. \end{cases}$$

We compute the contractive condition for the second iteration $\mathbb{T}^2(x)$ of the mapping:

Case 1: $0 < x, y < 2$ For $\mathbb{T}^2(x) = \frac{x}{4}$:

$$d(\mathbb{T}^2(x), \mathbb{T}^2(y)) = \left|\frac{x}{4} - \frac{y}{4}\right| = \frac{|x-y|}{4}.$$

The contraction constant is:

$$k = \frac{1}{4}.$$

Case 2: $2 \leq x, y \leq 5$ For $\mathbb{T}^2(x) = 2 + \frac{x-2}{9}$:

$$d(\mathbb{T}^2(x), \mathbb{T}^2(y)) = \left|\left(2 + \frac{x-2}{9}\right) - \left(2 + \frac{y-2}{9}\right)\right| = \frac{|x-y|}{9}.$$

The contraction constant is:

$$k = \frac{1}{9}.$$

Case 3: $x, y > 5$ For $\mathbb{T}^2(x) = 5 + \frac{x-5}{16}$:

$$d(\mathbb{T}^2(x), \mathbb{T}^2(y)) = \left|\left(5 + \frac{x-5}{16}\right) - \left(5 + \frac{y-5}{16}\right)\right| = \frac{|x-y|}{16}.$$

The contraction constant is:

$$k = \frac{1}{16}.$$

We compute the contractive condition for the third iteration $\mathbb{T}^3(x)$ of the mapping:

Case 1: $0 < x, y < 2$ For $\mathbb{T}^3(x) = \frac{x}{8}$:

$$d(\mathbb{T}^3(x), \mathbb{T}^3(y)) = \left|\frac{x}{8} - \frac{y}{8}\right| = \frac{|x-y|}{8}.$$

The contraction constant is:

$$k = \frac{1}{8}.$$

Case 2: $2 \leq x, y \leq 5$ For $\mathbb{T}^3(x) = 2 + \frac{x-2}{27}$:

$$d(\mathbb{T}^3(x), \mathbb{T}^3(y)) = \left|\left(2 + \frac{x-2}{27}\right) - \left(2 + \frac{y-2}{27}\right)\right| = \frac{|x-y|}{27}.$$

The contraction constant is:

$$k = \frac{1}{27}.$$

Case 3: $x, y > 5$ For $\mathbb{T}^3(x) = 5 + \frac{x-5}{64}$:

$$d(\mathbb{T}^3(x), \mathbb{T}^3(y)) = \left|\left(5 + \frac{x-5}{64}\right) - \left(5 + \frac{y-5}{64}\right)\right| = \frac{|x-y|}{64}.$$

The contraction constant is:

$$k = \frac{1}{64}.$$

Both $\mathbb{T}^2(x)$ and $\mathbb{T}^3(x)$ satisfy the contractive condition of Theorem. 4.17. The unique fixed point of $\mathbb{T}(x)$ is:

$$x^* = 2.$$

This fixed point is guaranteed by the contractiveness of $\mathbb{T}(x)$, $\mathbb{T}^2(x)$ and $\mathbb{T}^3(x)$ as required by Theorem. 4.17.

Theorem 4.18 [50] Point-wise Contraction (Sehgal's Theorem) *Let $\mathbb{T}$ be a continuous self-mapping on a complete metric space (X, d). If for each $x \in X$, there exists $n(x) \in \mathbb{N}$ such that:*

$$d(\mathbb{T}^{n(x)}(x), \mathbb{T}^{n(x)}(y)) \leq k \cdot d(x, y),$$

for some $k \in [0, 1)$ and all $y \in X$, then $\mathbb{T}$ has a unique fixed point x^*, and $\{\mathbb{T}^n x\}$ converges to x^*.

Example 4.4 Consider the metric space $X = [0, 1]$ with the standard metric $d(x, y) = |x - y|$, and define the mapping $\mathbb{T} : X \to X$ as:

$$\mathbb{T}(x) = \begin{cases} \frac{x}{2}, & \text{if } x \in [0, \frac{1}{2}], \\ \frac{x+1}{6}, & \text{if } x \in (\frac{1}{2}, 1]. \end{cases}$$

We aim to show that $\mathbb{T}(x)$ satisfies Theorem 4.18 by verifying:

1. $\mathbb{T}(x)$ is continuous on X.
2. $\mathbb{T}(x)$ satisfies the contractive condition for $n(x) = 2, 3$.
3. $\mathbb{T}(x)$ has a unique fixed point x^*, and the sequence $\mathbb{T}^n(x)$ converges to x^* for all $x \in X$.

Proof Now, we would like to check the continuity of $\mathbb{T}(x)$:

Case 1:
On $[0, \frac{1}{2}]$:
For $x \in [0, \frac{1}{2}]$, $\mathbb{T}(x) = \frac{x}{2}$, which is a linear function and therefore continuous on this interval.

Case 2:
On $(\frac{1}{2}, 1]$:
For $x \in (\frac{1}{2}, 1]$, $\mathbb{T}(x) = \frac{x+1}{6}$, which is also a linear function and hence continuous on this interval.

Case 3:
At $x = \frac{1}{2}$:

$$\text{Left-hand limit:} \quad \lim_{x \to \frac{1}{2}^-} \mathbb{T}(x) = \frac{\frac{1}{2}}{2} = \frac{1}{4},$$

$$\text{Right-hand limit:} \quad \lim_{x \to \frac{1}{2}^+} \mathbb{T}(x) = \frac{\frac{1}{2} + 1}{6} = \frac{\frac{3}{2}}{6} = \frac{1}{4},$$

$$\text{Value at } x = \frac{1}{2}\text{:} \quad \mathbb{T}\left(\frac{1}{2}\right) = \frac{\frac{1}{2}}{2} = \frac{1}{4}.$$

Since the left-hand limit, right-hand limit, and $T\left(\frac{1}{2}\right)$ agree, $\mathbb{T}(x)$ is continuous at $x = \frac{1}{2}$.

Thus, $\mathbb{T}(x)$ is continuous on $[0, 1]$.

The second iteration is:

$$\mathbb{T}^2(x) = \mathbb{T}(\mathbb{T}(x)) = \begin{cases} \frac{x}{4}, & \text{if } x \in [0, \frac{1}{2}], \\ \frac{x+1}{12}, & \text{if } x \in (\frac{1}{2}, 1]. \end{cases}$$

The third iteration is:

$$\mathbb{T}^3(x) = \mathbb{T}(\mathbb{T}^2(x)) = \begin{cases} \frac{x}{8}, & \text{if } x \in [0, \frac{1}{2}], \\ \frac{x+1}{24}, & \text{if } x \in (\frac{1}{2}, 1]. \end{cases}$$

The mappings $\mathbb{T}^2(x)$ and $\mathbb{T}^3(x)$ satisfy the contractive condition:

$$d(\mathbb{T}^{n(x)}(x), \mathbb{T}^{n(x)}(y)) \leq k \cdot d(x, y), \quad \forall x, y \in X,$$

with $k \in [0, 1)$ for $n(x) = 2, 3$.

For $\mathbb{T}^2(x)$:

- $x, y \in [0, \frac{1}{2}]$: $d(\mathbb{T}^2(x), \mathbb{T}^2(y)) = \frac{|x-y|}{4}$, with $k = \frac{1}{4}$.
- $x, y \in (\frac{1}{2}, 1]$: $d(\mathbb{T}^2(x), \mathbb{T}^2(y)) = \frac{|x-y|}{12}$, with $k = \frac{1}{12}$.
- $x \in [0, \frac{1}{2}]$, $y \in (\frac{1}{2}, 1]$: Bounded by $k = \frac{1}{4}$.

For $\mathbb{T}^3(x)$:

- $x, y \in [0, \frac{1}{2}]$: $d(\mathbb{T}^3(x), \mathbb{T}^3(y)) = \frac{|x-y|}{8}$, with $k = \frac{1}{8}$.
- $x, y \in (\frac{1}{2}, 1]$: $d(\mathbb{T}^3(x), \mathbb{T}^3(y)) = \frac{|x-y|}{24}$, with $k = \frac{1}{24}$.
- $x \in [0, \frac{1}{2}]$, $y \in (\frac{1}{2}, 1]$: Bounded by $k = \frac{1}{8}$.

To find the fixed point x^*, solve $\mathbb{T}(x^*) = x^*$:

- For $x^* \in [0, \frac{1}{2}]$:

$$\mathbb{T}(x^*) = \frac{x^*}{2} = x^* \implies x^* = 0.$$

- For $x^* \in (\frac{1}{2}, 1]$:

$$\mathbb{T}(x^*) = \frac{x^* + 1}{6} = x^* \implies 6x^* = x^* + 1 \implies 5x^* = 1 \implies x^* = \frac{1}{5}.$$

  Since $x^* = \frac{1}{5} \notin (\frac{1}{2}, 1]$, there is no fixed point in this interval.

Thus, the unique fixed point is:

$$x^* = 0.$$

Thus, the mapping $\mathbb{T}(x)$, and its iterations $\mathbb{T}^2(x)$ and $\mathbb{T}^3(x)$, satisfies Theorem 4.18.

Example 4.5 Let $X = [0, \infty)$ with the usual metric $d(x, y) = |x - y|$, and define $\mathbb{T} : X \to X$ by:

$$\mathbb{T}(x) = \begin{cases} \frac{x}{2}, & \text{if } x \leq 2, \\ 1 + \frac{x-2}{3}, & \text{if } 2 < x \leq 5, \\ 2 + \frac{x-5}{4}, & \text{if } x > 5. \end{cases}$$

Then the mapping $\mathbb{T}$ satisfies the Sehgal's Theorem 4.18.

Theorem 4.19 [22] (Guseman's Theorem) *Let* $\mathbb{T}$ *be a self-mapping on a complete metric space* (X, d)*. Assume there exists a subset* $B \subset X$ *such that:*

1. $\mathbb{T}(B) \subset B$,
2. *For some* $k \in [0, 1)$ *and every* $y \in B$*, there exists an integer* $n(y) \geq 1$ *such that:*

$$d(\mathbb{T}^{n(y)}(x), \mathbb{T}^{n(y)}(y)) \leq k \cdot d(x, y), \quad \forall x \in B,$$

3. *For some* $x_0 \in B$*, the closure of the set* $\{\mathbb{T}^n(x_0) : n \geq 1\}$ *is contained in* B.

Then for any $y_0 \in B$*, the sequence* $\mathbb{T}^n(y_0)$ *converges to a unique fixed point* $x^* \in B$*. Moreover, if* $d(\mathbb{T}^{n(x^*)}(x), \mathbb{T}^{n(x^*)}(x^*)) \leq k \cdot d(x, x^*)$ *for all* $x \in X$*, then* x^* *is the unique fixed point of* $\mathbb{T}$*, and* $\mathbb{T}^n(x_0) \to x^*$ *for every* $x_0 \in X$.

Theorem 4.20 [22] (Guseman's Uniqueness Theorem) *Let* $\mathbb{T}$ *be a self-mapping on a complete metric space* (X, d)*. Assume* $x^*, x_0 \in X$ *such that* $\mathbb{T}^n(x_0) \to x^*$*. If* $\mathbb{T}$ *satisfies the contractive condition at* x^**, then* x^* *is the unique fixed point of* $\mathbb{T}$*, and* $\mathbb{T}^n(y) \to x^*$ *for all* $y \in X$.

Theorem 4.21 [51] (Matkowski's Theorem) *Let* $\mathbb{T}$ *be a self-mapping on a complete metric space* (X, d)*. Assume there exists an auxiliary function* $\alpha : [0, \infty)^5 \to [0, \infty)$ *such that:*

1. $\gamma(t) = \alpha(t, t, t, 2t, 2t)$*, and* $\lim_{n\to\infty} \gamma^n(t) = 0$ *for all* $t > 0$,
2. $\lim_{t\to\infty} (t - \gamma(t)) = \infty$,
3. α *is nondecreasing in each argument,*
4. *For every* $x \in X$*, there exists* $n(x) \in \mathbb{N}$ *such that for all* $y \in X$:

$$d(\mathbb{T}^{n(x)}(x), \mathbb{T}^{n(x)}(y)) \leq \psi_{\mathbb{T}}(x, y),$$

where:

$$\psi_{\mathbb{T}}(x, y) = \alpha\big(d(x, \mathbb{T}^{n(x)}(x)), d(x, \mathbb{T}^{n(x)}(y)), d(x, y), d(\mathbb{T}^{n(x)}(x), y), d(\mathbb{T}^{n(x)}(y), y)\big).$$

Then for any $x \in X$*, the sequence* $\{\mathbb{T}^n(x)\}$ *converges to a unique fixed point* x^* *of* $\mathbb{T}$.

Theorem 4.22 [52] (Ray and Rhoades) *Let $\mathbb{T}$ and $\mathbb{S}$ be self-mappings on a complete metric space (X, d). Assume there exists a constant $k \in (0, 1)$ such that for each $x, y \in X$, there exist integers $n(x), m(y) \in \mathbb{N}$ satisfying:*

$$d(\mathbb{T}^{n(x)}(x), \mathbb{S}^{m(y)}(y)) \leq k.\max\left\{\begin{array}{l} d(x, y), d(x, \mathbb{T}^{n(x)}(x)), d(y, \mathbb{S}^{m(y)}(y)), \\ \frac{d(x,\mathbb{S}^{m(y)}(y))+d(y,\mathbb{T}^{n(x)}(x))}{2} \end{array}\right\}.$$

Then $\mathbb{T}$ and $\mathbb{S}$ have a unique common fixed point $x^* \in X$.

Theorem 4.23 [53] (Singh) *Let $\mathbb{T}$ be a self-mapping on a complete metric space (X, d). Assume there exists $n(x) \in \mathbb{N}$ for each $x \in X$, such that for every $y \in X$:*

$$\begin{aligned} d(\mathbb{T}^{n(x)}(x), \mathbb{T}^{n(x)}(y)) \leq & q(x, y) \cdot d(x, y) + r(x, y) \cdot d(x, \mathbb{T}^{n(x)}(x)) + s(x, y) \cdot d(y, \mathbb{T}^{n(x)}(y)) \\ & + t(x, y) \cdot d(x, \mathbb{T}^{n(x)}(y)) + m(x, y) \cdot d(y, \mathbb{T}^{n(x)}(x)), \end{aligned}$$

where q, r, s, t, m are nonnegative functions satisfying:

$$\sup_{x,y\in X} \left\{2(m(x, y) + s(x, y) + t(x, y)) + q(x, y) + r(x, y)\right\} < 1.$$

Then for any $x_0 \in X$, the sequence $\mathbb{T}^n(x_0)$ converges to a unique fixed point x^* of $\mathbb{T}$.

Theorem 4.24 [54] (Ćirić) *Let $\mathbb{T}$ be a self-mapping on a complete metric space (X, d). Assume there exists $k \in (0, 1)$ such that for every $x \in X$, there exists $n(x) \in \mathbb{N}$ satisfying:*

$$d(\mathbb{T}^{n(x)}(x), \mathbb{T}^{n(x)}(y)) \leq k \cdot \max\left\{d(x, y), d(x, \mathbb{T}(y)), d(y, \mathbb{T}(x)), \ldots, d(x, \mathbb{T}^{n(x)}(y)), d(y, \mathbb{T}^{n(x)}(x))\right\}.$$

Then for any $x \in X$, the sequence $\{\mathbb{T}^n(x)\}$ converges to a unique fixed point x^*.

Theorem 4.25 [55] (Kincses-Totik) *Let $\mathbb{T}$ be a continuous self-mapping on a compact metric space (X, d). Assume that for every $x \in X$, there exists $p(x) \in \mathbb{N}$ such that for all $y \in X$, $y \neq x$:*

$$d(\mathbb{T}^{p(x)}(x), \mathbb{T}^{p(x)}(y)) < \max\left\{d(x, \mathbb{T}^{p(x)}(y)), d(y, \mathbb{T}^{p(x)}(x)), d(x, y)\right\}.$$

Then for any $x \in X$, the sequence $\{\mathbb{T}^n(x)\}$ converges to a unique fixed point x^*.

Theorem 4.26 [55] (Kincses-Totik) *Let $\mathbb{T}$ be a self-mapping on a complete metric space (X, d). Assume there exists $\alpha \in (0, 1)$ such that for every $x \in X$, there exists $p(x) \in \mathbb{N}$ satisfying:*

$$d(\mathbb{T}^{p(x)}(x), \mathbb{T}^{p(x)}(y)) \leq \alpha \cdot \max\left\{d(x, \mathbb{T}^{p(x)}(y)), d(y, \mathbb{T}^{p(x)}(x)), d(x, y)\right\}.$$

Then for any $x \in X$, the sequence $\{\mathbb{T}^n(x)\}$ converges to a unique fixed point x^*.

Theorem 4.27 [55] (Kincses-Totik) *Let $\mathbb{T}$ be a self-mapping on a complete metric space (X, d). Assume there exists a monotonic decreasing function $\alpha : (0, \infty) \to [0, 1)$, and for every $x \in X$, there exists $n(x) \in \mathbb{N}$ such that for all $y \neq x$:*

$$d(\mathbb{T}^{n(x)}(y), \mathbb{T}^{n(x)}(x)) < \alpha(d(y, x)).$$

Then for any $x \in X$, the sequence $\{\mathbb{T}^n(x)\}$ converges to a unique fixed point x^*.

Theorem 4.28 [56] (Jachymski) *Let $\mathbb{T}$ be a self-mapping on a complete metric space (X, d). Assume there exists a function $p : X \to \mathbb{N}$ and a mapping $\psi : [0, \infty) \to [0, \infty)$ satisfying:*

1. $\psi(t) < t$, *for all* $t > 0$,
2. $\lim_{t\to\infty} (t - \psi(t)) = \infty$.

If for every $x, y \in X$:

$$d(\mathbb{T}^{p(x)}(x), \mathbb{T}^{p(x)}(y)) \leq \psi\left(\max\{d(x, \mathbb{T}^j(y)), d(\mathbb{T}^{p(x)}(x), \mathbb{T}^j(y)) \mid j = 0, 1, 2, \ldots, p(x)\}\right),$$

then $\mathbb{T}$ has a unique fixed point x^*, and $\mathbb{T}^n(x) \to x^*$ for any $x \in X$.

Corollary 4.1 *[56] Let $\mathbb{T}$ be a self-mapping on a complete metric space (X, d). Assume there exists a nondecreasing function $\psi : [0, \infty) \to [0, \infty)$ such that:*

1. $\psi(t) < t$, *for all* $t > 0$,
2. $\lim_{t\to\infty} (t - \psi(t)) = \infty$.

If for every $x, y \in X$:

$$d(\mathbb{T}(x), \mathbb{T}(y)) \leq \psi(d(x, y)),$$

then $\mathbb{T}$ has a unique fixed point x^*, and the sequence $\{\mathbb{T}^n(x)\}$ converges to x^* for any $x \in X$.

Corollary 4.2 *[56] Let $\mathbb{T}$ be a self-mapping on a complete metric space (X, d), and let $p : X \to \mathbb{N}$ be a function satisfying:*

1. *$p(x)$ is nondecreasing,*
2. *For every $t > 0$, $\psi(t) < t$,*
3. $\lim_{t\to\infty} (t - \psi(t)) = \infty$.

If for any $x, y \in X$:

$$d(\mathbb{T}^{p(x)}(x), \mathbb{T}^{p(x)}(y)) \leq \psi\big(\max\{d(x, \mathbb{T}^j(y)), d(\mathbb{T}^{p(x)}(x), \mathbb{T}^j(y)) \mid j = 0, 1, 2, \ldots, p(x)\}\big),$$

then $\mathbb{T}$ has a unique fixed point x^*, and for any $x \in X$, $\{\mathbb{T}^n(x)\}$ converges to x^*.

4.3 Selected Key Contractive Mappings in Modern Era

The given below provides a detailed overview of generalized contractions, separate contractions, large contractions and F-contractions, within the context of metric space (X, d).

Definition 4.1 Consider a complete metric space (X, d). A mapping $\mathbb{T} : X \to X$ is called:

- **Generalized contraction** [57] if there exists a function $\Upsilon : X \times X \to [0, \infty)$ in such a way that,

  (Υ1) $\forall\, 0 < a \leq b,\ \sup_{a \leq d(x,y) \leq b} \Upsilon(x, y) < 1$;
  (Υ2) $\forall\, x, y \in X,\ d(\mathbb{T}x, \mathbb{T}y) \leq \Upsilon(x, y)d(x, y)$.

  - The proposed structure is more versatile than the standard contraction since it allows the contraction degree to change with point distance, enabling additional mappings.

- **Separate contraction** [42] if there exist two functions $\varphi, \psi : \mathbb{R}^+ \to \mathbb{R}^+$ satisfying the below stated conditions,

  1. $\psi(0) = 0$ and ψ is strictly increasing,
  2. $\forall\, x, y \in X,\ d(\mathbb{T}(x), \mathbb{T}(y)) \leq \varphi(d(x, y))$,
  3. $\forall\, r > 0,\ \psi(r) \leq r - \varphi(r)$.

- **Large contraction** [42, 58], for every $\delta > 0$ there exists $\mu \in (0, 1)$ in such a way that $\forall x, y \in X, d(\mathbb{T}x, \mathbb{T}y) \leq \mu d(x, y)$ for all $d(x, y) \geq \delta$.

  - Burton [42] constructed *large contractions*, a particular category of generalized contractions that substantially reduce the distance between relatively far apart.

Definition 4.2 [59] An F**-contraction** is a mapping $\mathbb{T} : X \to X$ satisfying:

$$\tau + F(d(\mathbb{T}x, \mathbb{T}y)) \leq F(d(x, y))$$

$\forall\ x, y \in X$ with $\mathbb{T}x \neq \mathbb{T}y$, where $\tau > 0$ and $F : (0, \infty) \to \mathbb{R}$ gratifies the below stated assertions:

(F1) $\forall \theta_1, \theta_2 > 0, \theta_1 > \theta_2$ implies $F(\theta_1) > F(\theta_2)$.
(F2) For any $(\theta_n) \subset (0, \infty)$, $\theta_n \to 0$ if and only if $F(\theta_n) \to -\infty$.
(F3) There is $k \in (0, 1)$ satisfying $\lim_{\theta \to 0^+} \theta^k F(\theta) = 0$.

Theorem 4.29 [59] *Let (X, d) be a complete metric space and let $\mathbb{T} : X \to X$ be a F-contraction. Then $\mathbb{T}$ has a unique fixed point.*

The function F drives the contractive characteristic in a nonlinear structure, permitting complex connections among the points within the metric space. The fixed-point theory continues to grow by the fact that every F-contraction on the complete metric space possesses a unique fixed point.

Turinici [60] found that the constraint $(F2)$ can be dropped to:

$(F2')$ $\lim_{\boldsymbol{\theta} \to 0^+} F(\boldsymbol{\theta}) = -\infty$.

From this, the implication:

$(F2'')$ $F(\boldsymbol{\theta}_n) \to -\infty \implies \boldsymbol{\theta}_n \to 0$,

can be derived from $(F1)$.

Expanding upon these notions, D. Wardowski [61] have presented the notion of $(\boldsymbol{\varphi}, F)$-**contraction (or nonlinear F-contraction)**. These mappings improve F-contraction by putting in another function, $\boldsymbol{\varphi} : (0, \infty) \to (0, \infty)$, which additionally develops the contraction criteria.

Definition 4.3 [61] A mapping $\mathbb{T} : X \to X$ is a (φ, F)**-contraction (or nonlinear F-contraction)** if there exist functions $F : (0, \infty) \to \mathbb{R}$ and $\boldsymbol{\varphi} : (0, \infty) \to (0, \infty)$ satisfying:

(H1) F gratifies $(F1)$ and the relaxed condition $(F2')$.
(H2) $\liminf_{s \to \boldsymbol{\theta}^+} \boldsymbol{\varphi}(s) > 0$, for all $\boldsymbol{\theta} \geq 0$.
(H3) $\forall x, y \in X$, $\boldsymbol{\varphi}(d(x, y)) + F(d(\mathbb{T}x, \mathbb{T}y)) \leq F(d(x, y))$ in such a way that $\mathbb{T}x \neq \mathbb{T}y$.

Theorem 4.30 [61] *Let (X, d) be a complete metric space and let $\mathbb{T} : X \to X$ be a (φ, F)-**contraction (or nonlinear F-contraction)**. Then $\mathbb{T}$ has a unique fixed point.*

Define Φ as the collection of mappings $\theta : (0, \infty) \to (1, \infty)$ that fulfill the subsequent criteria:

(Φ_1) The function θ is monotonic in its domain;
(Φ_2) If a sequence $\{s_n\} \subset (0, \infty)$ satisfies $\lim_{n\to\infty} \theta(s_n) = 1$, then $\lim_{n\to\infty} s_n = 0^+$, and vice versa;
(Φ_3) There are constants $r \in (0, 1)$ and $\lambda \in (0, \infty]$ such that $\lim_{t\to 0^+} \frac{\theta(t)-1}{t^r} = \lambda$;
(Φ_4) The function θ is continuous.

In a recent development, Jleli and Samet [62] presented an extension of the Banach fixed-point theorem within the framework of Branciari metric spaces.

Definition 4.4 [62] A self-mapping $\mathbb{T}$ on Branciari metric space (X, d) is called a JS-**contraction** if there exists a function $\theta \in \Phi$ such that:

$$\forall\, x, y \in X,\ d(\mathbb{T}x, \mathbb{T}y) \neq 0 \Rightarrow \theta(d(\mathbb{T}x, \mathbb{T}y)) \leq [\theta(d(x, y))]^r .$$

Theorem 4.31 [62] (Jleli and Samet) *Assume (X, d) is a complete generalized metric space (g.m.s), and let $\mathbb{T} : X \to X$ be a mapping. Suppose there exists a function $\theta \in \Phi$ and a constant $k \in (0, 1)$ such that:*

$$\forall\, x, y \in X,\ d(\mathbb{T}(x), \mathbb{T}(y)) \neq 0 \implies \theta(d(\mathbb{T}(x), \mathbb{T}(y))) \leq [\theta(d(x, y))]^k .$$

Under these conditions, $\mathbb{T}$ has exactly one fixed point.

Definition 4.5 [63] Let A and B be two non-empty subsets within a metric space (X, d). A mapping $\mathbb{T} : A \cup B \to A \cup B$ is referred to as a **cyclic operator** if the image of A under $\mathbb{T}$ is entirely contained within B, and the image of B under $\mathbb{T}$ lies completely within A.

Definition 4.6 [63] For a metric space (X, d) and subsets A and B that are non-empty, a cyclic operator $\mathbb{T} : A \cup B \to A \cup B$ is called a **cyclic contraction** if there exists a constant k such that $0 < k < 1$, and for every pair $x \in A$ and $y \in B$, the inequality

$$d(\mathbb{T}x, \mathbb{T}y) \leq k \cdot d(x, y)$$

is satisfied.

Theorem 4.32 [63] *Let $\{A_i\}_{i=1}^{p}$ represent a collection of non-empty, closed subsets within a complete metric space. Suppose that a mapping*

$$\mathbb{T} : \bigcup_{i=1}^{p} A_i \to \bigcup_{i=1}^{p} A_i$$

satisfies the following conditions, where $A_{p+1} = A_1$:

- *The image of each set under $\mathbb{T}$ is entirely contained within the next set in the sequence:*

$$\mathbb{T}(A_i) \subseteq A_{i+1}, \quad \textit{for } 1 \leq i \leq p.$$

- *There exists a constant k such that $0 \leq k < 1$ and:*

$$d(\mathbb{T}x, \mathbb{T}y) \leq k \cdot d(x, y),$$

  for all $x \in A_i$ and $y \in A_{i+1}$, where $1 \leq i \leq p$.

Under these conditions, $\mathbb{T}$ possesses exactly one fixed point.

The hypothesis provides a comprehensive framework for understanding these contraction types, revealing that numerous classical contraction mappings are special cases within this expanded category. This perspective delivers a cohesive and adaptable basis for fixed point theory in metric spaces.

4.3.1 Nonlinear Contractive Mappings: A Few Non-implications

Example 4.6 The set $Z = \{z_n := \frac{1}{n} \ : n \in \mathbb{N}\} \cup \{0\}$ with the metric $d(z, z^*) = |z - z^*|, \ z, z^* \in Z$ is a complete metric space. Define the mapping $\mathbb{T} : Z \to Z$ as follows:

$$\mathbb{T}(z_n) := z_{n+1} \text{ and } \mathbb{T}(0) = 1 \quad \text{for all } n \geq 1.$$

Then $\mathbb{T}$ is both a (φ, F)-contraction and a generalized contraction.

For our convenience, on a contrary, let us to prove $\mathbb{T}$ is not a generalized contraction.

By a routine computation, we begin with considering the distances:

$$|z_{n+1} - z_n| = \left| \frac{1}{n+1} - \frac{1}{n} \right| = \frac{1}{n(n+1)}.$$

For large n:

$$\frac{1}{n(n+1)} < \frac{1}{n^2}.$$

Suppose there exist a function Υ satisfying $(\Upsilon 2)$:

$$\Upsilon(z_{n+1}, z_n) \geq \frac{|\mathbb{T}(z_{n+1}) - \mathbb{T}(z_n)|}{|z_{n+1} - z_n|}, \ \forall \ z_{n+1}, z_n \in Z.$$

For $n > 0$:

$$\Upsilon(z_{n+1}, z_n) \geq \frac{|z_{n+2} - z_{n+1}|}{|z_{n+1} - z_n|} = \frac{\frac{1}{(n+1)(n+2)}}{\frac{1}{n(n+1)}} = \frac{n}{n+2} \approx 1.$$

Therefore:

$$\sup_{0 < d(z, z^*) \leq \frac{1}{2}} \Upsilon(z, z^*) \geq 1.$$

This contradicts $(\Upsilon 1)$ since $\sup \Upsilon(z, z^*)$ cannot be greater than 1. Therefore, as per the routine computation, $\mathbb{T}$ is not a generalized contraction.

However, this is not a sufficient logic to conclude $\mathbb{T}$ is not a generalized contraction.

The definition of a generalized contraction requires the existence of an interval $[a, b] \subset (0, \infty)$ such that

$$\sup \left\{ \Gamma(z, z^*) : a \leq d(z, z^*) \leq b \right\} < 1$$

for a function Γ satisfying $d(\mathbb{T}z, \mathbb{T}z^*) \leq \Gamma(z, z^*)d(z, z^*)$. Thus, to disprove that $\mathbb{T}$ is a generalized contraction, it is necessary to show that no such interval exists, which the above routine argument does not do.

We proceed as follows:

Fix any $a > 0$, and choose $p \in \mathbb{N}$ such that $\frac{1}{p} \leq a$. Consider $z = \frac{1}{n}$, $z^* = \frac{1}{m} \in Z$ with $n < m$ and $d(z, z^*) \geq a$. Then $\frac{1}{n} - \frac{1}{m} \geq \frac{1}{p}$, which implies $n < p$. Hence,

$$\frac{d(\mathbb{T}z, \mathbb{T}z^*)}{d(z, z^*)} = \frac{nm}{(n+1)(m+1)} = \frac{n}{n+1} \cdot \frac{m}{m+1} < \frac{p}{p+1},$$

since $\frac{n}{n+1} < \frac{p}{p+1}$ and $\frac{m}{m+1} < 1$. Thus,

$$d(\mathbb{T}z, \mathbb{T}z^*) \leq \frac{p}{p+1} \cdot d(z, z^*)$$

whenever $d(z, z^*) \geq a$. Since $\frac{p}{p+1} < 1$, this verifies that $\mathbb{T}$ is indeed a generalized contraction.

The initial computation suggesting otherwise is insufficient, and a correct conclusion requires the type of argument provided above.

Next, we show that $\mathbb{T}$ is a (φ, F)-contraction.

Let $\varphi(t) = \frac{1}{t+1}$ and $F(t) = -\frac{1}{t}$ for all $t > 0$. For distinct elements $z = \frac{1}{n}$, $y = \frac{1}{m}$ with $n < m$, we have

$$d(z, z^*) = \left|\frac{1}{n} - \frac{1}{m}\right| = \frac{m-n}{nm}, \quad d(\mathbb{T}z, \mathbb{T}z^*) = \left|\frac{1}{n+1} - \frac{1}{m+1}\right| = \frac{m-n}{(n+1)(m+1)}.$$

Then,

$$F(d(z, z^*)) = -\frac{nm}{m-n}, \quad F(d(\mathbb{T}z, \mathbb{T}z^*)) = -\frac{(n+1)(m+1)}{m-n}, \quad \varphi(d(z, z^*)) = \frac{1}{\frac{m-n}{nm} + 1} = \frac{nm}{nm + (m-n)}.$$

Thus,

$$\varphi(d(z, z^*)) + F(d(\mathbb{T}z, \mathbb{T}z^*)) = \frac{nm}{nm + (m-n)} - \frac{(n+1)(m+1)}{m-n} \leq -\frac{nm}{m-n} = F(d(z, z^*)),$$

which confirms the (φ, F)-contraction condition. Thus, this example shows that $\mathbb{T}$ is a $(\boldsymbol{\varphi}, F)$-contraction for the functions $\varphi(t) = \frac{1}{t+1}$, $F(t) = -\frac{1}{t}$. Therefore, $\mathbb{T}$ is a $(\boldsymbol{\varphi}, F)$-contraction as well as a generalized contraction.

Example 4.7 [64] Let $\mathscr{M} = [0, \infty)$ be a metric space with the usual distance $\mathscr{D}(x, y) = |x - y|$. Define the function $\mathscr{H} : \mathscr{M} \to \mathscr{M}$ by

$$\mathscr{H}(x) = \frac{x^2}{1+x}.$$

We show that $\mathscr{H}$ is a separate contraction mapping, but not a large contraction mapping.

To verify that $\mathscr{H}$ is a separate contraction, we define suitable functions $\varphi, \psi : \mathbb{R}^+ \to \mathbb{R}^+$, depending on the minimum of the pair $\{x, y\}$.

For any $x, y \in \mathscr{M}$, let $r = \mathscr{D}(x, y)$ and $n \in \mathbb{N}$ be such that $n - 1 < \min\{x, y\} \leq n$. Then define

$$\varphi(r) = \begin{cases} \dfrac{r\left(n^2 + 2n + (n+1)r\right)}{(n+1)^2 + (n+1)r}, & \text{if } \min\{x, y\} \in (n-1, n], \\ \dfrac{r(3+2r)}{4+2r}, & \text{if } \min\{x, y\} = 0. \end{cases}$$

Then set

$$\psi(r) = r - \varphi(r) = \begin{cases} \dfrac{r}{(n+1)^2 + (n+1)r}, & \text{if } \min\{x, y\} \in (n-1, n], \\ \dfrac{r}{4+2r}, & \text{if } \min\{x, y\} = 0. \end{cases}$$

We observe that $\psi(0) = 0$, and ψ is strictly increasing. Also, for any $x, y \in \mathscr{M}$, a direct computation shows:

$$|\mathscr{H}(x) - \mathscr{H}(y)| \leq \varphi(|x - y|),$$

and hence,

$$\psi(|x - y|) \leq |x - y| - |\mathscr{H}(x) - \mathscr{H}(y)|.$$

Thus, $\mathscr{H}$ satisfies the conditions of a separate contraction mapping as given in Definition 2.5.

Next, we prove that $\mathscr{H}$ is not a large contraction mapping. Suppose, for contradiction, that it is. Then, for every $\delta > 0$, there exists $0 < k < 1$ such that

$$|\mathscr{H}(x) - \mathscr{H}(y)| \leq k|x - y|, \quad \text{for all } x, y \in \mathscr{M} \text{ with } |x - y| \geq \delta.$$

Set $y = 0$ and choose $x > \max\{\delta, \frac{k}{1-k}\}$. Then

$$|\mathscr{H}(x) - \mathscr{H}(0)| = \mathscr{H}(x) = \frac{x^2}{1+x} = x \cdot \frac{x}{1+x} > kx = k|x - y|,$$

which contradicts the above inequality. Therefore, $\mathscr{H}$ is not a large contraction.

Thus, the function $\mathscr{H}(x) = \frac{x^2}{1+x}$ is a separate contraction mapping on $[0, \infty)$, with explicitly defined comparison function

$$\psi(r) = \begin{cases} \dfrac{r}{(n+1)^2 + (n+1)r}, & \text{if } n - 1 < \min\{x, y\} \leq n, \\ \dfrac{r}{4+2r}, & \text{if } \min\{x, y\} = 0, \end{cases}$$

but it is not a large contraction mapping.

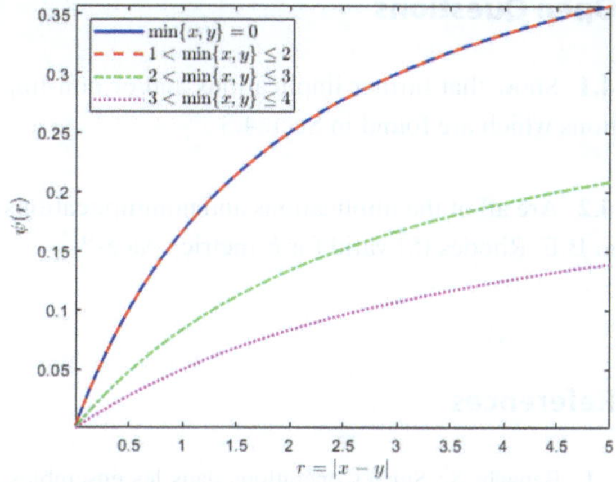

Fig. 4.1 Growth of $\psi(r)$ for various values of $\min\{x, y\}$

This example illustrates that the class of separate contractions properly includes the class of large contractions.

Figure 4.1 illustrates the growth of the function $\psi(r)$ as defined in Example 4.7 for different intervals of $\min\{x, y\}$:

- $\min\{x, y\} = 0$
- $\min\{x, y\} \in (1, 2]$
- $\min\{x, y\} \in (2, 3]$
- $\min\{x, y\} \in (3, 4]$

As observed:

- Each curve $\psi(r)$ is strictly increasing with respect to r.
- For a fixed r, the value of $\psi(r)$ decreases as $\min\{x, y\}$ increases.
- This behavior demonstrates that $\psi(r)$ satisfies the required conditions for a separate contraction mapping.

Thus, the plotted functions support the theoretical claim that $\mathscr{H}(x) = \frac{x^2}{1+x}$ is indeed a separate contraction mapping, although it is not a large contraction mapping.

Open Questions

4.1 Show that further implications and/or non-implications among the contractive conditions which are found in Sect. 4.3.

4.2 Are all of the implications and nonimplications of the contractive conditions described in B.E. Rhodes [8] valid for b-metric spaces?

References

1. Banach, S.: Sur les opérations dans les ensembles abstraits et leur application aux équations intégrales. Fund. Math. **3**, 133–181 (1922)
2. Bonsall, F.F.: Lectures on Some Fixed-point Theorems of Functional Analysis. Tata Inst. Fund. Res. Stud. Math, Mumbai (1962)
3. Brouwer, L.E.J.: Über Abbildungen von Mannigfaltigkeiten. Math. Ann. **71**, 97–115 (1912)
4. Palais, R.S.: A simple proof of the Banach contraction principle. J. Fixed Point Theory Appl. **2**, 221–223 (2007). https://doi.org/10.1007/s11784-007-0041-6
5. Bessaga, C.: On the converse of Banach "fixed-point principle". Colloq. Math. **7**(1), 41–43 (1959)
6. Meyers, P.R.: A converse to Banach's contraction theorem. J. Res. Natl. Bur. Stand. Sect. B Math. Math. Phys. **71B**, 2 and 3, 73 (1967)
7. Daskalakis, C., Tzamos, C., Zampetakis, M.: A Converse to Banach's fixed point theorem and its CLS-completeness. In: Proceedings of 50th Annual ACM SIGACT Symposium on the Theory of Computing (STOC'18). ACM, New York, NY, USA, 7 pages. https://doi.org/10.1145/3188745.3188968
8. Rhoades, B.E.: A comparison of various definitions of contractive mappings. Trans. Am. Math. Soc. **266**, 257–290 (1977). https://doi.org/10.1090/S0002-9947-1977-0433430-4
9. Rakotch, E.: A note on contractive mappings Proc. Am. Math. Soc. **13**, 459–465 (1962)
10. Edelstein, M.: On fixed and periodic points under contractive mappings. J. Lond. Math. Soc. **37**, 74–79 (1962)
11. Kannan, R.: Some results on fixed points-II. Am. Math. Mon. **76**, 405–408 (1969)
12. Bianchini, R.M.T.: Su un problema di S. Reich riguardante la teoria dei punti fissi Bollettino U.M.I. **5**(4), 103–108 (1972)
13. Reich, S.: Some remarks concerning contraction mappings. Canad. Math. Bull. **14**(1), 121–124 (1971)
14. Reich, S.: Kannan's fixed point theorem. Bollettino U.M.I. **4**(4), 1–11 (1971)
15. Sehgal, V.M.: On fixed and periodic points for a class of mappings. J. London Math. Soc. **5**(2), 571–576 (1972)
16. Chatterjea, S.K.: Fixed point theorems. Comptes rendus de l'Académie bulgare des Sciences **6**, 727–730 (1972)
17. Hardy, G.E., Rogers, T.D.: A generalisation of a fixed point theorem of Reich. Canad. Math. Bull. **16**(2), 201–206 (1973)
18. Zamfirescu, T.: Fix point theorems in metric spaces. Arch. Math. (Basel) **23**, 292–298 (1972)
19. Ćirić, L.B.: Generalized contractions and fixed-point theorems. Publ. Inst. Math. (Beograd) (N.S.) **12**(26), 19–26 (1971). MR 46 #8203

20. Ćirić, L.B.: A generalization of Banach's contraction principle. Proc. Am. Math. Soc. **45**, 267–273 (1974)
21. Bailey, D.F.: Some theorems on contractive mappings. J. London Math. Soc. **41**, 101–106 (1966). MR 32 #6434
22. Guseman, L.F., Jr.: Fixed point theorems for mappings with a contractive iterate at a point. Proc. Amer. Math. Soc. **26**, 615–618 (1970). MR 42 #919
23. Yen, C.L.: Remark on common fixed points. Tamkang J. Math. **3**, 95–96 (1972). MR 48 #1206
24. Singh, S.P.: Some results on fixed point theorems. Yokahama Math. J. **17**(6), 1–64 (1969). MR 41 #2488
25. Sehie, P., Rhoades, B.E.: Some Fixed Point Theorems Paper Presented at the 775th Meeting' of the American Mathematical Society, special session on Fixed Point Theory, Bloomington, Indiana (1980)
26. Rhoades, B.E.: Contractive definitions. In: Rassias, T.M. (ed.), Nonlinear Analysis, pp. 513–526. World Scientific Publishing Company, New Jersey (1988)
27. Kincses, J., Totik, V.: Theorems and counter examples on contractive mappings. Mathematica Balkanica, New Series, Vol. 4 (1990)
28. Collao, P., Silva, J.C.E.: A complete comparison of 25 contraction conditions. Nonlinear Anal. Theory Methods Appl. **30**(1), 471–476 (1997). https://doi.org/10.1016/S0362-546X(97)00353-2
29. Kannan, R.: Some results on fixed points. Bull. Calcutta Math. Soc. **60**, 71–76 (1968). MR 41 #2486
30. Ray, B.K.: On non-expansive mappings in a metric space. Nanta Math. **7**, 86–92 (1974)
31. Reich, S.: Fixed points of contractive functions. Boll. Unione. Mat. Ital. **5**, 26–42 (1972)
32. Rus, I.A.: On common fixed points. Studia Univ. Babes-Bolyai Ser. Math.-Mech. **18**, 31–33 (1973). MR 49 # 1502
33. Wong, C.S.: Common fixed points of two mappings. Pacific J. Math. **48**, 299–312 (1973). MR 48 #7245
34. Gupta, V.K., Srivastava, P.: A note on common fixed points. Yokahama Math. J. **19**, 91–95 (1971)
35. Karapınar, E., Agarwal, R.P.: Fixed Point Theory in Generalized Metric Spaces. Springer, Cham (2022)
36. Rakotch, E.: A note on contractive mappings. Proc. Am. Math. Soc. **13**, 459–465 (1962)
37. Browder, F.E.: On the convergence of successive approximations for nonlinear functional equations. Indag. Math. **30**, 27–35 (1968)
38. Boyd, D.W., Wong, J.S.W.: On nonlinear contractions. Proc. Am. Math. Soc. **20**, 458–464 (1969)
39. Matkowski, J.: Integrable solutions of functional equations. Diss. Math. 127, Warsaw (1975)
40. Meir, A., Keeler, E.: A theorem on contraction mappings. J. Math. Anal. Appl. **28**, 326–329 (1969)
41. Krasnosel'skii, M.A., Zabrejko, P.P.: Geometric Methods of Nonlinear Analysis. Nauka, Moscow (1975). (in Russia)
42. Burton, T.A.: Integral equations, implicit functions and fixed points. Proc. Am. Math. Soc. **124**, 2383–2390 (1996). https://doi.org/10.1090/S0002-9939-96-03533-2
43. Geraghty, M.: On contractive mappings. Proc. Am. Math. Soc. **40**, 604–608 (1973)
44. Dass, B.K., Gupta, S.: An extension of Banach contraction principle through rational expression. Indian J. Pure Appl. Math. **6**, 1455–1458 (1975)
45. Khan, M.S.: A fixed point theorem for metric spaces. Rend. Inst. Math. Univ. Trieste., Vol. VIII, Fase. 10, 1–4 (1976)
46. Alber, Y.I., Guerre-Delabriere, S.: Principle of weakly contractive maps in Hilbert spaces. In: Gohberg, I., Lybich, Y. (eds.) New Results in Operator Theory and its Applications, Operator Theory: Advances and Applications, vol. 98, pp. 7–22. Birkhäuser, Basel, Switzerland (1997)

47. Dutta, P.N., Choudhury, B.S.: A generalisation of contraction principle in metric spaces. Fixed Point Theory Appl. **2008**, 406368 (2008). https://doi.org/10.1155/2008/406368
48. Caristi, J.: Fixed point theorems for mappings satisfying inwardness conditions. Trans. Am. Math. Soc. **215**, 241–251 (1976)
49. Bryant, V.W.: A remark on a fixed point theorem for iterated mappings. Am. Math. Mon. **75**, 399–400 (1968)
50. Sehgal, V.M.: A fixed point theorem for mappings with a contractive iterate. Proc. Am. Math. Soc. **23**, 631–634 (1969)
51. Matkowski, J.: Fixed point theorems for mappings with a contractive iterate at a point. Proc. Am. Math. Soc. **62**(2) (1977)
52. Ray, B.K., Rhoades, B.E.: Fixed point theorems for mappings with a contractive iterate. Pac. J. Math. **71** (1977)
53. Singh, K.L.: Fixed-point theorems for contractive-type mappings. J. Math. Anal. Appl. **72**, 283–290 (1979). Ćirić, L.B.: On Sehgal's maps with a contractive iterate at a point. Publ. De L'Institut Mathématique. **33**(47), 59–62 (1983)
54. Kincses, J., Totik, V.: Theorems and counter examples on contractive type mappings. Math. Balkanica (N.S.) **4**(1), 69–90 (1990)
55. Ćirić, L.B.: On Sehgal's maps with a contractive iterate at a point. Publ. De L'Institut Mathématique. **33**(47), 59–62 (1983)
56. Jachymski, J.: Fixed points of maps with a contractive iterate at a point. Math. Balkanica, New Ser. **9**(4) (1995)
57. Jachymski, J.: Equivalent conditions for generalized contractions on (ordered) metric spaces. Nonlinear Anal. **74**(3), 768–774 (2011). https://doi.org/10.1016/j.na.2010.09.025
58. Liu, Y., Li, Z.: Schaefer type theorem and periodic solutions of evolution equations. J. Math. Anal. Appl. **316**(1), 237–255 (2006). https://doi.org/10.1016/j.jmaa.2005.04.045
59. Wardowski, D.: Fixed points of a new type of contractive mappings in complete metric spaces. Fixed Point Theory Appl. **94**, 6 pp. (2012). https://doi.org/10.1186/1687-1812-2012-94
60. Turinici, M.: Wardowski implicit contractions in metric spaces (2013). arxiv:1212.3164v2 [Math.GN]
61. Wardowski, D.: Solving existence problems via F-contractions. Proc. Am. Math. Soc. **146**(4), 1585–1598 (2018)
62. Jleli, M., Samet, B.: A new generalization of the Banach contraction principle. J. Inequal. Appl. **2014**(38) (2014). https://doi.org/10.1186/1029-242X-2014-38
63. Kirk, W.A., Srinivasan, P.S., Veeramani, P.: Fixed points for mapping satisfying cyclic contractive conditions. Fixed Point Theory **4**, 79–89 (2003)
64. Liu, Y.C., Li, Z.X.: Krasnoselskii type fixed point theorems and applications. Proc. Am. Math. Soc. **136**, 1213–1220 (2008)

Part II

Fixed-Point Theory for Modern Challenges

Control Process 5

Control systems are mechanisms or processes designed to regulate the behavior of other systems using control loops. These systems are pivotal in various fields such as engineering, robotics, and industrial automation.

In this chapter, we take a look back at the most important findings from the research on the system's control mechanism, specifically looking at how the contraction mapping theorem is applied to control systems. Thereafter, we aim to extend the result of [1] to metric fixed point theory context. Following that, our objective is to extend the finding of [1] to metric space while keeping the F-contraction mapping in mind.

5.1 Introduction to Control Systems

Before delve into this chapter, one should know that what is ***"Control Systems"*** mathematically?, one should know by defining what a system is and how it can be represented. A system, in general, is a process or a device that produces an output $y(t)$ in response to an input $u(t)$. This relationship can often be expressed mathematically as

$$y(t) = f(x(t), u(t), t),$$

where $x(t)$ represents the internal state variables that describe the system's dynamics.

S. K. Panda et al., *Fixed Point Theory from Early Foundations to Contemporary Challenges*, Synthesis Lectures on Mathematics & Statistics,
https://doi.org/10.1007/978-3-032-03585-1_5

In the field of control systems, the aim is to regulate or guide the behavior of such systems using a control input $u(t)$. A control system can operate in two modes: open-loop and closed-loop. In an open-loop system, the output $y(t)$ depends solely on the input $u(t)$, typically modeled in the Laplace domain as

$$Y(s) = G(s)U(s),$$

where $G(s)$ is the system's transfer function. On the other hand, a closed-loop system includes feedback, ensuring that the output follows a desired reference $r(t)$. The output in this case is given by

$$Y(s) = G(s)\frac{1}{1 + G(s)H(s)}R(s),$$

where $H(s)$ represents the feedback path.

Control systems are often modeled using differential equations in the time domain. For a linear time-invariant system, the governing equation is of the form

$$a_n\frac{d^n y(t)}{dt^n} + a_{n-1}\frac{d^{n-1} y(t)}{dt^{n-1}} + \cdots + a_1\frac{dy(t)}{dt} + a_0 y(t) = b_m\frac{d^m u(t)}{dt^m} + \cdots + b_0 u(t).$$

Taking the Laplace transform simplifies the analysis and yields

$$G(s) = \frac{Y(s)}{U(s)} = \frac{B(s)}{A(s)},$$

where $B(s)$ and $A(s)$ are polynomials of the Laplace variable s.

Another powerful representation is the state-space form, which describes the system in terms of first-order differential equations:

$$\dot{x}(t) = Ax(t) + Bu(t),$$

$$y(t) = Cx(t) + Du(t).$$

Here, $x(t)$ is the state vector, A is the system matrix, B is the input matrix, C is the output matrix, and D is the feedthrough matrix. These equations provide a compact way to analyze multi-variable systems.

The fundamental objectives of control are to ensure stability, controllability, and observability. Stability requires that the system's response remains bounded over time, which is analyzed by examining the poles of the transfer function or the eigenvalues of the system matrix A. Controllability implies that it is possible to drive the system from any initial state to any desired state, while observability ensures that the internal states can be inferred from the output.

5.2 Adaptive Control Systems

An adaptive control system is a type of control system that can modify its parameters automatically in response to changes in the environment or system dynamics. Unlike traditional control systems, adaptive systems are designed to handle uncertainties and varying operating conditions.

Key Features of Adaptive Control Systems:

- **Adaptation Mechanism**: Adjusts control parameters in real time based on system behavior.
- **Identification Model**: Monitors and estimates the system's parameters.
- **Feedback and Feedforward**: Combines feedback and feedforward control strategies for robustness.

Mathematical Representation:

In an adaptive control system, the control law $u(t)$ is adjusted dynamically. A typical representation includes:

$$\dot{x}(t) = A(t)x(t) + B(t)u(t),$$
$$y(t) = C(t)x(t),$$

where:

- $A(t), B(t), C(t)$: System matrices that vary with time.

The adaptive control system uses an **adaptive law** to update these parameters:

$$\dot{\theta}(t) = \gamma \cdot \phi(x, t),$$

where:

- $\theta(t)$: Parameter vector to be adapted,
- γ: Adaptation gain,
- $\phi(x, t)$: A function of the system's state and time.

A common example is the **Model Reference Adaptive Control (MRAC)**:

$$e(t) = x(t) - x_m(t),$$

$$\dot{\theta}(t) = -\Gamma e(t)\phi(x, t),$$

where:

- $x_m(t)$: Desired state from the reference model,
- Γ: Adaptation gain matrix.

Example 5.1 An adaptive control system can be explained through the simple example of a room heating system. The goal of this system is to maintain the room temperature ($\mathbf{T_{room}}$) at a desired level ($\mathbf{T_{desired}}$), regardless of external disturbances like changes in outdoor temperature or the presence of additional heat sources. In a traditional control system, the heater operates with fixed parameters.

Let us consider the heater's power (u) might be determined based on a constant proportional gain ($\mathscr{K}_p$), such that:

$$u = \mathscr{K}_p \cdot (\mathbf{T_{desired}} - \mathbf{T_{room}}).$$

- The difference "$\mathbf{T}_{\text{desired}} - \mathbf{T}''_{\text{room}}$ is referred to as the error.

This approach assumes that the system's dynamics, such as the insulation of the room and the efficiency of the heater, remain constant. Figure 5.1 shows a room heating system that uses adaptive control.

Nevertheless, these hypotheses often fail in practice. Since the weather is constantly changing and doesn't stay the same for very long, the outside environment can never be relied on to predict or control the inside environment. Current temperature control technologies are limited. One limitation is that the user must adjust the system when external conditions change. This is exhausting and ineffective for room temperature regulation. In practice, these speculations may often fail. So that, this could be managed by ***"adaptive control system"***

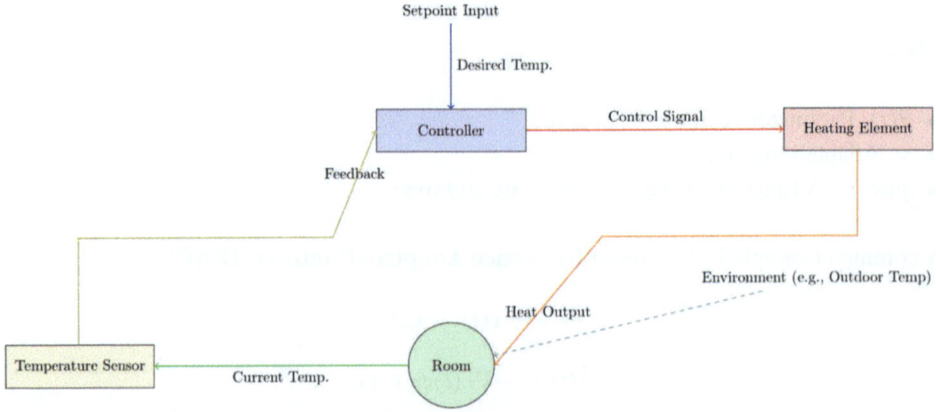

Fig. 5.1 Room heating system with adaptive control

that controls system methods and revises the parameters to maintain ***"stability"***. Moreover, this methodology considers continuous model dynamics such as, room thermal efficiency and heater effectiveness.

In this case, the proportional gain ($\mathscr{K}_p$) is no longer fixed but becomes time-dependent ($\mathscr{K}_p(t)$).

The control input now adapts in real-time:

$$u = \mathscr{K}_p(t) \cdot (\mathbf{T}_{\mathbf{desired}} - \mathbf{T}_{\mathbf{room}}).$$

The parameter $\mathscr{K}_p(t)$ is adjusted using an adaptive law, which modifies it based on the observed error and system behavior. For example, a simple adaptive law could be expressed as:

$$\dot{\mathscr{K}}_p(t) = \gamma \cdot (\mathbf{T}_{\mathbf{desired}} - \mathbf{T}_{\mathbf{room}}) \cdot \phi(\mathbf{state}).$$

- Here the term γ is the adaptation rate, ($\mathbf{T}_{\mathbf{desired}} - \mathbf{T}_{\mathbf{room}}$) is the error, and $\phi(\mathbf{state})$ is a function that reflects the system's current dynamics, such as the rate of temperature change.

When implemented, the adaptive control system works as follows. If the room experiences a disturbance, such as a window being opened, the system observes that the error is not reducing as expected. It reacts by increasing $\mathscr{K}_p(t)$, which amplifies the heater's power to compensate for the additional heat loss. Similarly, when the disturbance is removed (e.g., the window is closed), the system reduces $\mathscr{K}_p(t)$ to avoid overheating and minimize energy consumption.

Thus, we can easily conclude that:

1. The dynamic adjustment allows the adaptive control system to perform effectively under varying conditions.
2. Even with unanticipated climatic changes, this heating system's adaptive method keeps the space at the optimum temperature.
3. This improves comfort and energy efficiency, making the system more reliable and sustainable.

5.3 Utilization of Contraction Mapping Principle in Control Process

The idea of using ***"fixed-point theory"*** in control systems from its utility in solving equations and analyzing stability in nonlinear systems. Fixed-point theory provides a mathematical framework to study points that remain invariant under certain mappings, and this concept can be applied to design, analysis, and stability of control systems.

- **The contraction mapping principle** [2]: for a complete metric space $\mathscr{R}$ with metric d, a mapping $\mathbb{T}$ is called ***"contraction"*** if

$$d(\mathbb{T}(x), \mathbb{T}(y)) < \varkappa \cdot d(x, y), \quad 0 \leq \varkappa < 1.$$

This contraction mapping principle guarantees a unique fixed point $\mathbb{T}(x^*) = x^*$.

In [1], the author explores the application of the principle of contraction mapping in nonlinear and adaptive control systems.

The paper reviews the mathematical background of contraction mapping, its relevance in control theory, and its ability to ensure stability and convergence in nonlinear feedback systems.

The main application presented in the article [1] is, ***"aircraft-terrain"***, where the contraction mapping principle is utilized to produce a needed recursive control function. Thereafter, they presented a self-learning system for control processing associated with suitable parameters which attaching discrete iterative methods to continuous functions. Moreover, this article concludes by understanding the ***"contraction mapping principle"*** extendability in adaptive control systems as well as self-learning systems bringing its significance for stabilization and process of control in nonlinear models. The core idea is to employ a recursive relationship for the control function $\eta(x)$, defined as:

$$\eta_{i+1}(x) = -\gamma(x) + \eta_i(x) + \mathscr{H}[\eta_i(x)],$$

where $\gamma(x)$ represents the desired output, and $\mathscr{H}[\eta_i(x)]$ is the system's response to the current control input.

By repeatedly adjusting for changes in the system's response, this recursive connection guarantees that ***"the control function converges"*** to ***"the target output"***.

In order to scrutinize the analyze the convergence behaviour of this recursive procedure, the system applies to the contraction mapping principle. The updated recursive process is depicted as a mapping $\mathbb{T}$, defined as:

$$\mathbb{T}(\eta(x)) = -\gamma(x) + \eta(x) + \mathscr{H}[\eta(x)].$$

The ***"contraction mapping principle"***states that:

$$\|\mathbb{T}(\eta_1(x)) - \mathbb{T}(\eta_2(x))\| \leq \kappa \|\eta_1(x) - \eta_2(x)\|, \quad 0 \leq \kappa < 1,$$

where, $\eta_1(x)$ and $\eta_2(x)$ are control functions.

Thus, as per the contraction mapping theorem, $\{\eta_i(x)\}$ is converges.

The system response is often modeled by a differential equation such as:

$$\frac{d^3y}{dt^3} + a_2\frac{d^2y}{dt^2} + a_1\frac{dy}{dt} = b_2\frac{d^2\eta}{dt^2} + b_1\frac{d\eta}{dt} - b_0\eta,$$

where $y(t)$ represents the system state, and $\eta(t)$ is the control input. The operator $\mathscr{H}$ maps the control function $\eta(x)$ to the system's output response $\vartheta(x)$. By satisfying the contraction property, $\mathscr{H}$ ensures that the sequence of control updates converges to a fixed point, $\eta^*(x)$, that produces the desired output $\gamma(x)$. The solution is:

$$\mathscr{H}[\eta^*(x)] = \gamma(x),$$

indicating that the control function has successfully driven the system's output to the desired trajectory.

This approach have some highlights:

1. The iterative behaviour of this method constructs it explicitly matched for linear systems as well as nonlinear systems.
2. The operator $\mathscr{H}$ incorporates the system nonlinearities, and the recursive method succeeds when the ***"contraction mapping condition"*** is met. Thus, this method gives against system uncertainty and external disturbances. The convergence behaviour of the recursive process is characterized by the ***"contractive constant"*** with smaller values of κ pertaining to faster convergence.
3. This interplay between system dynamics, the operator $\mathscr{H}$, and the contraction constant highlights the mathematical elegance and practical relevance of the approach as in Fig. 5.2.

In [3], the author applies above principle to control systems, where input-output relationships are represented as $Y = MX$, with M being the transformation. By defining $\mathbb{T} = \mathcal{I} - \mathcal{M}$, iterative updates $X^{(k+1)} = \mathbb{T}X^{(k)}$ converge to a solution X^*, provided $\mathbb{T}$ is a contraction.

Further, the principle is used to compute Laplace transforms for systems governed by linear differential equations. The transformation

$$sL(X) = AX + G.$$

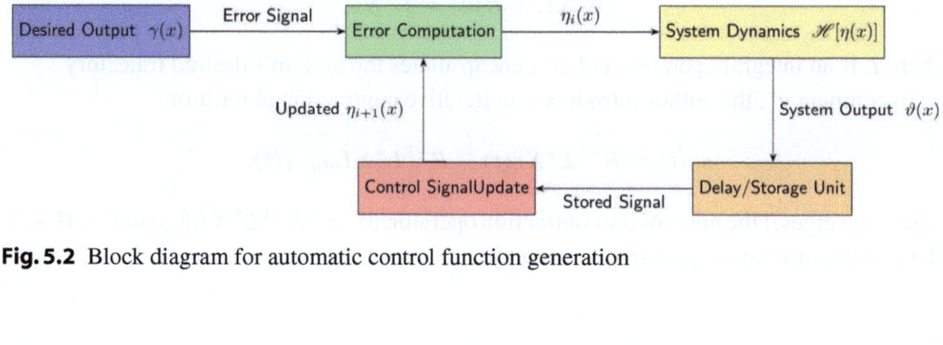

Fig. 5.2 Block diagram for automatic control function generation

can be solved iteratively using a defined metric, ensuring convergence if the mapping is a contraction. The synthesis of control matrices is also addressed, where C is designed to maintain $A = PC$ as a diagonal matrix under variable plant conditions. Successive approximations adjust C to meet the relationship $P_x C_x = P_o C_o$, preserving the desired response.

The work of [3] extends contraction mapping to variable transformations $\mathbb{T}_i$, showing that if these mappings commute ($\mathbb{T}_i \mathbb{T}_j = \mathbb{T}_j \mathbb{T}_i$), the system achieves uniform convergence in the metric space, making it robust against initial conditions. Lastly, self-organizing systems are modeled with iterative transformations

$$y^{(k+1)} = \mathscr{M}(y^{(k)}),$$

where M is designed as a contraction mapping. This ensures that subsystems converge to desired outputs while adapting to changes in inputs and environments.

In [4], the author introduces the optimization of time-varying linear multivariable control systems using the contraction-mapping algorithm (CMA). Existing methods, such as steepest descent and conjugate-gradient techniques, often face limitations in convergence rates and computational storage, particularly for complex systems. The author proposes the CMA, grounded in fixed-point principles, to overcome these challenges and ensure guaranteed convergence.

The optimization problem is formulated for a linear time-varying system described by the state equation:

$$\dot{x}(t) = A(t)x(t) + B(t)u(t),$$

and the output equation:

$$c(t) = C(t)x(t) + D(t)u(t).$$

The objective is to minimize a cost functional:

$$J = \int_{t_0}^{\mathbb{T}} \left[e^{\mathbb{T}}(t)X(t)e(t) + u^{\mathbb{T}}(t)R(t)u(t) \right] dt,$$

where $e(t) = c_d(t) - c(t)$ represents the error between the desired and actual outputs. The optimal control $u^*(t)$ satisfies:

$$(L^* X L + R)u^* = L^* X z,$$

where L is an integral operator, and $z(t)$ encapsulates the system's desired trajectory.

To compute u^*, the author introduces an iterative contraction algorithm:

$$u_n(t) = R^{-1} L^* X z(t) - R^{-1} L^* X L u_{n-1}(t),$$

which converges if the norm of the contraction operator, $M = \|R^{-1} L^* X L\|$, satisfies $M < 1$. The convergence rate is geometric, given by:

$$\|u^* - u_n\| \leq \frac{M^n}{1-M}\|R^{-1}L^*X\|\|z\|.$$

To address the issue of slow convergence when M approaches unity, the author employs perturbation techniques. By introducing a perturbation constant k, the algorithm transforms to:

$$(R + kI)^{-1}(L^*XL - kI)u^* = (R + kI)^{-1}L^*Xz.$$

The optimal k minimizes the contraction-operator norm N_g, expressed as:

$$N_g = \frac{M_L - m_L}{M_L + m_L + 2m_R},$$

where M_L and m_L are spectral bounds of L^*XL, and m_R is the smallest eigenvalue of R.

The work of [4] then compares the optimal CMA (OCMA) with traditional methods, including steepest descent, conjugate-gradient, and Riccati equation-based approaches. The OCMA demonstrates superior convergence rates due to its optimized geometric progression and requires fewer computational resources per iteration. For high-order systems, the OCMA is computationally efficient and achieves significant improvements over alternatives, particularly for systems with well-conditioned operators.

As a short description, the authors of [4] establishes the OCMA as a robust method for optimal control of linear time-varying systems. The algorithm guarantees convergence with minimal storage requirements and offers practical advantages in computational speed.

5.4 *F*-Contraction Mapping in Automatic Flight-Path Generator

Wardowski [5] generalized the Banach Contraction Principle in 2012 by constructing *F*-contractions in metric spaces. Throughout the years that followed, many scholars employed Wardowski's method, or some of its adaptations, to produce novel fixed-point solutions for single- and multivalued mappings in various spaces (refer to [6–17]).

Definition 5.1 ([5]) An *F*-**contraction** is a mapping $\mathbb{T} : X \to X$ satisfying:

$$\tau + F(d(\mathbb{T}x, \mathbb{T}y)) \leq F(d(x, y)),$$

$\forall\ x, y \in X$ with $\mathbb{T}x \neq \mathbb{T}y$, where $\tau > 0$ and $F : (0, \infty) \to \mathbb{R}$ gratifies the below stated assertions:

($F1$) $\forall \theta_1, \theta_2 > 0, \theta_1 > \theta_2$ implies $F(\theta_1) > F(\theta_2)$.
($F2$) For any $(\theta_n) \subset (0, \infty)$, $\theta_n \to 0$ if and only if $F(\theta_n) \to -\infty$.
($F3$) There is $k \in (0, 1)$ satisfying $\lim_{\theta \to 0^+} \theta^k F(\theta) = 0$.

Theorem 5.1 ([5]) *Let (X, d) be a complete metric space and let $\mathbb{T} : X \to X$ be a F-contraction. Then $\mathbb{T}$ has a unique fixed point.*

5.4.1 Automatic Flight Path Generation for Terrain-Following Aircraft Through *F*-Contraction

An airplane is following a terrain, and for simplicity, we consider a 2D scenario: horizontal flight distance and height, so the airplane flies in its pitch plane. The terrain is scanned by a radar, and the airplane has a radar altimeter. There are constraints on the airplane's acceleration, which can be simplified as limitations on the radius of curvature of the flight path at a given speed. Additionally, there may be a maximum allowable inclination of the flight to the horizontal. Given these factors and a desired ground clearance, it is possible to define the required flight path $\gamma(x)$, where x is the horizontal distance.

The task is to automatically generate the flight path that satisfies these constraints. We begin by defining:

- $\gamma(x)$: The required flight path function, which specifies the desired height above the ground at each horizontal distance x.
- $\eta(x)$: The control function, which corresponds to the setting of the flight controls, like the elevator control signal.
- $\rho(x)$: The flight path resulting from the control $\eta(x)$, which can represent the flight angle at each point along the path.
- $\vartheta(x)$: The actual height of the airplane at each point along the flight.

The system works by updating the control function $\eta(x)$ recursively. Starting from an initial control function $\eta_0(x)$, the control function is updated using feedback to match the required path $\gamma(x)$. The recursive update rule is given by:

$$\eta_{i+1}(x) = \mu_i(x) = -\gamma(x) + \eta_i(x) + \vartheta_i(x),$$

where $\vartheta_i(x)$ represents the actual height at the i-th iteration.

The height $\vartheta_i(x)$ is related to the control function $\eta_i(x)$ by:

$$\vartheta_i(x) = \mathscr{H} A \eta_i(x), \tag{5.1}$$

then

$$\eta_{i-1}(x) = \mu_i(x) = -\gamma(x) + (\mathscr{H} A + I)\eta_i(x), \tag{5.2}$$

where $\mathscr{H}$ is a scaling factor and A represents the dynamics of the aircraft.

For the recursive process to converge, a sufficient condition is that the right-hand side of the equation becomes a *F-contraction mapping*. This ensures that the system will eventually reach a stable solution, where the actual height $\vartheta(x)$ matches the required path $\gamma(x)$.

Thus, the solution to the system is:

$$\eta(x) = -\gamma(x) + (\mathscr{H}A + I)\eta(x).$$

Rearranging this equation gives:

$$\gamma(x) = \mathscr{H}A\eta(x) = \vartheta(x).$$

This means that the control function $\eta(x)$ is such that the actual position $\vartheta(x)$ equals the required path $\gamma(x)$, which is the desired outcome.

For an aircraft flying at a constant speed V, the height $\vartheta(x)$ above some reference point is given by:

$$\vartheta(x) = \vartheta(x_0) + V\int_{x_0}^{x} \tan(\rho)dx,$$

where $\rho(x)$ is the flight angle and x_0 is the initial horizontal distance. For simplicity, we assume:

$$\vartheta(x) = \mathscr{H}\rho(x). \tag{5.3}$$

This relationship allows us to describe the aircraft dynamics in terms of the flight angle $\rho(x)$.

The dynamics of the aircraft can be described by a third-order differential equation:

$$\frac{d^3\rho}{dt^3} + a_2\frac{d^2\rho}{dt^2} + a_1\frac{d\rho}{dt} = b_2\frac{d^2\eta}{dt^2} + b_1\frac{d\eta}{dt} - b_0\eta, \tag{5.4}$$

where $\rho(x)$ is the flight angle and $\eta(x)$ is the control input.

The solution to the system involves using Laplace transforms, and it can be expressed as:

$$f\{x(t)\} \equiv b_2\frac{d^2\eta}{dt^2} + b_1\frac{d\eta}{dt} - b_0\eta.$$

For a given speed, x is a function of time t, and hence both ρ and η are functions of time or distance.

From the differential equation Eq. 5.4, the solution for $\rho(x)$ can be written as:

$$\rho\{x(t)\} = \rho(x_0) + \int_{t_0}^{t} e^{\beta t_1}\int_{t_0}^{t_1} e^{-\beta t_2}e^{\alpha t_2}\int_{t_0}^{t_3} e^{-\alpha t_3} f\{x(t_3)\}\,dt_3\,dt_2\,dt_1,$$

which simplifies to:

$$\rho(x(t)) = A\eta(x),$$

where A and $\mathscr{H}$ are constants determined by the system dynamics.

The block diagram for the system is as shown in Fig. 5.3.

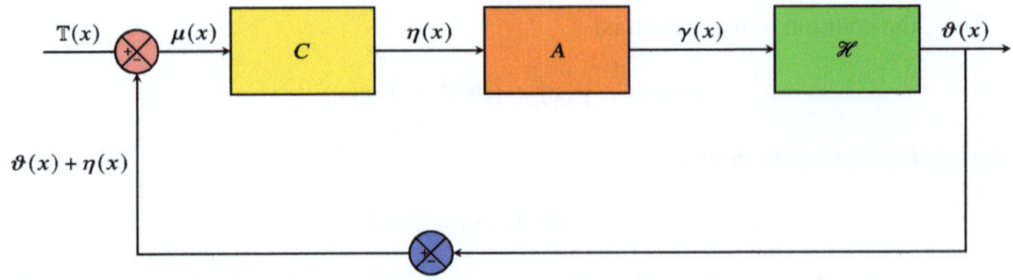

Fig. 5.3 Block diagram of the automatic flight-path generator

Theorem 5.2 *Suppose the conditions given below are satisfied.*

$(\boldsymbol{H}_1)$. $x - x_0 < \frac{1+e^{-\tau}}{Vm}, \tau > 0,$
$(\boldsymbol{H}_2)$. $\frac{\mathscr{H} A\eta_i(x) - \mathscr{H} A\eta_{i-1}(x)}{\eta_i - \eta_{i-1}} < -1 + e^{-\tau}.$

Then (5.2) converges.

Proof Let us define the metric

$$d(f(x), g(x)) = \max_x |f(x) - g(x)|.$$

If the right side of (5.2) satisfies the F-contraction, there exists one and only one solution such that $x^* = -\gamma + (\mathscr{H} A + I)x^*$ or $\mathscr{H} Ax^* = \gamma$.

Consider,

$$\begin{aligned} |\eta_{i+1}(x) - \eta_i(x)| =& |(\mathscr{H} A + I)\eta_i(x) - (\mathscr{H} A + I)\eta_{i-1}(x)| \\ =& \mathscr{H} A\eta_i(x) + \eta_i(x) - \mathscr{H} A\eta_{i-1}(x) - \eta_{i-1}(x)| \\ =& \mathscr{H} A\eta_i(x) - \mathscr{H} A\eta_{i-1}(x) + \eta_i(x) - \eta_{i-1}(x)| \\ =& \left\{\left|\frac{\mathscr{H} A\eta_i(x) - \mathscr{H} A\eta_{i-1}(x)}{\eta_i(x) - \eta_{i-1}(x)} + 1\right|\left|\eta_i(x) - \eta_{i-1}(x)\right|\right\}. \end{aligned} \tag{5.5}$$

We know that,

$$x - x_0 < \frac{1 + e^{-\tau}}{Vm}. \tag{5.6}$$

The required control function $\eta(x)$ can be automatically generated over the given range.

To better understand, note the following:

$$\frac{d(\tan\rho)}{d\eta} = \sec^2\rho\frac{d\rho}{d\eta} = (1 + \tan^2\rho)\frac{d\rho}{dt}\frac{dt}{d\eta}.$$

This can be rewritten as:

$$\left(1+\left(\frac{dh}{dx}\right)^2\right)\left(\frac{d\rho/dt}{d\eta/dt}\right).$$

Here, $\frac{dh}{dx}$ is limited by restrictions on the aircraft's acceleration, and $\frac{d\rho}{dt}$ and $\frac{d\eta}{dt}$ are related through Eq. (5.4), which accounts for the aircraft dynamics.

Therefore, m is determined by these parameters, and the horizontal distance $x - x_0$ can be computed using Eq. (5.6) for an appropriate value of m.

$$\Rightarrow Vm\,(x - x_0) < 1 + e^{-\tau}$$
$$\Rightarrow -Vm\,(x - x_0) > -1 - e^{-\tau}. \tag{5.7}$$

Let

$$\frac{d(\tan\rho)}{d\eta} \geq -m \quad \forall\, i, x, \tag{5.8}$$

where m is a positive constant, which has not been specified yet.

Now, consider the difference between $\tan\rho_i$ and $\tan\rho_{i-1}$:

$$\tan\rho_i - \tan\rho_{i-1} = \int_{\eta_{i-1}}^{\eta_i} \frac{d(\tan\rho)}{d\eta}\, d\eta \geq -m(\eta_i - \eta_{i-1}).$$

Therefore, we have:

$$\frac{V\int_{x_0}^{x}(\tan\rho_i - \tan\rho_{i-1})\,dx}{\eta_i - \eta_{i-1}} \geq -Vm(x - x_0).$$

From (5.7) and (5.8)

$$\frac{V\int_{x_0}^{x}(\tan\rho_i - \tan\rho_{i-1})\,dx}{\eta_i - \eta_{i-1}} \geqslant -V_m\,(x - x_0) > -1 - e^{-\tau}. \tag{5.9}$$

However

$$\mathscr{H}A\eta_i(x) - \mathscr{H}A\eta_{i-1}(x) = V\int_{x_0}^{x}(\tan\rho_i - \tan\rho_{i-1})\,dx.$$

Thus, (5.9) become

$$\frac{\mathscr{H}A\eta_i(x) - \mathscr{H}A\eta_{i-1}(x)}{\eta_i - \eta_{i-1}} > -1 - e^{-\tau}.$$

From physical considerations, when $\eta(x)$ increases $\rho(x)$ decreases, and vice versa. Hence, as per the assumption ($\mathbf{H}_2$),

$$\frac{\mathscr{H}A\eta_i(x) - \mathscr{H}A\eta_{i-1}(x)}{\eta_i - \eta_{i-1}} < -1 + e^{-\tau}.$$

Thus from the above two inequalities

$$-1-e^{-\tau} < \frac{\mathscr{H}A\eta_i(x) - \mathscr{H}A\eta_{i-1}(x)}{\eta_i - \eta_{i-1}} < -1+e^{-\tau}$$

$$-e^{-\tau} < \frac{\mathscr{H}A\eta_i(x) - \mathscr{H}A\eta_{i-1}(x)}{\eta_i - \eta_{i-1}} + 1 < e^{-\tau},$$

which yields that

$$\left|\frac{\mathscr{H}A\eta_i(x) - \mathscr{H}A\eta_{i-1}(x)}{\eta_i - \eta_{i-1}} + 1\right| < e^{-\tau} \text{ for all } i, x.$$

Thus, from Eq. 5.5

$$\begin{aligned} &|\eta_{i+1}(x) - \eta_i(x)| < e^{-\tau}\,|\eta_i(x) - \eta_{i-1}(x)| \\ \Rightarrow &\max_x |\eta_{i+1}(x) - \eta_i(x)| < e^{-\tau} \max_x |\eta_i(x) - \eta_{i-1}(x)|\,. \end{aligned}$$

Thus it satisfies the Theorem 5.1, when

$$F(\alpha) = \log \alpha.$$

Hence (5.2) converges if $\mathbf{H}_1 \& \mathbf{H}_2$ satisfies.

The utilisation of the F-contraction mapping within the realm of control theory is a relatively new development. Nevertheless, there are consequences concerning to adaptive control process, feedback control systems and stability analysis.

The study of control process through F-contraction mapping renders it a valuable domain for ongoing research endeavours.

Open Questions

5.1 What could be the suitable value of "m" if we take complex-valued b-metric space [18] instead of standard metric space?

5.2 Can we apply Bryant mapping theorem [19] in adaptive control systems? If "YES", explain it with real-world example.

5.3 Prove or disprove automatic flight path generation for terrain-following aircraft through "**Ray and Rhoades contraction mapping** [20]".

References

1. Leibovic, K.: The principle of contraction mapping in nonlinear and adaptive control systems. IEEE Trans. Autom. Control **9**(4), 393–398 (1964). https://doi.org/10.1109/TAC.1964.1105743
2. Banach, S.: Sur les opérations dans les ensembles abstraits et leur application aux équations intégrales. Fund. Math. **3**, 133–181 (1922)
3. Leibovic, K.: Contraction mapping with application to control processes. J. Electron. Control **15**(1), 81–96 (1963). https://doi.org/10.1080/00207216308937557
4. Freeman, E.A.: Optimisation of contraction-mapping algorithm for calculating optimal controls. Proc. Inst. Electr. Eng. **119**(9). https://doi.org/10.1049/piee.1972.0270
5. Wardowski, D.: Fixed points of a new type of contractive mappings in complete metric spaces. Fixed Point Theory Appl. **2012**, 94 (2012). https://doi.org/10.1186/1687-1812-2012-94
6. Hussain, N., Salimi, P.: Suzuki–Wardowski type fixed point theorems for α-GF-contractions. Taiwan. J. Math. **18**(6), 1879–1895 (2014). https://doi.org/10.11650/tjm.18.2014.4462
7. Wardowski, D.: Solving existence problems via f-contractions. Proc. Am. Math. Soc. **146**(4), 1585–1598 (2018)
8. Panda, S.K., Atangana, A., Abdeljawad, T.: Existence results and numerical study on novel coronavirus 2019-nCoV/Sars-CoV-2 model using differential operators based on the generalized Mittag-Leffler kernel and fixed points. Fractals **30**(08), 2240214 (2022)
9. Mınak, G., Helvacı, A., Altun, I.: Ciric type generalized f-contractions on complete metric spaces and fixed point results. Filomat **28**(6), 1143–1151 (2014). https://doi.org/10.2298/FIL1406143M
10. Piri, H., Kumam, P.: Some fixed point theorems concerning f-contraction in complete metric spaces. Fixed Point Theory Appl. **2014**, 210 (2014). https://doi.org/10.1186/1687-1812-2014-210
11. Secelean, N.A.: Weak f-contractions and some fixed point results. Bull. Iran. Math. Soc. **42**(3), 779–798 (2016)
12. Agarwal, R.P., et al.: F-contraction mappings on metric-like spaces in connection with integral equations on time scales. Revista de la Real Academia de Ciencias Exactas, FíÃ‚Âsicas y Naturales. Serie A. Matemáticas **114**(3), 147 (2020)
13. Wardowski, D., Van Dung, N.: Fixed points of f-weak contractions on complete metric spaces. Demonstr. Math. **47**(1), 146–155 (2014). https://doi.org/10.2478/dema-2014-0012
14. Panda, S.K.: Applying fixed point methods and fractional operators in the modelling of novel coronavirus 2019-nCoV/SARS-CoV-2. Results Phys. **19**, 103433 (2020)
15. Kumari, P.S., Zoto, K., Panthi, D.: d-Neighborhood system and generalized F-contraction in dislocated metric space. SpringerPlus **4**, 1–10 (2015)
16. Panda, S.K., Karapınar, E., Atangana, A.: A numerical schemes and comparisons for fixed point results with applications to the solutions of Volterra integral equations in dislocatedextendedb-metricspace. Alex. Eng. J. **59**(2), 815–827 (2020)
17. Karapınar, E., Fulga, A., Agarwal, R.P.: A survey: f-contractions with related fixed point results. J. Fixed Point Theory Appl. **22**, 69 (2020). https://doi.org/10.1007/s11784-020-00803-7
18. Rao, K.P.R., Swamy, P.R., Prasad, J.R.: A common fixed point theorem in complex valued b-metric spaces. Bull. Math. Stat. Res. **1**(1), 1–8 (2013)
19. Bryant, V.W.: A remark on a fixed point theorem for iterated mappings. Am. Math. Mon. **75**, 399–400 (1968)
20. Ray, B.K., Rhoades, B.E.: Fixed point theorems for mappings with a contractive iterate. Pac. J. Math. **71** (1977)

Equilibrium Points, Dynamics and Synchronization of Neural Networks 6

Neural networks (NNs) are computer models based on the human brain, which include input, hidden, and output layers with interconnected nodes or neurons. Artificial intelligence includes NNs as one of its subclasses of deep learning technologies. Continuously expanding this mode of thought is essential to keep up with contemporary technology and science. Mcculloch and Pitts initially explored artificial neural networks (ANNs) in [1]. They investigated how such neurons work. Their analysis suggests various valuable applications for this method in the context of their work. Such types of applications can be found in many different domains, including statistical modelling, data processing, circuit theory, and others [2–7].

The input layer of a neural network receives unstructured information, with each neuron encoding a dataset feature. Hidden layers between the input and output layers operate complex computation involving weighted sums and activation functions to gather and enhance data patterns and characteristics. The final estimates or categorisations are made by the output layer, which may have numerous neurons for identification or one for regression. By progressive learning and abstraction, these layers help the network turn raw input into understandable output as in the Fig. 6.1 [8]. In NNs, an activation function is a mathematical function that is introduced to the output of every neuron, inside the network. Its main function is to incorporate nonlinearity within the model, thereby enabling the network to learn complex patterns and correlations in the data. We can observe numerous activation functions in the existing literature from Table 6.1.

In real-valued neural networks (RVNNs), weighted, biases, and activations are interpreted as real values $w, b, a(x) \in \mathbb{R}$. The network's operation includes computational functions like $f(x) = \theta(w \cdot z + b)$, where θ acts as an activation function. RVNNs succeed at tasks that need real data and connections, such as picture categorisation and regression.

S. K. Panda et al., *Fixed Point Theory from Early Foundations to Contemporary Challenges*, Synthesis Lectures on Mathematics & Statistics,
https://doi.org/10.1007/978-3-032-03585-1_6

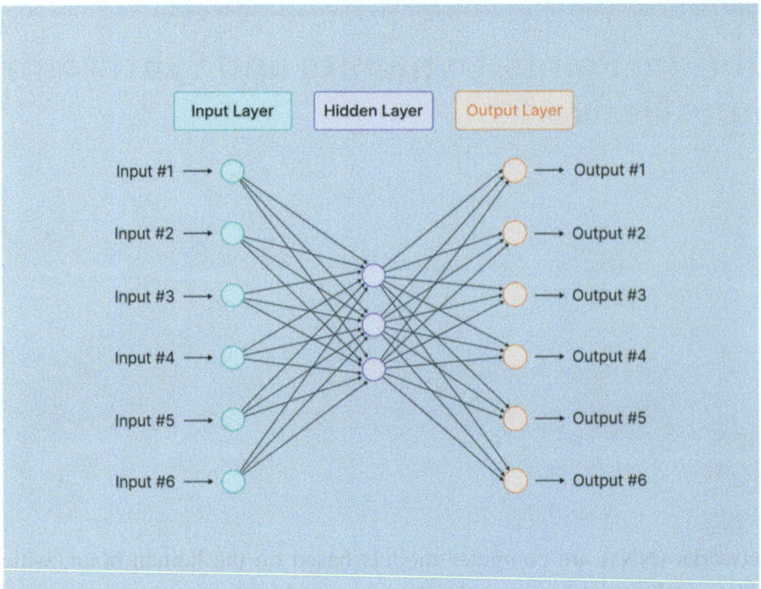

Fig. 6.1 Structure of a neural network with input, hidden, and output layers [8]

Table 6.1 Common activation functions

Activation function			
Sigmoid function	Hyperbolic tangent	Rectified linear unit (ReLU)	Softmax
$\theta(x) = \frac{1}{1+e^{-x}}$	$\text{Tanh}(x) = \frac{e^x - e^{-x}}{e^x + e^{-x}}$	$f(x) = \max(0, x)$	$\text{Softmax}\ (z_i) = \frac{e^{z_i}}{\sum_j e^{z_j}}$
Range: (0, 1)	Range: $(-1, 1)$	Range: $[0, \infty)$	Range: (0, 1)

In complex-valued neural networks (CVNNs), weighted, biases, and activations are interpreted as complex values $w, b, a(z) \in \mathbb{C}$, where $z = x + iy$ and the operations involve real parts as well as imaginary parts. The associated networks functions as $f(z) = \theta(w \cdot z + b)$, where $z \in \mathbb{C}$ is complex input, and θ is a complex activation function. CVNNs are exceptionally benit for medical imaging, where magnitude and phase argument function of z require to be modeled constructively. As a result, CVNNs grab substantial information in data as compared to RVNNs.

In this instance, the only neural networks that we examine and/or focus on specifically in this chapter are fractional-order neural networks (FoNNs). Additionally, we discuss how the different types of FoNNs continue to expand and develop. This study examines three different types of FoNNs: general class, complex-valued, and discrete. There is a short overview of the other types of FoNNs, including memristive, quaternion-valued, and others. In addition, specific stability hypothesis results, synchronization, and sufficient criteria for the existence and uniqueness of the nontrivial solutions are addressed.

6.1 Dynamics of Fractional-Order Neural Networks

FoNNs are an effective and flexible subclass of regular NNs, in which the derivatives utilised in their dynamic systems are of fractional-order rather than integer-order. This modification deals with the system's ability to capture memory and hereditary characteristics; in this context, FoNNs are more useful than traditional NNs in a variety of domains.

A typical method for controlling classical NNs is to use integer-order differential equations, with the current input and weight updates serving as the basis for each system state. Nevertheless, in contrast to previous models, *fractional calculus* provides dependence for the system's whole history, not just the current state. This is a very useful advantage for activities where **memory** or **long-term dependencies** are important.

The classical form of a fractional derivative of order α, where $\alpha \in (0, 1)$, is given by:

$$\mathscr{D}^{\alpha} f(t) = \frac{1}{\Gamma(1-\alpha)} \int_0^t \frac{f'(\tau)}{(t-\tau)^{\alpha}} d\tau,$$

where, Γ is the Gamma function, and $\mathscr{D}^{\alpha} f(t)$ represents the fractional derivative of the function $f(t)$. The integral based definition shows that the present state $f(t)$ is governed by the past values $f(\tau)$ through the integral term.

Classical NNs are more essential for tasks with short-term dependencies due to their dependence on integer-order derivatives. However, they lack inherent memory unless they are specifically constructed (see, for example, Recuucent Neural Networks). Instead, FoNNs make use of fractional derivatives, which reliably capture long-term dependencies through the memory effect. Smoother convergence and improved stability are provided by this method's capacity to support more robust and adaptive learning dynamics. Thus, FoNNs outperform standard models in tasks that require a balance of short- and long-term data.

For example, in standard gradient descent, the weight update rule is as follows:

$$w_{t+1} = w_t - \eta \frac{\partial L}{\partial w},$$

where w_t represents the weights at time step t, η is the learning rate, and L is the loss function.

In fractional-order gradient descent, the weight update rule is generalized to:

$$w_{t+1} = w_t - \eta \, {}_0D_t^{\alpha} \frac{\partial L}{\partial w},$$

where ${}_0D_t^\alpha$ is the fractional derivative operator of order α, and $0 < \alpha < 1$. This operator incorporates the history of the weight changes, allowing for long-term dependencies in the learning process.

Hopfield FoNNs were introduced in [9], substituting fractance components for the traditional capacitors. The initial approach to mathematically characterize Hopf bifurcation phenomenon in FoNNs was established in [10]. The same authors of [10] have substituted fractance in the conventional capacitor of the Hopfield neural network, resulting in a well-known Hopfield FoNNs. In 2012, they proposed an analytical stability analysis based on characteristic features for the Hopfield-type FoNNs [11].

The FoNNs of Hopfield type [10, 11] is as follows:

$$ {}^{c}\mathscr{D}^{\alpha} x_i(t) = -a_i x_i(t) + \sum_{j=1}^{n} T_{ij} g_j\left(x_j(t)\right) + I_i; \text{ for all } i = \overline{1, n}, \text{ and for all } t > 0, \quad (6.1) $$

where $\alpha \in (0, 1)$, $a_i > 0$ are the self-regulating parameters of the neurons, $T = \left(T_{ij}\right)_{n\times n}$ is the interconnection matrix, $g_i : \mathbb{R} \to \mathbb{R}$ are the neuron input-output activation functions and I_i denotes the external inputs.

The system (6.1) can be expressed in vector form as follows:

$$ {}^{c}\mathscr{D}^{\alpha} \mathbf{x}(t) = -A\mathbf{x}(t) + T\mathbf{g}(\mathbf{x}(t)) + \mathbf{I}, \quad (6.2) $$

where $A = \operatorname{diag}(a_1, a_2, \ldots, a_n) \in \mathbb{R}^{n\times n}$, $\mathbf{g}(\mathbf{x}) = (g_1(x_1), g_2(x_2), \ldots, g_n(x_n))^T$ and $\mathbf{I} = (I_1, I_2, \ldots, I_n)^T \in \mathbb{R}^n$. In light of the Caputo fractional derivative's characteristics, it follows that, $x^\star \in \mathbb{R}^n$ is a steady state of (6.1)–(6.2), that are to the input $\mathbf{I}$, if and only if

$$ -A\mathbf{x}^\star + T\mathbf{g}\left(\mathbf{x}^\star\right) + \mathbf{I} = 0. $$

It is evident that $\mathbf{x}^\star \in \mathbb{R}^n$ is a steady state of the fractional-order neural network (6.1)–(6.2) if and only if it is a steady state of the integer-order neural network

$$ \dot{\mathbf{x}}(t) = -A\mathbf{x}(t) + T\mathbf{g}(\mathbf{x}(t)) + \mathbf{I}. \quad (6.3) $$

The existence, uniqueness, or multiplicity of steady states in FoNNs is the same as for integer-order neural networks. However, the study of Eva Kaslik and Seenith Sivasundaram [10, 11] explores the dynamics of FoNNs of the Hopfield type, including stability, multi-stability, bifurcations, and chaos.

Based on a mathematical point of view, the Hopfield type FoNNs model has various advantages, mainly because it incorporates the Caputo fractional derivative. The following are some benefits:

- The neural network experiences a memory effect when a fractional-order derivative ${}^{C}\mathscr{D}^{\alpha}$ is used, meaning that the system's current state is dependent on both its past and present states. This contextual scenario is especially valuable for biomedical and mathematical modeling systems because neurons exhibit memory behavior, which fractional calculus may represent better than integer-order models.

- The model generalizes the classical Hopfield neural network by allowing the order of the derivative α to be $0 < \alpha < 1$. When $\alpha = 1$, the system reduces to the widely recognized integer-order model, however, when α is fractional, the model reflects richer and more complex dynamics.
- The model reveals that the steady state conditions for the fractional-order system (see, (6.1)–(6.2)) are the same as those of the systems (6.3). This scenario is beneficial since current results for integer-order Hopfield neural networks, which include the existence, uniqueness, and multiplicity of steady states, simply apply to the fractional-order system. This simplifies analysis and allows results to be easily extended to fractional-order frameworks.

Remark 6.1 In the work of [10], a comprehensive analysis of stability and bifurcation for a two-dimensional neural network was conducted. This network can be viewed as a simplified structure consisting of a core neuron and one connected peripheral neuron. Specifically, the system is described by the equations:

$$\begin{cases} {}^{c}\mathscr{D}^{\alpha} x_1(t) = -a x_1(t) + T_{11} g_1(x_1(t)) + T_{12} g_2(x_2(t)) + I_1 \\ {}^{c}\mathscr{D}^{\alpha} x_2(t) = -b x_2(t) + T_{21} g_1(x_1(t)) + T_{22} g_2(x_2(t)) + I_2. \end{cases}$$

When studying the stability of the steady state $\mathbf{x}^{\star}$, the resulting characteristic equation is:

$$\varrho^2 - (\rho + \omega)\varrho + \rho\omega - \vartheta = 0.$$

Here, $\rho = -a + T_{11} g_1'(x_1^{\star})$, $\omega = -b + T_{22} g_2'(x_2^{\star})$, and $\vartheta = T_{12} T_{21} g_1'(x_1^{\star}) g_2'(x_2^{\star})$. Similar to previous cases, the steady state $\mathbf{x}^{\star}$ of this system is asymptotically stable if and only if the below stated conditions fulfilled.

1. $\rho\omega - \vartheta > 0$;
2. $\rho + \omega < 2\cos\left(\frac{\alpha\pi}{2}\right)\sqrt{\rho\omega - \vartheta}$.

The discrete FoNNs is dealt with using a fuzzy lyapunov synthesis technique, as stated in [12]. To utilise this suggested NNs in simulating a heat transfer, a nonlinear discrete FoNNs methodology has been developed in this particular instance. A class of FoNNs with delay is discussed in the work in [13]. Considering a delay, this study provides a sufficient requirement for the uniform stability associated with these FoNNs. Furthermore demonstrated the stability, existence, and uniqueness of the equilibrium point. Additionally, in [14], the study looks into the finite-time stability of a class of FoNNs. Necessary criteria have been established for the time-stability of FoNNs through Caputo fractional derivatives. A class of FoNNs with delayed systems has had its finite-time stability examined in [15]. The innovative needed criteria that are obtained by the inequality technique have been demonstrated to guarantee the stability associated with these FoNNs. These criteria are dependent on delay. Additionally, the Cauchy-Schwartz inequality has been used to develop and validate the Caputo derivative. A class of NNs having fractional-order derivatives is examined in [16]. To guarantee the uniqueness and existence of the non-trivial solution in this case, several novel conditions that involve Krasnosel'skiĭ's fixed point theorem and inequality technique are used. A class of fractional-order chaotic neural networks has been provided to be synchronized using the Mittag-Leffler function in [17]. The acquired conceptual insights are implemented in a Hopfield FoNNs model. When the complex dynamical behaviors of these networks were examined within the same year, it quickly became apparent that a wide variety of fascinating dynamical phenomena existed.

6.1.1 Existence, Uniqueness, and the Uniform Stability of the Solution

In 2013, Liping Chen et al. [13], a necessary condition has been set out for uniform stability, and a class of fractional-order neural networks with delay is described. Its equilibrium point's existence, uniqueness, and stability are also demonstrated. To show the applicability and viability of the suggested outcomes, a numerical example is provided in their work.

The following differential equation can be used to describe how a continuous fractional-order delayed neural network evolves over time:

$$\mathscr{D}^{\alpha} x_i(t) = -c_i x_i(t) + \sum_{j=1}^{n} a_{ij} f_j(x_j(t)) + \sum_{j=1}^{n} b_{ij} g_j(x_j(t-\tau)) + I_i, \tag{6.4}$$

$$i \in \mathbb{N},\ t \in [0, T],\ T < \infty.$$

This can also be expressed in a more compact form as

$$\mathscr{D}^{\alpha} x(t) = -Cx(t) + Af(x(t)) + Bg(x(t-\tau)) + I,$$

where $0 < \alpha < 1$ and n represents the total number of neurons in the network. The vector $x(t) = (x_1(t), \ldots, x_n(t))^T \in \mathbb{R}^n$ denotes the state of the system at time t.

The functions $f(x(t)) = (f_1(x_1(t)), f_2(x_2(t)), \ldots, f_n(x_n(t)))^T$ and $g(x(t)) = (g_1(x_1(t)), g_2(x_2(t)), \ldots, g_n(x_n(t)))^T$ represent the neurons' activation functions. The matrices C, A, B are constants, with $C = \text{diag}(c_i > 0)$ indicating the rate at which each unit i returns to its resting state when isolated from both the network and any external stimuli. The elements a_{ij} and b_{ij} describe the influence of neuron j on neuron i at time t and $t - \tau$, respectively, where τ is a non-negative constant representing the delay in transmission. The vector $I = (I_1, I_2, \ldots, I_n)^T$ acts as an external bias input.

The initial conditions associated with this system (6.4) take the form

$$x_i(t) = \phi_i(t), \quad \phi_i(0) = 0, \quad t \in [-\tau, 0], \quad i \in \mathbb{N},$$

where it is generally assumed that $\phi_i(s) \in C([-\tau, 0], \mathbb{R})$ for $i \in N$. The norm of the function space $C([-\tau, 0], \mathbb{R})$ is given by

$$\|\phi(t)\| = \sum_{i=1}^{n} \sup_{t\in(-\tau,0]} \left\{e^{-t} |\phi_i(t)|\right\}.$$

Definition 6.1 The solution to system (6.4) is called stable if, for any $\varepsilon > 0$, there exists a $\delta(t_0, \varepsilon) > 0$ such that for $t \geq t_0 \geq 0$ and $\|\varphi(t) - \phi(t)\| < \delta$, it follows that

$$\|y(t, t_0, \varphi) - x(t, t_0, \phi)\| < \varepsilon,$$

holds for any two solutions $x(t, t_0, \phi)$ and $y(t, t_0, \varphi)$. This system is said to be uniformly stable if the δ can be chosen independently of t_0.

To get the principal outcomes, they have taken into account the following assumptions.

Assumption A: The activation functions of the neurons, denoted as f_j and g_j, exhibit Lipschitz continuity. This means there are positive constants F_j and G_j for $j = 1, 2, \ldots, n$ such that the following inequalities hold for all $u, v \in \mathbb{R}$:

$$\left|f_j(u) - f_j(v)\right| < F_j|u - v|, \quad \left|g_j(u) - g_j(v)\right| < G_j|u - v|.$$

Assumption B: The parameters c_i, a_{ij}, b_{ij}, F_j, and G_j are constrained by the condition:

$$\left\|A^*\right\| + \left\|B^*\right\| < \bar{c},$$

where:

$$\left\|A^*\right\| = \sum_{i=1}^{n} |a_i^*| = \sum_{i=1}^{n} \max_j \left\{|a_{ij}|L_j\right\}, \quad \bar{c} = \min\left\{1 - c_{\max}, c_{\min}\right\},$$

$$c_{\max} = \max_i \left\{c_i\right\}, \quad c_{\min} = \min_i \left\{c_i\right\},$$

$$\left\|B^*\right\| = \sum_{i=1}^{n} |b_i^*| = \sum_{i=1}^{n} \max_j \left\{|b_{ij}|G_j\right\}.$$

Theorem 6.1 ([13]) *The system 6.4 is uniformly stable if Assumptions A and B are satisfied.*

Theorem 6.2 ([13]) *If Assumptions A and B are met, system (6.4) has a unique equilibrium point that is uniformly stable.*

Example 6.1 ([13]) The two-state fractional-order delayed Hopfield neural network is defined by the following equations:

$$\begin{aligned}\mathscr{D}^{\alpha} x_1(t) &= -0.25x_1(t) - 0.05f(x_1(t)) + 0.01f(x_2(t)) \\ &\quad - 0.01f(x_1(t-\tau)) + 0.02f(x_2(t-\tau)) - 0.1, \\ \mathscr{D}^{\alpha} x_2(t) &= -0.2x_2(t) + 0.02f(x_1(t)) - 0.01f(x_2(t)) \\ &\quad + 0.02f(x_1(t-\tau)) + 0.01f(x_2(t-\tau)) + 0.4,\end{aligned}$$

where $\alpha = 0.8$ and the piecewise linear activation function is $f\left(x_j(t)\right) = 0.5\,(|x+1| - |x-1|)$ for $j = 1, 2$.
Parameters are defined as:

$$C = \operatorname{diag}(0.25, 0.2), \quad A = \begin{bmatrix} -0.05 & 0.01 \\ 0.02 & -0.01 \end{bmatrix}, \quad B = \begin{bmatrix} -0.01 & 0.02 \\ 0.02 & 0.01 \end{bmatrix}, \quad I = \begin{bmatrix} -0.1 \\ 0.4 \end{bmatrix}, \quad \tau = 0.1.$$

With $L_j = G_j = 1$, Assumption B is satisfied, leading to a unique equilibrium point $x^* = \left(x_1^*, x_2^*\right)^T$ given by:

$$\begin{aligned}-0.25x_1^* - 0.06f\left(x_1^*\right) + 0.03f\left(x_2^*\right) - 0.1 = 0, \\ -0.2x_2^* + 0.04f\left(x_1^*\right) + 0.4 = 0.\end{aligned}$$

The solution is $x^* = (-0.2258, 1.9548)$, which is uniformly stable according to Theorem 6.2.

Now, let's consider another numerical example to further illustrate the results mentioned above.

Example 6.2 Consider a two-state fractional-order delayed Hopfield neural network model described by the following equations:

$$\begin{cases}\mathscr{D}^{\alpha} x_1(t) = -0.3x_1(t) - 0.05g(x_1(t)) + 0.02g(x_2(t)) \\ -0.01g(x_1(t-\tau)) + 0.03g(x_2(t-\tau)) - 0.1, \\ \mathscr{D}^{\alpha} x_2(t) = -0.25x_2(t) + 0.03g(x_1(t)) - 0.02g(x_2(t)) \\ +0.01g(x_1(t-\tau)) + 0.02g(x_2(t-\tau)) + 0.5,\end{cases}$$

where $\alpha = 0.8$ and the activation function is defined by the sigmoid function

$$g(x) = \frac{1}{1+e^{-x}}.$$

The system uses a diagonal matrix

$$C = \operatorname{diag}(0.3, 0.25),$$

matrix

$$A = \begin{bmatrix} -0.05 & 0.02 \\ 0.03 & -0.02 \end{bmatrix},$$

and matrix

$$B = \begin{bmatrix} -0.01 & 0.03 \\ 0.01 & 0.02 \end{bmatrix}.$$

The input vector is given by

$$I = \begin{bmatrix} -0.1 \\ 0.5 \end{bmatrix}.$$

To find the equilibrium point

$$x^* = \begin{bmatrix} x_1^* \\ x_2^* \end{bmatrix},$$

the following equations must be satisfied:

$$\begin{cases} -0.3x_1^* - 0.05g(x_1^*) + 0.02g(x_2^*) - 0.1 = 0, \\ -0.25x_2^* + 0.03g(x_1^*) + 0.5 = 0. \end{cases}$$

By solving these equations, Assumption B is satisfied, and we can determine that the unique equilibrium point is approximately

$$x^* \approx \begin{bmatrix} -0.2 \\ 1.8 \end{bmatrix}.$$

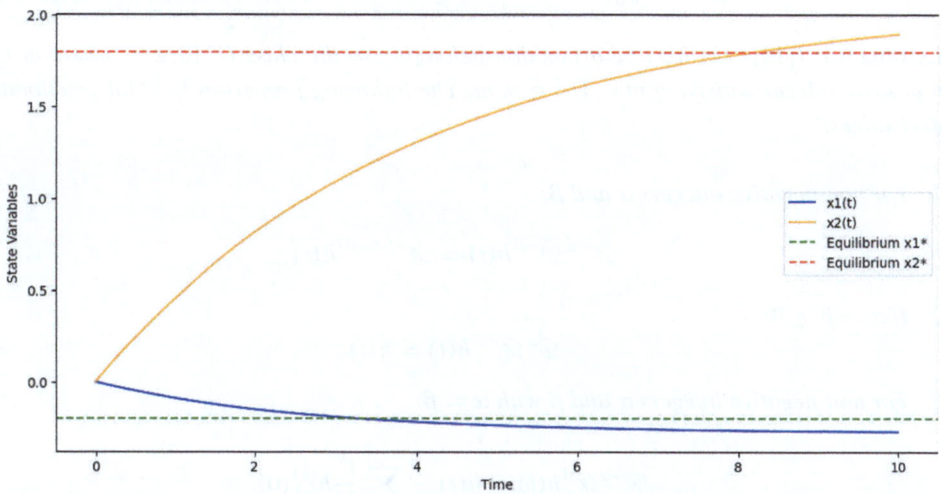

Fig. 6.2 Dynamics of two-state fractional-order delayed hopfield neural network

The stability analysis indicates that the equilibrium point x^* is uniformly stable, meaning that the system's trajectory will converge to this equilibrium point over time as in Fig. 6.2.

Chao Song and Jinde Cao [16] explored a broad category of FoNNs in 2014. To guarantee the existence and uniqueness of the nontrivial solution, several new suitable requirements are developed by applying the inequality approach, the Krasnosel'skiĭ fixed point theorem, and the contraction mapping principle. Additionally, it is suggested that FoNNs have uniform stability over predetermined time intervals. Lastly, a few instances are provided to show how useful the theoretical findings are.

A general class of FoNNs is examined in [16] and is defined as follows:

$$\begin{cases} {}^{c}\mathscr{D}^{\alpha} x_i(t) & = -c_i x_i(t) + \sum_{j=1}^{n} a_{ij} f_j(x_j(t)) + I_i, \quad t \geq 0, \\ x_i(0) & = x_{i0}, \end{cases} \tag{6.5}$$

where $0 < \alpha < 1$, and ${}^{c}\mathscr{D}^{\alpha}$ represents the Caputo fractional derivative. Here, n indicates the number of neurons in the network, $x_i(t)$ denotes the state of the i-th neuron at time t, and $f_j(x_j(t))$ reflects the activation response of the j-th neuron to its inputs at time t. The constant $c_i > 0$ (for $i = 1, \ldots, n$) is a positive coefficient, I_i is the constant input vector, and a_{ij} represents the strength of the synaptic connection between neurons in the network.

Definition 6.2 ([18]) The Caputo fractional derivative of a function $h(t)$ for an order $\alpha > 0$ is expressed as:

$${}^{c}\mathscr{D}^{\alpha} h(t) = \frac{1}{\Gamma(n-\alpha)} \int_0^t (t-s)^{n-\alpha-1} h^{(n)}(s)\, ds, \quad n-1 < \alpha < n,\ n \in \mathbb{N}^{+}.$$

Lemma 6.1 ([19]) *Let $h(t)$ be a function belonging to the class $C^m[0, \infty)$, where m is a positive integer satisfying $m - 1 < \alpha < m$. The following properties hold for fractional derivatives:*

1. *For non-negative integers α and β:*

$$\mathscr{D}^{-\alpha} \mathscr{D}^{-\beta} h(t) = \mathscr{D}^{-(\alpha+\beta)} h(t).$$

2. *If $\alpha = \beta \geq 0$:*

$$\mathscr{D}^{\alpha} \mathscr{D}^{-\beta} h(t) = h(t).$$

3. *For non-negative integers α and β with $\alpha = \beta$:*

$$\mathscr{D}^{-\alpha} \mathscr{D}^{\beta} h(t) = h(t) - \sum_{k=0}^{m-1} \frac{t^k}{k!} h^{(k)}(0).$$

Lemma 6.2 ([20]) *For $\alpha > 0$, the differential equation ${}^{c}\mathscr{D}^{\alpha} y(t) = h(t)$ has the solution:*

$$y(t) = I^{\alpha} h(t) + c_0 + c_1 t + c_2 t^2 + \cdots + c_{n-1} t^{n-1},$$

where $c_i \in \mathbb{R}$ *and* $n = [\alpha] + 1$.

Lemma 6.3 ([21] (Krasnosel'skiĭ's Fixed Point Theorem)): *Let D be a closed, convex, and nonempty subset of a Banach space X. Suppose two operators* ϕ_1 *and* ϕ_2 *satisfy the following conditions:*

- $\phi_1 x + \phi_2 y \in D$ *for all* $x, y \in D$;
- ϕ_1 *is compact and continuous;*
- ϕ_2 *is a contraction mapping.*

Then, there exists $x \in D$ *such that* $\phi_1 x + \phi_2 x = x$.

Following Lemma 6.2, the authors of [16] define the solution to system (6.5) as follows.

Definition 6.3 [16] A continuous function $x_i(t)$ is considered a solution of system (6.5) if it satisfies the following integral equation:

$$x_i(t) = x_{i0} + \int_0^t \frac{(t-s)^{\alpha-1}}{\Gamma(\alpha)} \left[-c_i x_i(s) + \sum_{j=1}^{n} a_{ij} f_j(x_j(s)) + I_i \right] ds.$$

Theorem 6.3 ([16]) *Consider the fractional-order neural network system* (6.5) *under the assumption that the neuron activation functions* f_i *(for* $i = 1, \ldots, n$*) are Lipschitz continuous, meaning that there exist non-negative constants* l_i *such that for any* $u, v \in \mathbb{R}$,

$$|f_i(u) - f_i(v)| \leq l_i |u - v| \leq l_0 |u - v|,$$

where $l_0 = \max_{1 \leq i \leq n} l_i$. *Given this assumption, the system* (6.5) *has a unique solution on the interval* $[0, T]$, *provided there exists a real number* $p > 1$ *such that*

$$c_0 T^{\alpha} n^{1/p} + l_0 T^{\alpha} \left(\sum_{i=1}^{n} \xi_i^p \right)^{1/p} < \Gamma(\alpha + 1), \tag{6.6}$$

where $\xi_i = \left(\sum_{j=1}^{n} |a_{ij}|^{p/(p-1)} \right)^{(p-1)/p}$ *and* $c_0 = \max_{1 \leq i \leq n} c_i$.

Sketch of the proof. The proof is based on the application of Banach's Fixed-Point Theorem to demonstrate the existence and uniqueness of the solution for the fractional-order neural network system under the assumption of Lipschitz continuity. We start by considering the fractional-order neural network system given by:

$$\mathscr{D}^{\alpha}x_i(t) = -c_i x_i(t) + \sum_{j=1}^{n} a_{ij} f_j(x_j(t)), \quad i = 1, \dots, n,$$

where $\mathscr{D}^{\alpha}$ is the Caputo fractional derivative, $x_i(t)$ represents the state of the neurons, and f_j are the neuron activation functions. Under the assumption that each activation function f_j satisfies a Lipschitz condition, i.e., there exists a constant $l_j > 0$ such that

$$|f_j(u) - f_j(v)| \leq l_j|u - v| \quad \forall u, v \in \mathbb{R},$$

we define $l_0 = \max_{1 \leq j \leq n} l_j$.

By utilizing the properties of the Caputo derivative, the system can be rewritten as an equivalent Volterra-type integral equation:

$$x_i(t) = x_i(0) + \frac{1}{\Gamma(\alpha)} \int_0^t (t-s)^{\alpha-1} \left(-c_i x_i(s) + \sum_{j=1}^{n} a_{ij} f_j(x_j(s)) \right) ds,$$

where $\Gamma(\alpha)$ is the Gamma function. This integral equation allows us to define a mapping Φ in a Banach space of continuous functions such that

$$\Phi(x_i(t)) = x_i(0) + \frac{1}{\Gamma(\alpha)} \int_0^t (t-s)^{\alpha-1} \left(-c_i x_i(s) + \sum_{j=1}^{n} a_{ij} f_j(x_j(s)) \right) ds.$$

Next, we show that this mapping Φ is a contraction. For two solutions $x(t)$ and $y(t)$, we estimate the difference:

$$|\Phi(x_i(t)) - \Phi(y_i(t))| \leq \frac{1}{\Gamma(\alpha)} \int_0^t (t-s)^{\alpha-1} \left(c_0|x_i(s) - y_i(s)| + l_0 \sum_{j=1}^{n} |a_{ij}||x_j(s) - y_j(s)| \right) ds.$$

Using the Lipschitz condition and appropriate bounds, we obtain:

$$\|\Phi(x) - \Phi(y)\| \leq \left(c_0 T^{\alpha} n^{1/p} + l_0 T^{\alpha} \left(\sum_{i=1}^{n} \xi_i^p \right)^{1/p} \right) \|x - y\|,$$

where $\xi_i = \left(\sum_{j=1}^{n} |a_{ij}|^{p/(p-1)} \right)^{(p-1)/p}$. To ensure that Φ is a contraction, the constant multiplying $\|x - y\|$ must be less than 1, which leads to the condition

$$c_0 T^{\alpha} n^{1/p} + l_0 T^{\alpha} \left(\sum_{i=1}^{n} \xi_i^p \right)^{1/p} < \Gamma(\alpha + 1).$$

Under this condition, Banach's fixed-point theorem guarantees the existence and uniqueness of the solution to the fractional-order system. Thus, the theorem is proved.

Theorem 6.4 ([16]) *If the system (6.5) satisfies below condition:*

- *for $i = 1, \dots, n$, there exists $M > 0$ such that for $u \in \mathbb{R}$, $|f_i(u)| \le M$.*

Then the system has at least one solution on $[0, T]$ if there exists a real number $p > 1$ such that

$$c_0 T^\alpha n^{1/p} < \Gamma(\alpha + 1).$$

Sketch of the proof. We will show that the system (6.5) has at least one solution on $[0, T]$ by constructing appropriate operators and invoking Krasnosel'skiĭ's fixed point theorem (Lemma 6.3).

Define operators L and N on the ball $B_\delta = \{X \in \mathbb{R}^n : \|X\| \le \delta\}$. The operator L is given by

$$(Lx)(t) = \left(x_{i0} + \int_0^t \frac{(t-s)^{\alpha-1}}{\Gamma(\alpha)} [-c_i x_i(s) + I_i] ds \right)_{i=1}^n,$$

and the operator N is defined as

$$(Nx)(t) = \left(\int_0^t \frac{(t-s)^{\alpha-1}}{\Gamma(\alpha)} \sum_{j=1}^n a_{ij} f_j(x_j(s)) ds \right)_{i=1}^n.$$

First, we demonstrate that $Lx + Ny \in B_\delta$ for any $x, y \in B_\delta$. Using the Minkowski inequality, we find that

$$\|Lx + Ny\| \le \left[\sum_{i=1}^n |x_{i0}|^p \right]^{1/p} + \frac{c_0 T^\alpha n^{1/p}}{\Gamma(\alpha+1)} \delta + \frac{I_0 T^\alpha n^{1/p}}{\Gamma(\alpha+1)} + \frac{M T^\alpha \left(\sum_{i=1}^n \xi_i^p \right)^{1/p}}{\Gamma(\alpha+1)},$$

which, by our choice of δ, shows that $Lx + Ny \in B_\delta$.

Next, we show that L is a contraction mapping. For any $x, y \in B_\delta$, we obtain

$$\|Lx - Ly\| \le \frac{c_0 T^\alpha n^{1/p}}{\Gamma(\alpha+1)} \|x - y\|.$$

This implies that L is a contraction.

We also demonstrate that N is compact. Since f_j are continuous functions, N is continuous. We find that

$$\|(Nx)(t)\| \le \frac{M T^\alpha \left(\sum_{i=1}^n \xi_i^p \right)^{1/p}}{\Gamma(\alpha+1)},$$

showing that N is uniformly bounded. To establish equicontinuity, we analyze

$$|(Nx)(t_1) - (Nx)(t_2)| \le \frac{M T^\alpha \left(\sum_{i=1}^n \xi_i^p \right)^{1/p}}{\Gamma(\alpha+1)} |2(t_1 - t_2)^\alpha + t_2^\alpha - t_1^\alpha|.$$

As $t_2 \to t_1$, the right-hand side tends to zero, indicating that $N(B_\delta)$ is relatively compact. By the Arzelá-Ascoli theorem, we conclude that N is compact.

Given that L is a contraction mapping and N is compact, we can apply Krasnosel'skiĭ's Fixed Point Theorem (Lemma 6.3), which guarantees the existence of at least one solution to the system (6.5) on the interval $[0, T]$. Thus, we conclude that the system has at least one solution.

Theorem 6.5 ([16]) *Consider the fractional-order neural network system* (6.5) *under the assumption that the neuron activation functions* f_i *(for* $i = 1, \ldots, n$*) are Lipschitz continuous, meaning that there exist non-negative constants* l_i *such that for any* $u, v \in \mathbb{R}$*,*

$$|f_i(u) - f_i(v)| \leq l_i|u - v| \leq l_0|u - v|,$$

where $l_0 = \max_{1 \leq i \leq n} l_i$. *By considering this assumption and the inequality 6.6 the solution of the system* (6.5) *is uniformly stable on* $[0, 1]$.

Sketch of the proof. To show that the solution of system (6.5) is uniformly stable on $[0, T]$, we consider two solutions $x_i(t)$ and $y_i(t)$ with initial conditions $x_i(0) = x_{i0}$ and $y_i(0) = y_{i0}$, where the initial states are close in the sense that

$$\left(\sum_{i=1}^{n} |x_{i0} - y_{i0}|^p\right)^{1/p} \leq \delta.$$

The solutions can be expressed as follows:

$$x_i(t) = x_{i0} + \int_0^t \frac{(t-s)^{\alpha-1}}{\Gamma(\alpha)} \left[-c_i x_i(s) + \sum_{j=1}^{n} a_{ij} f_j(x_j(s)) + I_i\right] ds,$$

$$y_i(t) = y_{i0} + \int_0^t \frac{(t-s)^{\alpha-1}}{\Gamma(\alpha)} \left[-c_i y_i(s) + \sum_{j=1}^{n} a_{ij} f_j(y_j(s)) + I_i\right] ds.$$

To analyze the stability, we calculate the distance between the two solutions:

$$\|x - y\| = \sup_{0 \leq t \leq T} \left(\sum_{i=1}^{n} |x_i(t) - y_i(t)|^p\right)^{1/p}.$$

By employing Gronwall's inequality and the properties of the integral, we can derive an inequality that relates $\|x - y\|$ to δ:

$$\|x-y\| \le \delta + \sup_{0\le t\le T}\left[\sum_{i=1}^{n}\left(\int_0^t \frac{c_i(t-s)^{\alpha-1}}{\Gamma(\alpha)}|x_i(s)-y_i(s)|ds\right)^p\right]^{1/p}$$
$$+\sup_{0\le t\le T}\left[\sum_{i=1}^{n}\left(\int_0^t \sum_{j=1}^{n}\frac{(t-s)^{\alpha-1}|a_{ij}|}{\Gamma(\alpha)}|f_j(x_j(s))-f_j(y_j(s))|ds\right)^p\right]^{1/p}.$$

We ultimately arrive at the inequality:

$$\|x-y\| \le \frac{\Gamma(\alpha+1)}{\Gamma(\alpha+1)-c_0T^\alpha n^{1/p}-l_0T^\alpha\left(\sum_{i=1}^{n}\xi_i^p\right)^{1/p}}\delta,$$

where c_0 and l_0 are constants derived from the system parameters.

To ensure stability, we select δ such that

$$\delta < \frac{\Gamma(\alpha+1)-c_0T^\alpha n^{1/p}-l_0T^\alpha\left(\sum_{i=1}^{n}\xi_i^p\right)^{1/p}}{\Gamma(\alpha+1)} > 0,$$

which implies that for any $\varepsilon > 0$, there exists a $\delta > 0$ such that $\|x-y\| \le \varepsilon$.

Thus, we conclude that the solution of system (6.5) is uniformly stable on $[0, T]$.

Moreover, without taking into account the nontrivial solution of FoNNs, the authors of [13, 22] examined the existence and uniqueness of equilibrium points for a class of FoNNs with or without time delay, accordingly.

In [23], the investigation focuses on the existence, uniqueness, and stability of solutions for a certain class of fractional-order neural networks. By applying the Banach fixed-point theorem, a sufficient condition is established to guarantee both the existence and uniqueness of solutions, as well as the uniform stability, for neural networks with fractional orders, time-varying coefficients, and multiple delays. Moreover, the authors of [23] considered the space $\Omega = (C([0,T],\mathbb{R}^n), \|\cdot\|)$, where $C([0,T],\mathbb{R}^n)$ represents the set of continuous functions that map into $\mathbb{R}^n$ as column vectors. For any $\xi \in C([0,T],\mathbb{R}^n)$, the norm is specified as

$$\|\xi\| = \sum_{i=1}^{n}\sup_t\left\{e^{-Nt}|\xi_i(t)|\right\}.$$

Additionally, for a matrix $A = \left(a_{ij}(t)\right)_{n\times n}$, the norm is defined as

$$\|A\| = \sum_{i=1}^{n}\sup_{t,\forall j}\left|a_{ij}(t)\right|.$$

In [23], the authors analyzed a neural network system characterized by time-varying coefficients and multiple delays. The system is described by the following equation:

$$\mathscr{D}^{\alpha} x_i(t) = -c_i(t)x_i(t) + \sum_{j=1}^{n} a_{ij}(t) f_j(x_j(t)) + \sum_{j=1}^{n} b_{ij}(t) f_j\left(x_j(t-\tau_{ij})\right) \\ + I_i(t), \quad t \in [0, T]. \tag{6.7}$$

Here, $T < +\infty$, and $\mathscr{D}^{\alpha}$ represents the Caputo fractional derivative of order α with $0 < \alpha < 1$. The variable i ranges from 1 to n, where n is the total number of neurons in the network, and $x_i(t)$ denotes the state of the i-th neuron at time t. The function $f_j(x_j(t))$ is the activation function for the j-th neuron.

The parameter $c_i(t) > 0$ reflects the rate at which neuron i reverts to its resting state when isolated from both the network and external input at time t. The terms $a_{ij}(t)$ and $b_{ij}(t)$ denote the synaptic strength of the connection from neuron j to neuron i at times t and $t - \tau_{ij}$, respectively. The variable $I_i(t)$ represents the external input received by neuron i at time t. The delay along the axon of neuron j is indicated by τ_{ij}, where $0 \leq \tau_{ij} \leq \tau = \max\left\{\tau_{ij} \mid i, j = 1, 2, \ldots, n\right\}$.

The initial conditions corresponding to Eq. 6.7 are expressed as:

$$x_i(t) = \phi_i(t), \quad t \in [-\tau, 0], \quad i = 1, 2, \ldots, n, \tag{6.8}$$

where $\phi_i(t) \in C([-\tau, 0], \mathbb{R})$. The norm of an element in $C([-\tau, 0], \mathbb{R}^n)$ is given by $\|\phi\| = \sum_{i=1}^{n} \sup_{t\in[-\tau,0]} \left\{e^{-Nt}|\phi_i(t)|\right\}$.

The following assumptions were made throughout their analysis:

- **Assumption 1**: The functions $c_i(t)$, $a_{ij}(t)$, $b_{ij}(t)$, and $I_i(t)$ are continuous on the interval $[0, T]$.
- **Assumption 2**: The activation functions f_j are Lipschitz continuous, meaning that for some positive constants L_j, we have:

$$|f_j(u) - f_j(v)| \leq L_j|u - v|, \tag{6.9}$$

for all $u, v \in \mathbb{R}$ and for each $j = 1, 2, \ldots, n$. To simplify further references, the following notations were stated for the model in Eq. 6.7:

$$\|A\| = \sum_{i=1}^{n} \sup_{t,\forall j} |a_{ij}(t)|, \quad \|B\| = \sum_{i=1}^{n} \sup_{t,\forall j} |b_{ij}(t)|,$$

$$c^* = \max\left\{\sup_t |c_1(t)|, \sup_t |c_2(t)|, \ldots, \sup_t |c_n(t)|\right\},$$

$$\text{and } L = \max\{L_1, L_2, \ldots, L_n\}.$$

Theorem 6.6 ([23]) *Given that Assumptions 1 and 2 are satisfied, the system described by Eq. 6.7 admits a unique solution*

$$x(t) = (x_1(t), x_2(t), \ldots, x_n(t))^T \in C\left([0, T], \mathbb{R}^n\right),$$

which adheres to the initial condition specified in Eq. 6.7.

Definition 6.4 ([23]) A solution of system (6.7) is said to be stable if, for any $\varepsilon > 0$ and $t_0 \geq 0$, there exists a corresponding $\delta(\varepsilon, t_0) > 0$ such that

$$\|y(t, t_0, \varphi) - x(t, t_0, \phi)\| < \varepsilon,$$

for all $t \geq t_0$, provided the initial conditions satisfy

$$\|\varphi(t) - \phi(t)\| < \delta(\varepsilon, t_0).$$

The solution is considered uniformly stable if δ can be chosen independently of t_0, i.e., $\delta(\varepsilon, t_0) \equiv \delta(\varepsilon)$.

Theorem 6.7 ([23]) *Provided that Assumptions 1 and 2 hold, the solution to the system defined by Eq. 6.7, which satisfies the initial condition in Eq. 6.8, exhibits uniform stability.*

Example 6.3 ([23]) Consider a fractional-order delayed neural network represented by the following differential equations:

$$\mathscr{D}^{0.7}x_1(t) = -2x_1(t) + 0.75f_1(x_1(t)) - 0.4f_2(x_2(t)) - 0.15f_1(x_1(t-0.01)) + 0.1f_2(x_2(t-0.01)) - 1.7,$$
$$\mathscr{D}^{0.7}x_2(t) = -x_2(t) - 0.25f_1(x_1(t)) + 0.6f_2(x_2(t)) - 0.12f_1(x_1(t-0.01)) - 0.7f_2(x_2(t-0.01)) + 1.2.$$

Here, the activation functions are defined as $f_1(x) = f_2(x) = |x + 0.6| - |x - 0.5|$. Although the parameters do not meet all conditions of Assumption B of Liping Chen et al. [13], Sierociuk and Petráš [12], the system still possesses a unique, uniformly stable solution. This solution corresponds to a fixed point defined by:

$$-2x_1^* + 0.6f_1(x_1^*) - 0.3f_2(x_2^*) - 1.7 = 0,$$
$$-x_2^* - 0.37f_1(x_1^*) - 0.1f_2(x_2^*) + 1.2 = 0.$$

The fixed point is $x^* = (-1.345, 1.497)^T$. Figure 6.3 illustrates how the solution converges to this fixed point over time.

In fixed-point theory, traditional approaches like *contraction mapping*, *Schauder*, and *Krasnosel'skiĭ* fixed-point results are commonly used to prove the existence of solutions and stability results in FoNNs. However, Panda et al. [24] proposed an alternative approach employing nonlinear F-contraction. Nonlinear F-contraction offers greater flexibility than

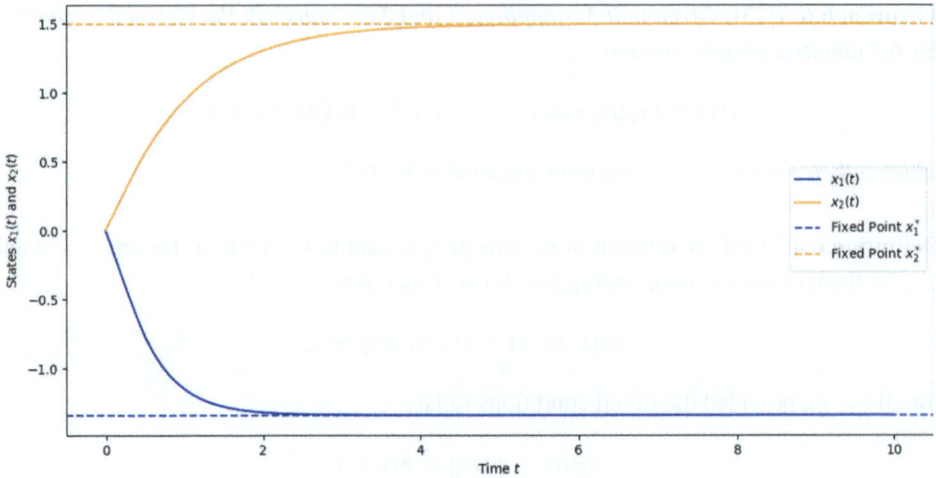

Fig. 6.3 Convergence of neural network solutions to fixed points

contractive mappings, which require a fixed and uniform contraction rate. These nonlinear contractions were used to study FoNNs. New necessary conditions ensure the existence and uniqueness of solutions for neural network systems with variable coefficients and numerous time delays. Additionally, they investigated FoNNs system uniform stability with time delay.

Definition 6.5 ([25]) An *F*-**contraction** is a mapping $f : \mathscr{M} \to \mathscr{M}$ satisfying:

$$\tau + F(\mathscr{D}(fx, fy)) \leq F(\mathscr{D}(x, y)),$$

$\forall\, x, y \in \mathscr{M}$ with $fx \neq fy$, where $\tau > 0$ and $F : (0, \infty) \to \mathbb{R}$ satisfies the below stated assertions:

$(F1)$ $\forall t_1, t_2 > 0$, $t_1 > t_2$ implies $F(t_1) > F(t_2)$.
$(F2)$ For any $(t_n) \subset (0, \infty)$, $t_n \to 0$ if and only if $F(t_n) \to -\infty$.
$(F3)$ There is $k \in (0, 1)$ satisfying $\lim_{t \to 0^+} t^k F(t) = 0$.

The function F drives the contractive characteristic in a nonlinear structure, permitting complex connections among the points within the metric space. The fixed-point theory continues to grow by the fact that every F-contraction on the complete metric space possesses a unique fixed point.

Turinici [26] found that the constraint $(F2)$ can be dropped to:

$(F2')$ $\lim_{t \to 0^+} F(t) = -\infty$.

From this, the implication:

$(F2'')$ $F(t_n) \to -\infty \implies t_n \to 0.$

can be derived from $(F1)$.

Expanding upon these notions, Wardowski [27] have presented the notion of (φ, F)-contraction (or nonlinear F-contraction). These mappings improve F-contraction by putting in another function, $\varphi : (0, \infty) \to (0, \infty)$, which additionally develops the contraction criteria.

Definition 6.6 ([27]) A mapping $f : \mathcal{M} \to \mathcal{M}$ is a (φ, F)-contraction (or nonlinear F-contraction) if there exist functions $F : (0, \infty) \to \mathbb{R}$ and $\varphi : (0, \infty) \to (0, \infty)$ gratifying:

(H1) F satisfies $(F1)$ and the relaxed condition $(F2')$.
(H2) $\liminf_{s \to t^+} \varphi(s) > 0$ for all $t \geq 0$.
(H3) $\forall x, y \in \mathcal{M}$, $\varphi(\mathcal{D}(x, y)) + F(\mathcal{D}(fx, fy)) \leq F(\mathcal{D}(x, y))$ in such a way that $fx \neq fy$.

The given hypothesis outlines the broader setting for these types of contractions and mentions that various known contractions can be derived as special cases of these generalized forms, providing a more comprehensive framework for fixed point theory in metric spaces.

Theorem 6.8 ([24]) *Let $(\mathcal{M}, \mathcal{S})$ be a complete suprametric space and let $f : \mathcal{M} \to \mathcal{M}$ be a (φ, F)-contraction. Then f possess a unique fixed point.*

Theorem 6.9 ([24]) *The problem of multiple time delays and variable coefficients in the FoNNs system (6.7) has a unique solution satisfying the initial condition (6.8), provided the following assertions hold.*

$(\mathcal{A}_1)$. *$c_i(t), a_{ij}(t), b_{ij}(t)$ and $I_i(t)$ are continuous on $[0, T]$.*
$(\mathcal{A}_2)$. *The activation functions f_j are Lipschitz continuous, i.e., there exist positive constants L_j such that:*

$$|f_j(\gamma) - f_j(\mu)| \leq L_j |\gamma - \mu|,$$

for all $\gamma, \mu \in \mathbb{R}$, where $j = 1, 2, \ldots n$.
$(\mathcal{A}_3)$. *There exists a strictly increasing sequence $(\flat_n)_{n \in \mathbb{N} \cup \{0\}}$ satisfying,*

$$\flat_0 = 0,\ \flat_n \geq 1,\ \flat_n - \flat_{n-1} \leq 1 \text{ for all } n \in \mathbb{N}.$$

$(\mathcal{A}_4)$. $\left(\frac{c^*}{N^\alpha} + \frac{(\|A\| + \|B\|)L}{N^\alpha} \right) < \frac{1}{2 + 3\flat_n^2}.$

In order to obtain the uniform stability, they define $\mathcal{M} = C([0, T], \mathbb{R}^n)$ the space of all continuous functions. Let $\mathcal{M} = \{x / x$ is column n-vector function $\}$; i.e., $\mathcal{M} =$

$(x_1, x_2, \dots x_n)^T$ where $x_i(t) \in C([0, T], \mathbb{R})$. We know that $(\mathscr{M}, \mathscr{S})$ is a suprametric space equipped with the norm $\|x\|_p = \sup_{0 \leqslant t \leqslant T} \left(\sum_{i=1}^n |x_i(t)|^p\right)^{\frac{1}{p}}$; where $p > 1$, and the associated suprametric is $\mathscr{S}(x(t), y(t)) = \sup_{0 \leqslant t \leqslant T} \left(\sum_{i=1}^n |x_i(t) - y_i(t)|^p\right)^{\frac{1}{p}}$ for $\rho > 0$.

The initial conditions associated with the system (6.7) are of the form:

$$x_i(t) = \phi_i(t), t \in [-\tau, 0], i = 1, 2 \dots n,$$

where $\phi_i(t) \in C([-\tau, 0], \mathbb{R})$, and the norm of an element in $C([-\tau, 0], \mathbb{R}^n)$ is

$$||\phi|| = \sup_{0 \leq t \leq T} \left(\sum_{i=1}^{n} |\phi_i(t)| \right).$$

Theorem 6.10 ([24]) *If Assumptions 1 and 2 are met, the FoNNs system solution* (6.7) *meets the initial condition* (6.8) *and is uniformly stable.*

Example 6.4 Consider the following delayed FoNNs described by the differential equation system:

$$\begin{cases} \mathscr{D}^\alpha y_1(t) &= -3y_1(t) + 0.5 f_1(y_1(t)) + 0.2 f_2(y_2(t)) - 0.1 f_1(y_1(t-\tau)) + 0.05 f_2(y_2(t-\tau)) - 2.5 \\ \mathscr{D}^\alpha y_2(t) &= -2y_2(t) + 0.3 f_1(y_1(t)) - 0.4 f_2(y_2(t)) - 0.2 f_1(y_1(t-\tau)) + 0.15 f_2(y_2(t-\tau)) + 0.8. \end{cases} \tag{6.10}$$

where the fractional order $\alpha = 0.6$, the activation functions are given by $f_1(x) = \tanh(x + 0.5)$ and $f_2(x) = \sin(x - 0.4)$, and the time delay $\tau = 0.02$.

To find the fixed point (y_1^*, y_2^*), we set $\mathscr{D}^\alpha y_1(t) = 0$ and $\mathscr{D}^\alpha y_2(t) = 0$:

$$\begin{cases} -3y_1^* + 0.5 f_1(y_1^*) + 0.2 f_2(y_2^*) - 2.5 = 0 \\ -2y_2^* + 0.3 f_1(y_1^*) - 0.4 f_2(y_2^*) + 0.8 = 0. \end{cases} \tag{6.11}$$

By using python software, we find that the fixed point (y_1^*, y_2^*) is approximately:

$$y_1^* \approx -0.90, \quad y_2^* \approx 0.35 \text{ as shown in Fig. 6.4.}$$

The authors of [28] investigated Caputo fractional-order BAM neural networks with delays in the leakage terms, utilizing fractional calculus, the contraction mapping principle, Krasnoselskii fixed-point theorem, and inequality techniques. They established new sufficient conditions for the existence and uniqueness of nontrivial solutions and proposed uniform stability within fixed time intervals. An illustrative example was provided to demonstrate the effectiveness of the results.

Very recently, the authors of [29] introduced a fractional-order neural network (FoNN) for system identification, incorporating the Grünwald-Letnikov fractional derivative in the learning algorithm. The proposed approach was validated by identifying two benchmark systems and one real-world system. Comparisons with other models showed that the FoNN achieved better accuracy with fewer parameters.

In the study of [30] addressed the uniform stability of fuzzy fractional-order genetic regulatory networks with random discrete delays, distributed delays, and parameter uncertainties. Unlike previous works that focused on fractional orders in (1, 2), this research explored stability for fractional orders within (0, 1). By employing fixed point theory, the authors established the uniqueness of the solution and derived a novel sufficient condition for the uniform stability of these networks. An example was provided to validate the effectiveness of the results.

6.1.2 Finite-Time Stability Results Under Caputo-Katugampola Derivative

A class of FoNNs with delayed systems and orders $\alpha \in (0, 0.5]$ and $\alpha \in (0.5, 1)$ were examined for their finite-time stability in [15]. On the other hand, the authors constructed sufficient conditions that ensure the stability of FoNNs for a certain amount of time. It is demonstrated that the acquired limitations are less conservative than those found in earlier studies. They provided two numerical examples to support their conclusions.

In the quest for finite-time stability, the study in [31] examines the stability characteristics of a certain class of fractional-order neural networks. Using tools such as the Laplace transform, the generalized Gronwall inequality, and bounds on Mittag–Leffler functions, the authors derive sufficient conditions that guarantee finite-time stability for these neural

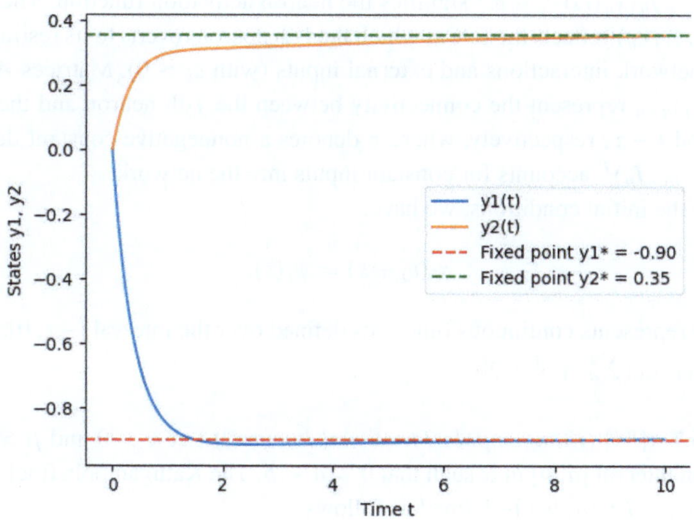

Fig. 6.4 The dynamic behaviour of system (6.11)

systems governed by the Caputo fractional derivatives. In addition to these findings, [31] also provides insights into the asymptotic stability of fractional-order neural networks. The authors of [32] investigated proportional delays in FoNNs. Using appropriate inequality techniques, they obtained adequate requirements to guarantee the stability of continuous FoNNs during a finite time interval. The investigators of [32] supply numerical examples to support their theoretical findings. The results of this innovative work enhance recent advancements in the field of FoNNs.

In the study of [33] proposed a novel scheme to analyze the finite-time stability, existence, and uniqueness of the equilibrium point in Caputo–Katugampola FoNNs with time delay. The scheme leverages the recently introduced Caputo–Katugampola fractional derivative. The effectiveness of the theoretical results was demonstrated through simulations of two numerical examples. The authors of [33] delve into the study of fractional-order neural networks characterized by time delays, which are represented mathematically as follows:

$$ {}^{C}\mathscr{D}_{t_0}^{\alpha,\beta} x_i(t) = -c_i x_i(t) + \sum_{j=1}^{n} a_{ij} f_j(x_j(t)) + \sum_{j=1}^{n} b_{ij} f_j(x_j(t-\tau)) + I_i, \quad i = 1, 2, \ldots, n. \tag{6.12} $$

In vector form, this can be expressed as:

$$ {}^{C}\mathscr{D}_{t_0}^{\alpha,\beta} x(t) = -Cx(t) + Af(x(t)) + Bf(x(t-\tau)) + I, \tag{6.13} $$

for $t \in [t_0, t_f]$, where ${}^{C}\mathscr{D}_{t_0}^{\alpha,\beta}$ denotes the Caputo-Katugampola fractional derivative. The parameters of the derivative are defined as $0 < \alpha < 1$ and $\beta > 0$. Here, $x(t) = (x_1(t), x_2(t), \ldots, x_n(t))^T \in \mathbb{R}^n$ represents the state vector, and $f(x(t)) = (f_1(x_1(t)), f_2(x_2(t)), \ldots, f_n(x_n(t)))^T \in \mathbb{R}^n$ signifies the neuron activation function. The matrix $C = \mathrm{diag}(c_1, c_2, \ldots, c_n)$ reflects the rate at which the i-th neuron reverts to its resting state in the absence of network interactions and external inputs (with $c_i > 0$). Matrices $A = (a_{ij})_{n\times n}$ and $B = (b_{ij})_{n\times n}$ represent the connectivity between the j-th neuron and the i-th neuron at times t and $t - \tau$, respectively, where τ denotes a nonnegative constant delay. Finally, $I = (I_1, I_2, \ldots, I_n)^T$ accounts for constant inputs into the network.

To define the initial conditions, we have:

$$ x_i(t_0 + s) = \psi_i(s), \tag{6.14} $$

where $\psi_i(s)$ represents continuous functions defined over the interval $[-\tau, 0]$. It holds that $\|\psi\| = \sup_{s\in[-\tau,0]} \sum_{i=1}^{n} |\psi_i(s)|$.

Definition 6.7 ([34] *(Katugampola Fractional Integral)*) Let $\alpha > 0$ and $\beta > 0$ be given, along with an interval $[a, b]$ in $\mathbb{R}$ such that $0 < a < b$. The Katugampola fractional integral of a function $x \in L^1([a, b])$ is defined as follows:

$$I_{a^+}^{\alpha,\beta} u(t) = \frac{\beta^{1-\alpha}}{\Gamma(\alpha)} \int_a^t \frac{s^{\beta-1} u(s)}{(t^\beta - s^\beta)^{1-\alpha}} ds,$$

where Γ denotes the gamma function.

Definition 6.8 ([35] *(Katugampola Fractional Derivative)*) For $0 < \alpha < 1$ and $\beta > 0$ over an interval $[a, b]$ of $\mathbb{R}$ with $0 < a < b$, the Katugampola fractional derivative is expressed as:

$$\mathscr{D}_{a^+}^{\alpha,\beta} u(t) = \frac{\beta^\alpha}{\Gamma(1-\alpha)} t^{1-\beta} \frac{d}{dt} \int_a^t \frac{s^{\beta-1} u(s)}{(t^\beta - s^\beta)^{1-\alpha}} ds.$$

Definition 6.9 ([35] *(Caputo-Katugampola Fractional Derivative)*) For $0 < \alpha < 1$ and $\beta > 0$ defined on an interval $[a, b]$ of $\mathbb{R}$ where $0 < a < b$, the Caputo-Katugampola fractional derivative is given by:

$$\begin{aligned} {}^C\mathscr{D}_{a^+}^{\alpha,\beta} u(t) &= {}^C\mathscr{D}_{a^+}^{\alpha,\beta} u(t)[u(t) - u(a)] \\ &= \frac{\beta^\alpha}{\Gamma(1-\alpha)} t^{1-\beta} \frac{d}{dt} \int_a^t \frac{s^{\beta-1}[u(s) - u(a)]}{(t^\beta - s^\beta)^{1-\alpha}} ds. \end{aligned}$$

Definition 6.10 ([35] *(Caputo-Katugampola Fractional Derivative)*) For $0 < \alpha < 1$ and $\beta > 0$ defined on an interval $[a, b]$ of $\mathbb{R}$ where $0 < a < b$, the Caputo-Katugampola fractional derivative is given by:

$$\begin{aligned} {}^C\mathscr{D}_{a^+}^{\alpha,\beta} u(t) &= {}^C\mathscr{D}_{a^+}^{\alpha,\beta} u(t)[u(t) - u(a)] \\ &= \frac{\beta^\alpha}{\Gamma(1-\alpha)} t^{1-\beta} \frac{d}{dt} \int_a^t \frac{s^{\beta-1}[u(s) - u(a)]}{(t^\beta - s^\beta)^{1-\alpha}} ds. \end{aligned}$$

Definition 6.11 ([35]) An equilibrium point $x^* = (x_1^*, x_2^*, \ldots, x_n^*)^T$ of system (1) is considered finite-time stable with respect to the parameters $\{t_0, \delta, \varepsilon, \Theta, \tau\}$, where $0 < \delta < \varepsilon$ and $\delta, \varepsilon \in \mathbb{R}$, and $\Theta = [t_0, t_0 + t_f]$. This stability condition holds if, for any solution $x(t) = (x_1(t), x_2(t), \ldots, x_n(t))^T$ of system (1) under the initial conditions (3), the following implication is satisfied:

$$\|\psi - x^*\| < \delta \implies \|x(t) - x^*\| < \varepsilon, \quad \forall t \in \Theta,$$

where

$$\|\psi - x^*\| = \sup_{s\in[-\tau,0]} \sum_{i=1}^n |\psi_i(s) - x_i^*|, \quad \|x(t) - x^*\| = \sum_{i=1}^n |x_i(t) - x_i^*|.$$

Lemma 6.4 ([36]) *For $n \in \mathbb{N}$ and nonnegative real numbers $a_1, a_2, \ldots, a_n$, it holds that for $l > 1$:*

$$\left(\sum_{i=1}^{n} a_i\right)^l \leq n^{l-1} \sum_{i=1}^{n} a_i^l.$$

Lemma 6.5 ([37]) *Let $p, q > 1$ such that $\frac{1}{p} + \frac{1}{q} = 1$. If $f_1 \in L^p([a,b])$ and $f_2 \in L^q([a,b])$, then the product $f_1 f_2 \in L^1([a,b])$ and satisfies:*

$$\int_a^b |f_1(x) f_2(x)|\, dx \leq \left(\int_a^b |f_1(x)|^p\, dx\right)^{\frac{1}{p}} \left(\int_a^b |f_2(x)|^q\, dx\right)^{\frac{1}{q}},$$

where $L^p([a,b])$ denotes the Banach space of all Lebesgue measurable functions $f : [a,b] \to \mathbb{R}$ for which $\int_a^b |f(x)|^p\, dx < \infty$. Specifically, if $p = q = 2$, this reduces to the Cauchy-Schwarz inequality:

$$\int_a^b |f_1(x) f_2(x)|\, dx \leq \left(\int_a^b |f_1(x)|^2\, dx\right)^{\frac{1}{2}} \left(\int_a^b |f_2(x)|^2\, dx\right)^{\frac{1}{2}}.$$

Before presenting their primary findings, the authors established the following hypotheses:

(H1) The functions f_i are Lipschitz continuous. There exist positive constants $F_i > 0$ such that for any $x, y \in \mathbb{R}$ (where $i = 1, 2, \ldots, n$), the following inequality holds:

$$\|f_i(x) - f_i(y)\| \leq F_i \|x - y\|.$$

(H2) Given the parameters a_{ij}, b_{ij}, c_i and constants F_j, the following condition is satisfied:

$$\sum_{i=1}^{n} \max_{1 \leq j \leq n} \left\{ \left|a_{ij}\right| F_j + \left|b_{ij}\right| F_j \right\} < \min_{1 \leq i \leq n} \{c_i\}.$$

Theorem 6.11 ([33]) *Assuming that conditions (H1) and (H2) are fulfilled and that $0 < \alpha \leq \frac{1}{2}$, if the following criterion is met:*

$$\sqrt[q]{3^{q-1} + N A_2^q \left[(t_0+\tau)^\beta - t_0^\beta\right]} \exp\left[t_f^\beta + N\left(\frac{A_1^q}{q}\left(t_f^\beta - t_0^\beta\right) + \frac{A_1^q}{q}\left((t_f+\delta)^\beta - (t_0+\delta)^\beta\right)\right)\right] < \frac{\varepsilon}{\delta}, \tag{6.15}$$

where $N = \frac{3^{q-1} p^{-q\alpha}}{(\Gamma(\alpha))^q} \left(\frac{\Gamma(p(\alpha-1)+1)}{p^{\alpha p} - p + 1}\right)^{\frac{q}{p}}$, $p = 1 + \alpha$, and $q = \frac{\alpha+1}{\alpha}$. Here, $A_1 = \max_{1 \leq i \leq n}(c_i) + \sum_{i=1}^{n} \max_{1 \leq j \leq n}\left(\left|a_{ij}\right| F_j\right)$, and $A_2 = \sum_{i=1}^{n} \max_{1 \leq j \leq n}\left(\left|b_{ij}\right| F_j\right)$. Then, a unique equilibrium point $x^* = (x_1^*, x_2^*, \ldots, x_n^*)^T$ exists for system 6.12, which is stable in finite time with respect to the parameters $\{t_0, \delta, \varepsilon, \Theta, \tau\}$, where $0 < \delta < \varepsilon$ and $\delta, \varepsilon \in \mathbb{R}$, and $\Theta = [t_0, t_0 + t_f]$.

Theorem 6.12 ([33]) *Let α be in the interval $\left(\frac{1}{2}, 1\right)$, and consider a fractional-order system 6.12 that adheres to the initial condition $x\,(t_0 + s) = \varphi(s)$ for $-\tau < s < 0$. For the system to exhibit finite-time stability, the following inequality must hold:*

$$\sqrt{3 + 2MA_2^2\left((t_0+\tau)^\beta - t_0^\beta\right)\exp\left(T^\beta + M\left(A_1^2\left(t_f^\beta - t_0^\beta\right) + A_2^2\left((t_f+\tau)^\beta - (t_0+\tau)^\beta\right)\right)\right)} < \frac{\varepsilon}{\delta}, \tag{6.16}$$

where M is defined as

$$M = \frac{3\beta^{-2\alpha}\Gamma(2\alpha - 1)}{4^\alpha(\Gamma(\alpha))^2}.$$

Under these conditions, the system represented by (6.12) *is considered to be finite-time stable concerning the parameters $\{t_0, \delta, \varepsilon, \Theta, \tau\}$ with the constraint that $\delta < \varepsilon$.*

In their exploration of fractional-order neural networks with time delays, the authors of [33] introduce a sophisticated framework for assessing finite-time stability, as well as the existence and uniqueness of equilibrium points. Central to this investigation is the Caputo–Katugampola fractional derivative, which is pivotal for capturing the dynamics of such systems. The mathematical representation of these neural networks is articulated through a fractional-order differential equation, expressed in both component and vector forms, effectively delineating the interrelationships among neurons influenced by time-varying delays. Initial conditions are carefully stipulated, ensuring the continuity of the functions that govern the state of the system. Moreover, the authors of this article discuss fractional-integrals and derivatives, pointing-out the impact of the parameters α and β on derivative operations. Key conditions (H1) and (H2) are posited to facilitate the verification of stability, ultimately leading to the derivation of significant theorems. These theorems stipulate conditions under which a unique equilibrium point exists, ensuring its finite-time stability, thus providing critical insights into the dynamics of fractional-order systems. The validity of the theoretical assertions is substantiated through the implementation of numerical simulations, reinforcing the practical applicability of their findings.

The proofs of Theorems 6.11 and 6.12 are constructed upon the foundational results presented in the authors–of [33] preceding work [38]. The inherent complexities associated with the simultaneous demonstration of both theorems under the constraint $0 < \alpha \leq 1$ necessitated a strategic division of the analysis. By delineating the proofs into two distinct theorems, the authors enhance the clarity and rigor of their argumentation. This bifurcation allows for a more focused examination of the underlying mathematical principles, ensuring that the intricacies of each theorem are thoroughly addressed without the conflation of their respective proofs. Consequently, this methodical approach not only fortifies the validity of the results but also elucidates the relationships between the theorems in the context of fractional-order dynamics.

Remark 6.2 When $\beta = 1$, system 6.12 simplifies to a fractional-order neural network governed by the Caputo derivative. In this specific scenario, the analysis regarding the existence of a singular equilibrium point, along with the conditions for finite-time stability, has been extensively explored in references [39, 40]. These studies provide the necessary theoretical framework to establish the uniqueness of the equilibrium point and confirm the system's stability behavior within a finite time stability under the Caputo derivative formulation.

The article of [33] represents the pioneering work addressing the problem of finite-time stability in Caputo–Katugampola fractional-order neural networks with time delay. No prior studies have specifically focused on this problem within the framework of Caputo–Katugampola fractional derivatives.

In the study of [41], the authors investigated the finite-time stability of Caputo–Katugampola fractional-order projection neural networks with time delay. Utilizing the Sadovskii fixed point theorem, the Banach fixed point theorem, and the generalized Gronwall inequality, the authors established existence and boundedness theorems for solutions of these neural networks. Additionally, by applying the proposed theorems and inequality techniques, they derived conditions ensuring the finite-time stability of the equilibrium point for the system. The effectiveness of their theoretical findings was further demonstrated through simulations on a numerical example.

However, very recently, [42] introduced a novel fractional-order Gronwall integral inequality using the beta function to explore the existence, uniqueness, and finite-time stability of the zero solution for a class of Caputo–Katugampola fractional-order neural networks with time delay. Based on this inequality, the authors derived a new condition to guarantee the finite-time stability of the system with fractional orders between 0 and 1. Under specific conditions, the results were shown to be less conservative compared to existing literature. Finally, two numerical examples were provided to demonstrate the validity of the proposed results.

Remark 6.3 If $\rho = 1$, condition (H3) of [42] becomes a particular example of Theorem 3.5 of [43].

Remark 6.4 Theorem 3.2 of [42] is less conservative than the corresponding result in [33] due to its broader applicability and less restrictive conditions. This is theoretically obvious when considering each Theorems' fixed-time estimates. The finite-time stability parameters in [33] needs stronger time-delays and fractional-order parameters, which leads to, in a lower estimate of the time T^* to settle. Nevertheless, Theorem 3.2 permits for a longer duration T, enable the system to sustain stability for longer under the same conditions.

This difference arises from the fact that Theorem 3.2 uses a time-delay-independent criterion, which enables the time T to be maximized over a broader parameter space. As a result, the fixed time T in Theorem 3.2 is consistently larger than T^* in [33], indicating that the system can remain stable for a longer duration before reaching the specified error bounds. This makes Theorem 3.2 less conservative in practice, as demonstrated by the numerical comparisons where larger T values are obtained.

Remark 6.5 Theorem 3.2 [42] of has been numerically demonstrated to be less conservative than the corresponding result in [33]. Through two numerical examples, it has been shown that Theorem 3.2 provides a larger estimate for the fixed time T, meaning that the system maintains stability for a longer duration compared to the time T^* derived in [33]. This is primarily due to the less restrictive nature of condition (H3) in [42], which does not require the time delay to be sufficiently small. The numerical comparisons clearly highlight that Theorem 3.2 offers a more flexible and efficient criterion for stability, as evidenced by the larger T values obtained in the simulations.

6.2 Dynamics of Fractional-Order Complex-Valued Neural Networks

Fractional-order Complex-valued Neural Networks (FoCvNNs) represent an intriguing blend of neural network theory and fractional calculus, providing a powerful framework for modeling dynamic systems that exhibit memory and hereditary properties. These networks incorporate complex-valued signals, which allow for richer representations of information, especially in fields like signal processing and quantum computing. When fractional-order derivatives are applied to such systems, they provide more flexibility due to the capability of the network behavior to be driven by the previous states in a continuous time scale and not on a discrete time scale as it is usually in the traditional models.

FoCvNNs can be mathematically described by the system of differential equations containing a fractional derivative, usually in the Caputo or Riemann-Liouville sense. The ordinary neuron in the network satisfies an equation of the type

$$\mathscr{D}_t^{\alpha} z(t) = -\lambda z(t) + f(Wz(t) + b), \tag{6.17}$$

where $\mathscr{D}_t^{\alpha}$ represents the fractional derivative of order α, $z(t)$ is the complex-valued state of the neuron, λ is a decay constant, and f is a complex activation function. The fractional-order α offers a degree of freedom which can minutely adjust the temporal response of the network, and thus FoCvNNs are expected to be very surpassing at learning long-term dependencies in intricate data. Such a combination of complex-valued and fractional-order components provides a flexible solution to real-life problems, which are multi-dimensional and dynamic in nature.

Compared to the classical FoNNs, FoCvNNs have the following key benefits since they process complex-valued signals: they can represent more complex dynamics; they can process data with a magnitude and a phase part. Mathematically, a neuron's state in an FoCvNN is represented as a complex number $z(t) = x(t) + iy(t)$, where $x(t)$ and $y(t)$ are real and imaginary parts, respectively. This is because it can process two-dimensional information in parallel, which is important in signals processing where the phase and amplitude are important. The use of activation function in FoCvNNs, on complex numbers, also increases the ability of the network to learn non-linearities and phase-magnitude relationships. One important characteristic of such networks is that they use fractional-order derivatives, namely the Caputo derivative $\mathscr{D}_t^\alpha$, which adds a memory effect, whose duration extends in time, enabling the network to remember long-term dependencies. The dynamics of the complex-valued neuron state are governed by the equations such as (6.17), with the fractional-order α allowing more compliant dynamics. In comparison with FONNs, where the data is restricted to real values, FoCvNNs offer improved stability and convergence properties, and thus are well-suited to problems exhibiting an oscillatory solution, or where it is needed to model a complex time-dependence. Such a combination of fractional-order dynamics and complex-valued representations makes FoCvNNs very useful in a broad variety of complex and multi-dimensional tasks.

6.2.1 Various Stability Results Under Caputo Derivative

This contribution discusses the stability of FoCvNNs with time delays and highlights the necessary requirements for uniform stability. Furthermore, some helpful conclusions previously published in the literature are presented regarding the existence and uniqueness of FoCvNNs with time delays.

In [44], the authors examined a continuous fractional-order complex-valued neural network model with time delays, expressed as:

$$\mathscr{D}^\alpha z_j(t) = -c_j z_j(t) + \sum_{k=1}^{n} a_{jk} f_k(z_k(t)) + \sum_{k=1}^{n} b_{jk} g_k(z_k(t-\tau)) + U_j. \tag{6.18}$$

Alternatively, in vector notation, it can be written as:

$$\mathscr{D}^\alpha z(t) = -Cz(t) + Af(z(t)) + Bg(z(t-\tau)) + U. \tag{6.19}$$

Here, $\mathscr{D}^\alpha$ represents the Caputo fractional-order derivative of order α where $0 < \alpha < 1$, and $z = (z_1, z_2, \ldots, z_n)^T \in \mathbb{C}^n$ is the state vector. The matrix $C = \text{diag}(c_1, c_2, \ldots, c_n) \in \mathbb{R}^{n\times n}$, where $c_j > 0$ for $j = 1, 2, \ldots, n$, is the self-feedback connection weight matrix. The matrices $A = (a_{jk})_{n\times n} \in \mathbb{C}^{n\times n}$ and $B = (b_{jk})_{n\times n} \in \mathbb{C}^{n\times n}$ represent the connection weights without and with time delays, respectively. The activation functions $f(z(t)) = (f_1(z_1(t)), f_2(z_2(t)), \ldots, f_n(z_n(t)))^T$ and $g(z(t-\tau)) = (g_1(z_1(t-\tau_1)), g_2(z_2(t-\tau_2)), \ldots, g_n(z_n(t-\tau_n)))^T$ are vector-valued, mapping $\mathbb{C}^n \longrightarrow \mathbb{C}^n$, and involve complex-valued nonlinear functions. The constant τ denotes the time delay, and $U = (U_1, U_2, \ldots, U_n)^T \in \mathbb{C}^n$ is the external input vector.

The initial conditions corresponding to equation (4) are given as:

$$z_i(s) = \varphi_i(s) + \mathbf{i}\phi_i(s), \quad s \in [-\tau, 0], \quad i \in n. \tag{6.20}$$

It is commonly assumed that $\varphi_i(s), \phi_i(s) \in C([-\tau, 0], \mathbb{R}^n)$ for $i \in n$, where $C([-\tau, 0], \mathbb{R}^n)$ refers to the Banach space of continuous real vector functions defined on the interval $[-\tau, 0]$. The norms of these functions are defined as $\|\varphi(t)\| = \sum_{i=1}^{n} \sup_{t\in(-\tau,0]}\{e^{-t}|\varphi_i(t)|\}$ and $\|\phi(t)\| = \sum_{i=1}^{n} \sup_{t\in(-\tau,0]}\{e^{-t}|\phi_i(t)|\}$.

Definition 6.12 ([44]) A solution to Eq. 6.18 is considered stable if, for any chosen value of $\varepsilon > 0$, there exists a corresponding $\delta(t_0, \varepsilon) > 0$ such that, for $t \geq t_0 \geq 0$, if $\|\chi(t) - \psi(t)\| < \delta$, then it follows that $\|z_1(t, t_0, \chi) - z(t, t_0, \psi)\| < \varepsilon$. This applies to any pair of solutions, $z_1(t, t_0, \chi)$ and $z(t, t_0, \psi)$. When this condition holds irrespective of t_0, the solution is said to be uniformly stable.

They considered $f_j(\cdot)$ as a collection of complex-valued functions where $j = 1, 2, \ldots, n$. These functions fall into two categories of complex activation functions that meet the requirements of Assumptions 3 and 4. To derive core results, they considered the following assumptions.

Assumption 3: The values of $c_i, a_{jk}, b_{jk}, \lambda_k$, and μ_k satisfy the following constraints:

$$\sum_{j=1}^{4} \|p_j^*\| + \sum_{j=1}^{4} \|q_j^*\| + \sum_{j=1}^{4} \|s_j^*\| + \sum_{j=1}^{4} \|t_j^*\| < \bar{c},$$

where:

$$\bar{c} = \min\left(|1 - c_{\max}|, c_{\min}\right), \quad c_{\max} = \max_{\forall j}\{c_j\}, \quad c_{\min} = \min_{\forall j}\{c_j\}.$$

Additionally, they defined the norms as follows:

$$\|p^*\| = \sum_{j=1}^{n} |p_j^*| = \sum_{j=1}^{n} \max_{\forall k}\left\{|a_{jk}|\lambda_k\right\}, \quad \|q^*\| = \sum_{j=1}^{n} |q_j^*| = \sum_{j=1}^{n} \max_{\forall k}\left\{|b_{jk}|\mu_k\right\}.$$

We break down each of the terms for clarity:

$$\|p_1^*\| = \sum_{j=1}^{n} |p_{1j}^*| = \sum_{j=1}^{n} \max_{\forall k}\left\{|a_{jk}^R|\lambda_k^{RR}\right\}.$$

$$\|p_2^*\| = \sum_{j=1}^{n} |p_{2j}^*| = \sum_{j=1}^{n} \max_{\forall k}\left\{|a_{jk}^R|\lambda_k^{RI}\right\}.$$

$$\|p_3^*\| = \sum_{j=1}^{n} |p_{3j}^*| = \sum_{j=1}^{n} \max_{\forall k}\left\{|a_{jk}^I|\lambda_k^{IR}\right\}.$$

$$\|p_4^*\| = \sum_{j=1}^{n} |p_{4j}^*| = \sum_{j=1}^{n} \max_{\forall k} \left\{ |a_{jk}^I| \lambda_k^{\mathrm{II}} \right\}.$$

Similarly, the terms for q_j^* are defined as:

$$\|q_1^*\| = \sum_{j=1}^{n} |q_{1j}^*| = \sum_{j=1}^{n} \max_{\forall k} \left\{ |b_{jk}^R| \mu_k^{\mathrm{RR}} \right\}.$$

$$\|q_2^*\| = \sum_{j=1}^{n} |q_{2j}^*| = \sum_{j=1}^{n} \max_{\forall k} \left\{ |b_{jk}^R| \mu_k^{\mathrm{RI}} \right\}.$$

And it continues similarly for q_3^*, q_4^*, and so on for s_j^* and t_j^*. Each term relates to a specific combination of values of a, b, λ, and μ with their real and imaginary parts indicated by R and I subscripts. For more info, the reader can refer [44].

Assumption 4: Define $z = x + \mathbf{i}y$, where $\mathbf{i}$ represents the imaginary unit, which means $\mathbf{i} = \sqrt{-1}$. The functions $f_j(z)$ and $g_j(z(t-\tau))$ can be split into their real and imaginary parts as follows:

$$f_j(z) = f_j^R(x, y) + \mathbf{i} f_j^I(x, y),$$
$$g_j(z(t-\tau)) = g_j^R(x(t-\tau), y(t-\tau)) + \mathbf{i} g_j^I(x(t-\tau), y(t-\tau)),$$

where $f_j^R(\cdot,\cdot) : \mathbb{R}^2 \to \mathbb{R}$, $f_j^I(\cdot,\cdot) : \mathbb{R}^2 \to \mathbb{R}$, $g_j^R(\cdot,\cdot) : \mathbb{R}^2 \to \mathbb{R}$, and $g_j^I(\cdot,\cdot) : \mathbb{R}^2 \to \mathbb{R}$. For simplicity, x_τ and y_τ to refer to $x(t-\tau)$ and $y(t-\tau)$, respectively.

(1) The partial derivatives of $f_j(\cdot,\cdot)$ with respect to x and y exist and are continuous. These include $\partial f_j^R/\partial x$, $\partial f_j^R/\partial y$, $\partial f_j^I/\partial x$, and $\partial f_j^I/\partial y$. Similarly, the partial derivatives of $g_j(\cdot,\cdot)$ with respect to x and y, such as $\partial g_j^R/\partial x$, $\partial g_j^R/\partial y$, $\partial g_j^I/\partial x$, and $\partial g_j^I/\partial y$, also exist and are continuous.

(2) The partial derivatives of f_j^R and f_j^I with respect to x and y are bounded. Specifically, there exist constants λ_j^{RR}, λ_j^{RI}, λ_j^{IR}, and λ_j^{II} such that:

$$\left|\frac{\partial f_j^R}{\partial x}\right| \le \lambda_j^{\mathrm{RR}}, \quad \left|\frac{\partial f_j^R}{\partial y}\right| \le \lambda_j^{\mathrm{RI}},$$
$$\left|\frac{\partial f_j^I}{\partial x}\right| \le \lambda_j^{\mathrm{IR}}, \quad \left|\frac{\partial f_j^I}{\partial y}\right| \le \lambda_j^{\mathrm{II}}.$$

(3) Likewise, the partial derivatives of g_j^R and g_j^I with respect to x and y are also bounded. There are constants μ_j^{RR}, μ_j^{RI}, μ_j^{IR}, and μ_j^{II} such that:

$$\left|\frac{\partial g_j^R}{\partial x}\right| \le \mu_j^{\mathrm{RR}}, \quad \left|\frac{\partial g_j^R}{\partial y}\right| \le \mu_j^{\mathrm{RI}},$$
$$\left|\frac{\partial g_j^I}{\partial x}\right| \le \mu_j^{\mathrm{IR}}, \quad \left|\frac{\partial g_j^I}{\partial y}\right| \le \mu_j^{\mathrm{II}}.$$

Applying the mean value theorem for functions of several variables, for any $x, x', y, y' \in \mathbb{R}^n$:

$$\begin{aligned}
\left|f_j^R(x', y') - f_j^R(x, y)\right| &\le \lambda_j^{\mathrm{RR}} \left|x' - x\right| + \lambda_j^{\mathrm{RI}} \left|y' - y\right|, \\
\left|f_j^I(x', y') - f_j^I(x, y)\right| &\le \lambda_j^{\mathrm{IR}} \left|x' - x\right| + \lambda_j^{\mathrm{II}} \left|y' - y\right|, \\
\left|g_j^R(x'_\tau, y'_\tau) - g_j^R(x_\tau, y_\tau)\right| &\le \mu_j^{\mathrm{RR}} \left|x'_\tau - x_\tau\right| + \mu_j^{\mathrm{RI}} \left|y'_\tau - y_\tau\right|, \\
\left|g_j^I(x'_\tau, y'_\tau) - g_j^I(x_\tau, y_\tau)\right| &\le \mu_j^{\mathrm{IR}} \left|x'_\tau - x_\tau\right| + \mu_j^{\mathrm{II}} \left|y'_\tau - y_\tau\right|.
\end{aligned} \tag{6.21}$$

Assumption 5: The functions $f_j(\cdot)$ and $g_j(\cdot)$ are assumed to satisfy the Lipschitz condition over the complex plane. That is, for any $u, v \in \mathbb{C}$, there exist positive constants λ_j and μ_j such that the following holds:

$$\begin{aligned}
\|f_j(u) - f_j(v)\| &\le \lambda_j \|u - v\|, \\
\|g_j(u) - g_j(v)\| &\le \mu_j \|u - v\|.
\end{aligned} \tag{6.22}$$

Utilising the aforementioned hypothesis along with related assumptions 3, 4, and 5, the authors of [44] deduced an adequate prerequisite for the uniform stability of a class of fractional-order complex-valued neural networks with time delays. Thereafter, the existence and uniqueness findings were obtained by applying the Banach fixed point theorem.

Theorem 6.13 ([44]) *If Assumptions 3 and 4 hold, then* (6.18) *is uniformly stable.*

Theorem 6.14 ([44]) *If Assumptions 3 and 5 hold, there exists a unique equilibrium point in* (6.18)*, which is uniformly stable.*

To prove above theorems and establish the uniform stability of the system, the assumptions play a critical role, particularly Assumptions 3, 4, and 5. The system breakdown into its real and imaginary components, as expressed in Eqs. (9) and (10) of [44]. This allows the complex-valued system to be treated as two coupled real-valued systems, simplifying the analysis.

Assumption 5 is crucial because it ensures that the functions f_j and g_j, which govern the system's activation, satisfy the Lipschitz condition. The Lipschitz condition provides bounds on the growth of the functions, meaning that for any two distinct points in the complex plane, the difference between the function values is linearly bounded by a constant multiplied by the distance between the points. This was denoted by the constants λ_j and μ_j. Such bounds guarantee that a relatively small change in the input results in a comparatively small change in the output, which is essential to being able to control the system behavior and to avoid instability.

Assumption 3 gives further restrictions on the system parameters, namely the connection weights, the self-feedback coefficients and the time-delay effects. These make sure that the system reacts in a controlled way. Likewise, Assumption 4 linearises the non-linear functions into real and imaginary parts, thus it is less involving to use the Lipschitz conditions and regulate the discrepancies between the solutions.

The proof continues with estimation of the difference between two solutions of the system, $z_1(t) = x_1(t) + iy_1(t)$ and $z_2(t) = x_2(t) + iy_2(t)$. Using the Lipschitz conditions of Assumption 5 we can get bounds on the differences $|x_1(t) - x_2(t)|$ and $|y_1(t) - y_2(t)|$, namely that the distance between any two solutions decreases with time. The introduction of the multiplication by e^{-t} on both sides is done in the proof of uniform stability property with the aim of simplifying and bounding the solution behaviour in time. To be more specific, the method is useful in addressing the exponential terms which come up in the integral equations of the system behavior. The decay of the terms in time can be better managed by multiplying by e^{-t} and then the analyst can gain uniform bounds on the difference of the solution which is critical in establishing stability.

Banach Contraction Mapping Theorem is then invoked, using the fact that the system happens to satisfy the Lipschitz conditions, which essentially means that it is a contraction mapping on a complete metric space (space of bounded continuous functions). It follows from the contraction mapping theorem that the unique fixed point of a contraction is attractive: that is, given any two solutions of the system, in the sense of any two initial conditions, the solutions will approach a common solution as time tends to infinity. This culminates into the uniqueness and existence of the solution of the system.

Eventually, uniform stability of the system is determined. The Lipschitz estimates ensure that the gap between two solutions grows exponentially with no dependence on the initial time. This shows that the system's stability is uniform, as the rate of convergence to the equilibrium does not depend on when the system starts. In conclusion, Assumptions 3, 4, and 5 work together to ensure that the system's behavior is bounded and controlled, leading to the uniform stability of the system as stated in Theorem 6.13 and Theorem 6.14.

As an alternative to the long-established method in the literature, the authors of [45] relied on topological notions like compactness and closed convex sets, and others. They investigated the uniform stability problem of FoCvNNs with time delay and obtained equilibrium point results for impulsive FoCvNNs with time delay using $(\varepsilon - \mu)$-uniformly locally contractive mappings in complete suprametric ε-chainable space. To demonstrate the accuracy and viability of the findings, a few numerical results are given. Moreover, the authors of [45] considered another assumption as below:

Assumption 6:

1. $\sum_{j=1}^{n} |U_j| < \omega$;
2. $\sup_{\mathcal{U}\in\mathscr{B}_r} \sup_t |\mathcal{U}(t)| < \infty$;
3. $|\mathcal{U}_n'(t)| < \mathcal{M}(\mathcal{M} > 0)$;
4. $\overline{o} = ||p^*|| + ||q^*||$.

Theorem 6.15 ([45]) *If Assumptions 3, 5 and 6 are hold, then FoCVNNs model 6.18 has an equilibrium point.*

Example 6.5 ([45]) Consider the following two-neuron FoCvNNs with time delay:

$$\mathscr{D}^{\alpha} z(t) = -Cz(t) + Af(z(t)) + Bg(z(t-\tau)) + U,$$

where $\alpha = 0.9,\ \tau = 0.5,\ z(t) = [z_1(t)\ z_2(t)]^T,\ C = diag(2, 1),\ U = [-0.2 + \mathbf{i}\ \ 1 - \mathbf{i}]^T$,

$$A = \begin{bmatrix} -1 + 3\mathbf{i} & 4 - \mathbf{i} \\ 2 - \mathbf{i} & 3 + 2\mathbf{i} \end{bmatrix}, \quad B = \begin{bmatrix} 1 + \mathbf{i} & 2 - \mathbf{i} \\ -3 - \mathbf{i} & 4 + 2\mathbf{i} \end{bmatrix},$$

and

$$f_j(z_j) = \frac{2}{1 + \exp(-2x_j - 2y_j)} - 1, \quad j = 1, 2,$$

$$g_j(z_j) = \frac{1}{1 + \exp(-x_j - y_j)}, \quad j = 1, 2.$$

To solve the provided system, we select various values for the initial conditions as follows:

$$\begin{bmatrix} 3 - \mathbf{i} & 2 - 3\mathbf{i} \\ 2 - 2\mathbf{i} & -2 + \mathbf{i} \\ 0.1 - 3\mathbf{i} & 0.2 - 3\mathbf{i} \\ -2.5 + 0.5\mathbf{i} & -1 - 0.5\mathbf{i} \\ -1 + 0.5\mathbf{i} & 1 - 3\mathbf{i} \end{bmatrix},$$

where each row of this vector represents an initial condition for the states $z_j,\ i = 1, 2$. Figure 6.5 illustrates the real and imaginary components of the neural network states $z_j,\ i = 1, 2$, with parameters $\alpha = 0.9$ and $\tau = 0.5$. It is clear that the solutions for Example 6.5, given the selected initial conditions, converge to the equilibrium point (−2.4546, −0.4683, −4.0838, −2.0612), which aligns with the theoretical outcome presented in Theorem 6.20.

In the work of [46] examined the global asymptotic stability of impulsive fractional-order complex-valued neural networks with time delay. Using the contraction mapping principle, comparison theorem, and inequality scaling techniques, they established sufficient conditions for the existence, uniqueness, and global asymptotic stability of the equilibrium point. A numerical example was provided to demonstrate the validity and practicality of the

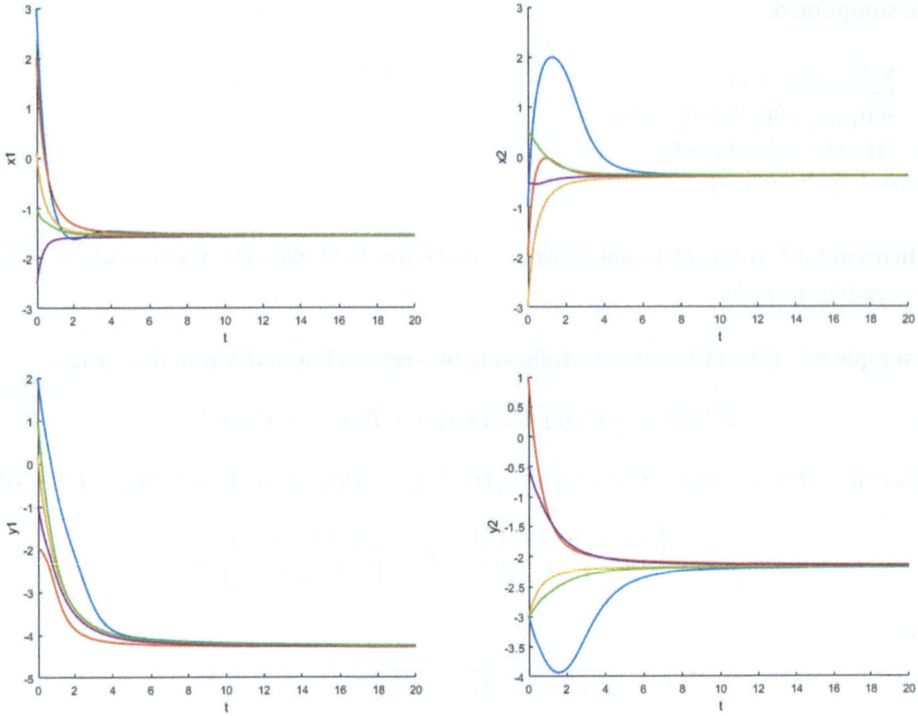

Fig. 6.5 The trajectories of the components states z_j, $j = 1, 2$, for Example 6.5 with $\alpha = 0.9$, and $\tau = 0.5$

results. As a generalization of [46], Sumati et al. [45] obtained the equilibrium point results for impulsive FoCvNNs with time delay using Ćirić-Riech-Rus type $(\varepsilon - \mu)$-uniformly locally contractive mappings in complete suprametric ε-chainable space.

The study conducted by [47] focused on examining the behavior of fractional-order neural networks with complex-valued dynamics. Their research emphasized the conditions that led to both stability and instability in the network, as well as the emergence of Hopf bifurcations. [47] established criteria to ensure whether the steady state of the network would be asymptotically stable or unstable, taking into account the system's complex parameters and fractional order, particularly in simplified network configurations like hub and ring structures. In certain cases, they identified critical thresholds for the fractional order that indicated the occurrence of Hopf bifurcations. To support their theoretical findings, [47] also performed computational simulations, further analyzing the stability and characteristics of the limit cycles that emerged from these bifurcations.

The FoCvNNs with linear impulses and fixed time delays was discussed by [48]. Utilizing the sign function, the Banach fixed-point theorem, and two classes of activation functions, they derived several criteria for the uniform stability of the solution, as well as conditions for the existence and uniqueness of the equilibrium solution. Finally, three experimental simulations were presented to demonstrate the accuracy and effectiveness of the results obtained in the study.

The study conducted by [49] examined the stability of complex-valued fractional-order neural networks with uncertain parameters and neutral-type delay. By extending reciprocally convex inequalities and integral inequalities to the complex domain, the authors established conditions for existence, uniqueness, and both delay-independent and delay-dependent robust stability using the CV homeomorphism theorem and a Lyapunov–Krasovskii functional. The effectiveness of these findings was demonstrated through two simulation examples.

The authors of [50] addressed the uniform stability of fractional-order fuzzy impulsive complex-valued neural networks with mixed delays in infinite dimensions. By applying fixed-point theory, differential inclusion, and set-valued mappings, they derived the uniqueness of the solution and established stability criteria without the need for a complex Lyapunov function, simplifying the computational process. An example was provided to validate the results.

In [51], the authors were investigated the dynamic behaviors of fractional-order complex-valued Hopfield neural networks (CVHNNs) with time delay, focusing on local asymptotic stability and Hopf bifurcation. For networks with hub and ring structures, they analyzed the stability of equilibrium states using eigenvalue analysis of the characteristic matrix and Laplace transformations for Caputo-fractional derivatives. Sufficient conditions for the uniqueness of the equilibrium point and the occurrence of Hopf bifurcation were established. Numerical examples were provided to validate the results.

The authors of [52] explored the stability of FoCvNNs within complex-valued rectangular b-metric spaces. By applying the generalized contraction principle, they derived conditions for uniform stability and established the existence and uniqueness of the equilibrium point. Numerical results were provided to demonstrate the feasibility and accuracy of the findings.

In [53], the authors were examined the drive-response synchronization of fractional-order uncertain complex-valued neural networks (FOUCNNs) with mixed time-varying delays and impulses. Using the contraction mapping principle, robust analysis techniques, and the Riemann-Liouville (R-L) derivative, they established novel sufficient conditions for the existence and uniqueness of the equilibrium point. Through the Lyapunov functional approach, they also ensured the global stability of the equilibrium solutions. Additionally, they addressed the synchronization of FOUCNNs using an adaptive error feedback control strategy. Two examples with simulations were provided to confirm the effectiveness of the results.

6.2.2 Uniform Stability Results Under Caputo-Katugampola Derivative

Very recently, in 2024, the authors of [54] extensively examined the dynamics and existence results of generalized Caputo fractional derivatives. However, they highlighted that uniform stability and equilibrium in fractional-order neural networks with generalized Caputo derivatives in real-valued settings had not been thoroughly explored. Unlike earlier works, they were the first to investigate the uniform stability and equilibrium properties for complex-valued neural networks within the framework of generalized Caputo fractional derivatives. Additionally, they explored the intermittent behavior of complex-valued neural networks in these fractional-order contexts, providing numerical results to validate the accuracy of their findings. The article concluded with open questions, encouraging future research in this area. As per our convenience, we rewrite the complex-valued neural networks within the framework of generalized Caputo fractional derivatives:

$$ {}^{C}\mathscr{D}_{t_0}^{\gamma,\mu} x_i(t) = -c_i x_i(t) + \sum_{j=1}^{n} a_{ij} f_j(x_j(t)) + \sum_{j=1}^{n} b_{ij} f_j(x_j(t-\tau)) + I_i, \quad i = 1, 2, \ldots, n. \tag{6.23} $$

In vector form, this can be expressed as:

$$ {}^{C}\mathscr{D}_{t_0}^{\gamma,\mu} x(t) = -Cx(t) + Af(x(t)) + Bf(x(t-\tau)) + I. \tag{6.24} $$

The initial conditions associated with Eq. 6.23 are articulated through the representation

$$ x_i(\mathfrak{q}) = \Psi_i(\mathfrak{q}) + \mathrm{i}\Theta_i(\mathfrak{q}), \quad \mathfrak{q} \in [-\tau, 0], \quad i = 1, 2, 3, \ldots, n, $$

where the functions $\Psi_i(\mathfrak{q})$ and $\Theta_i(\mathfrak{q})$ are elements of the complex-valued metric space $\mathbb{C}([-\tau, 0], \mathbb{R}^n)$. This space encompasses continuous functions that produce n-dimensional real vector outputs defined over the interval $[-\tau, 0]$. The norms for these functions are computed using the supremum of a weighted sum, defined as

$$ ||\Psi(t)|| = \sup_{t\in(-\tau,0]} \sum_{i=1}^{n} e^{t(i-1)} |\Psi_i(t)|, $$

and

$$ ||\Theta(t)|| = \sup_{t\in(-\tau,0]} \sum_{i=1}^{n} e^{t(i-1)} |\Theta_i(t)|. $$

These mathematical expressions provide a comprehensive framework for understanding the behaviors of the complex functions within the specified domain, highlighting the intricate relationship between the components Ψ and Θ as they contribute to the overall solution of the problem at hand.

Theorem 6.16 ([54]) *If Assumptions 3 and 4 hold, then the system 6.23 is uniformly stable.*

Let $\mathscr{M} = C([-\tau, 0], \mathbb{C}^n)$, where C denotes the collection of continuous functions over the interval $[-\tau, 0]$.

Define a distance function $d : \mathscr{M} \times \mathscr{M} \to \mathbb{C}$ by

$$d(x(t), x'(t)) = |x(t) - x'(t)| + \mathbf{i}|x(t) - x'(t)|.$$

Clearly, $(\mathscr{M}, d)$ forms a complete complex-valued metric space. Next, define an operator $\mathcal{F} : \mathscr{M} \to \mathscr{M}$ by

$$\mathcal{F}_i(u) = \sum_{j=1}^{n} a_{ij} f_j \left(\frac{u_j}{c_j}\right) + \sum_{j=1}^{n} b_{ij} f_j \left(\frac{u_j}{c_j}\right) + I_i, i = 1, 2, \ldots, n,$$

where $i = 1, 2, 3, \ldots, n$, and $u = (u_1, u_2, \ldots u_n)^T$.

Theorem 6.17 ([54]) *A unique equilibrium point exists for the system described by 6.23, provided the following conditions are satisfied, ensuring uniform stability:*

(1) Assumptions 3, 4 and 5 hold;
(2) $||p^*|| + ||q^*|| < c_{\min}$.

Remark 6.6 The recent literature has significantly contributed to understanding stability results for fractional-order neural networks, particularly within Banach spaces.

- Numerous authors have effectively utilized contractive mappings to establish various stability results. The work of [54], presented a distinct approach by employing a complex-valued metric space, which enables to demonstration of uniform stability for the problem under consideration.
- Additionally, while exploring the stability of equilibrium points and their existence and uniqueness concerning generalized Caputo fractional-order systems, earlier studies, primarily focused on real-valued neural networks. In contrast, the authors of [54] advances the field by first utilizing complex-valued neural networks in the setting of generalized Caputo fractional-order systems. This approach not only broadens the analytical framework but also provides new insights into the stability of these networks, thereby addressing challenges associated with fractional-order dynamics.

Example 6.6 Consider a fractional-order complex-valued neural network (CVNN) with time delays, described by the following equation:

$$ {}^{C}\mathscr{D}_0^{\gamma,\mu} x_i(t) = -c_i x_i(t) + \sum_{j=1}^{2} a_{ij} f_j(x_j(t)) + \sum_{j=1}^{2} b_{ij} f_j(x_j(t-\tau)) + I_i, \quad i = 1, 2. \tag{6.25} $$

Let us assume the parameters of the system are chosen as follows:

$$ \tau = 1,\ c_1 = 0.15,\ c_2 = 0.14,\ a_{11} = -0.03 - 0.04i,\ a_{12} = 0.05 - 0.04i, $$
$$ a_{21} = 0.04 + 0.007i,\ a_{22} = -0.006 + 0.02i, $$

$$ b_{11} = 0.03 - 0.08i,\ b_{12} = 0.2 - 0.02i,\ b_{21} = 0.02 + 0.03i,\ b_{22} = -0.02 + 0.06i, $$
$$ I_1 = 0.002 - 0.003i,\ I_2 = -0.001 + 0.002i,\ x_i(t) = \Psi_i(t) + \mathbf{i}\Theta_i(t), $$
$$ \varphi_i(x_i) = 0.2\tanh(\Psi_i) + 0.2\tanh(\Theta_i)\mathbf{i}, \quad i = 1, 2. $$

The function in terms of the delay is defined as:

$$ \varphi_i(x_i) = 0.1\tanh(\Psi_i) + 0.1\tanh(\Theta_i)\mathbf{i}, \quad i = 1, 2. $$

The system was analyzed using various values for γ and μ, along with two initial conditions:

$$ \vartheta_1 = \begin{bmatrix} -0.005 + 0.002i \\ 0.001 \end{bmatrix}, \quad \vartheta_2 = \begin{bmatrix} 0.003 + 0.004i \\ 0.003 - -0.005i \end{bmatrix}. $$

Figure 6.6 illustrates the time trajectories of system (6.25) for various values of γ and μ with the initial condition ϑ_1. It is evident that the solutions for all given values of γ and μ converge to an equilibrium point, consistent with the theoretical results.

Table 6.2 presents the equilibrium points at different values of γ and μ for $x_i(t),\ i = 1,\ 2$. Additionally, Fig. 6.7 displays the time trajectories for the components of $x_i(t),\ i = 1,\ 2$, with two initial conditions ϑ_1 and ϑ_2 at $\gamma = \mu = 0.95$.

The equilibrium points of the system (6.25) at different values of γ and μ highlight an interesting pattern in the behavior of the fractional-order CVNN. As γ and μ decrease, the real and imaginary components of the equilibrium points for x_1 and x_2 also show a gradual decline. This suggests that the dynamics of the system become less responsive as the fractional-order parameters reduce, with smaller magnitudes in both the real and imaginary parts.

For instance, at $\gamma = 0.95$ and $\mu = 0.95$, the equilibrium points for x_1 and x_2 are $(0.0120, -0.0172)$ and $(-0.0059, 0.0117)$, respectively. However, when γ and μ decrease to 0.8 and 0.55, the equilibrium points shift to $(0.0077, -0.0122)$ and $(-0.0042, 0.0085)$. This demonstrates a consistent shrinking in magnitude for both variables.

Table 6.2 serves as a clear illustration of how fractional-order parameters influence the system's stability and equilibrium. The decreasing values of γ and μ lead to more compressed equilibrium points, suggesting a tighter range of oscillatory behavior in the network's response to different fractional orders. Furthermore, it reveals that as the fractional

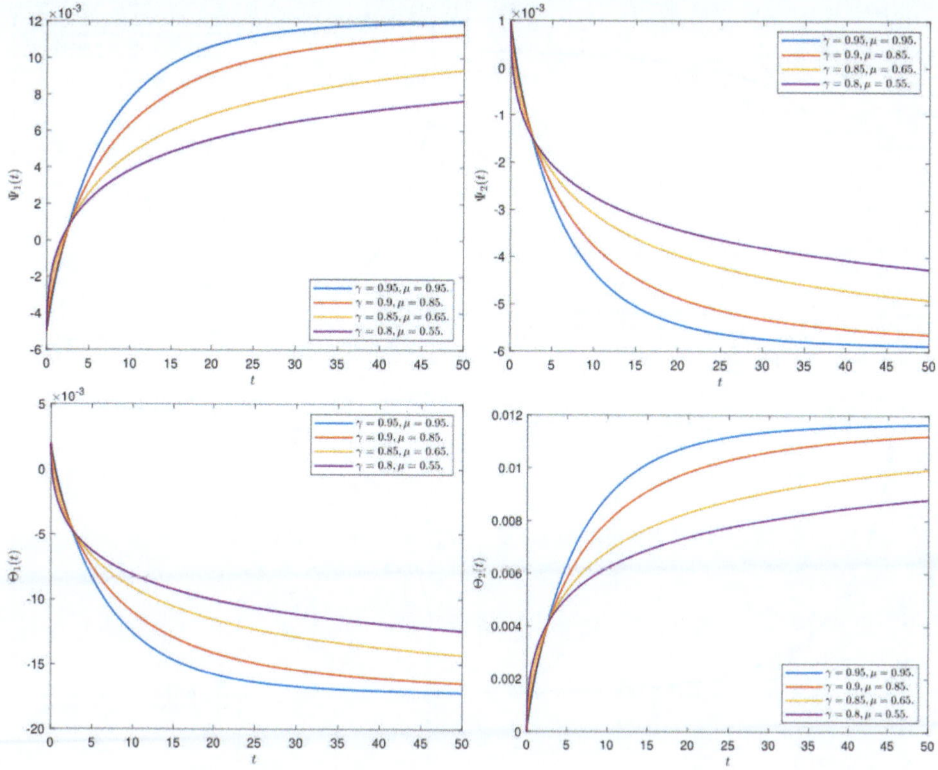

Fig. 6.6 The trajectories of the real and imaginary parts of x_i, $i = 1, 2$ for Example 6.6 with the initial condition ϑ_1 and various values of γ and μ

Table 6.2 The equilibrium points for system (6.25) at various values of γ and μ

γ	μ	Equilibrium points as real and imaginary parts for x_1 and x_2, respectively
0.95	0.95	(0.0120, –0.0172) and (–0.0059, 0.0117)
0.9	0.85	(0.0113, –0.0165) and (–0.0056, 0.0112)
0.85	0.65	(0.0077, –0.0124) and (–0.0043, 0.0088)
0.8	0.55	(0.0077, –0.0122) and (–0.0042, 0.0085)

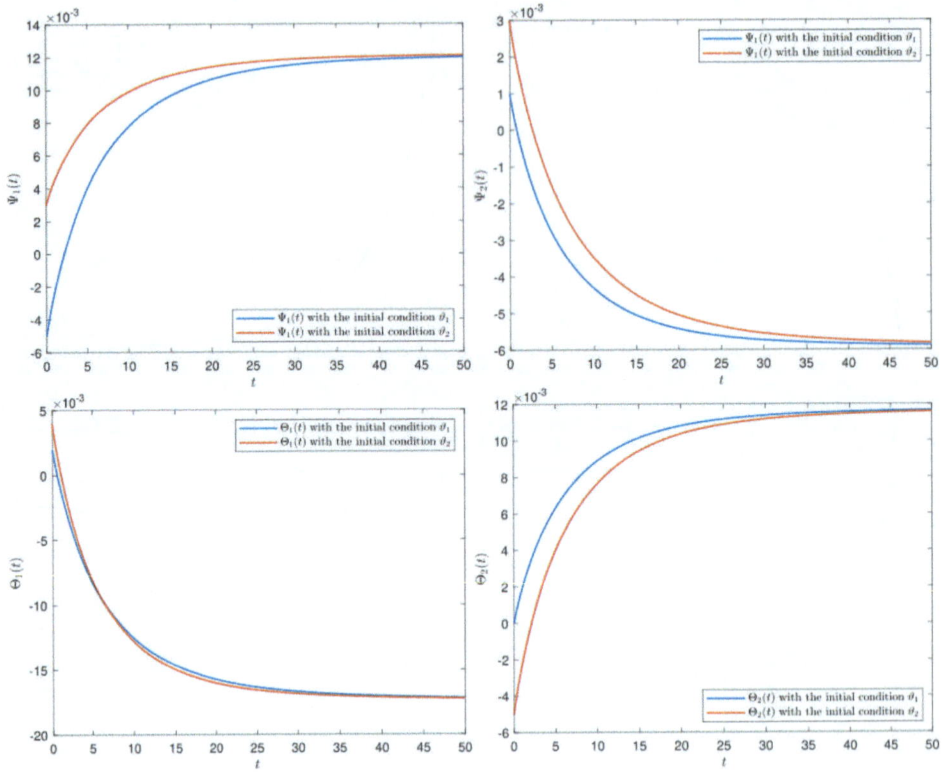

Fig. 6.7 The trajectories of the real and imaginary parts of x_i, $i = 1, 2$ for Example 6.6 with the initial conditions ϑ_1 and ϑ_2 at $\gamma = \mu = 0.95$

orders reduce, the system converges to equilibrium points closer to zero, which could imply that lower fractional orders result in a slower rate of change in the network's dynamics.

6.3 Discrete Fractional-Order Complex-Valued Neural Networks

Discrete fractional differential equations are an extension of the classical discrete difference equations, in which a more general model of dynamic systems with memory effects can be studied. These equations are the generalisation of the difference operator Δ^n (with integer n) to Δ^α with non-integer α. Most used discrete fractional difference operators are Grumwald-Letnikov and Caputo definitions, and they are given by:

$$\Delta^\alpha y(k) = \sum_{i=0}^{k} (-1)^i \binom{\alpha}{i} y(k-i),$$

where $\binom{\alpha}{i}$ is the generalized binomial coefficient. This operator is the accumulation of historical values with fractional weights, therefore it embodies memory effects. Equations of this form appear importantly in fields such as control theory and time series analysis, where the systems under study evolve in discrete time, but are driven by their previous states in a non-integer way. The solution to the equations is usually through special techniques such as discrete Laplace transforms or generating functions, and gives intimate information on the stability and long-time behavior of the system.

The concept of discrete fractional differential equations in neural networks brings forward a fascinating idea of modeling dynamics of neural systems having memory and hereditary characteristics. Traditional neural networks Difference equations describing the updates of neuron states with respect to discrete time steps are often expressed in integer order. Using fractional-order differences, noted as Δ^α with $0 < \alpha < 1$, neural network models may be made to exhibit the effect of past states in a non-integer manner which is more flexible and precise way of modeling neuron behavior.

In this context the dynamics of the state $y(k)$ of a neuron at discrete time k may be described by a fractional difference equation of the form:

$$\Delta^\alpha y(k) = \sum_{i=0}^{k} (-1)^i \binom{\alpha}{i} y(k-i) + f\,(Wx(k))\,.$$

In this case Δ^α is the discrete fractional difference operator, W is the weight matrix, $x(k)$ is the input at time step k, and $f(\cdot)$ is the activation function. Here the symbol $\binom{\alpha}{i}$ denotes the generalized binomial coefficient, which controls the fractional accumulation of past states.

These FoNNs are most applicable in modelling biological neural systems in which memory, delay and long-term dependencies are important. Also in fields such as deep learning and time series forecasting, they are being investigated, because the fractional dynamics can learn more complex patterns in data, and thus generalize better and learn more efficiently than integer-order models.

When using fractional calculus with neural networks, it is usually necessary to use modified learning algorithms, e.g. fractional gradient descent, to train the network parameters taking into consideration the non-local dependence induced by the fractional operators. This enriches the traditional neural network framework, enabling a deeper understanding of complex temporal dynamics in various real-world applications.

6.3.1 General Class

We begin by presenting some preliminary concepts:

Definition 6.13 ([55, 56])

1. Let e be a natural number and $t \in \mathbb{R}$. The *e-ascending function* of t is defined as:

$$t^{\overline{e}} = \prod_{k=0}^{e-1}(t+k), \quad t^{\overline{0}} = 1.$$

2. For a real number $\alpha \in \mathbb{R}$, the *α-ascending function* is defined as:

$$t^{\overline{\alpha}} = \frac{\Gamma(t+\alpha)}{\Gamma(t)}, \quad t \in \mathbb{R} \setminus \{-2, -1, 0\}, \quad 0^{\overline{\alpha}} = 0.$$

This provides a generalized formulation for ascending sequences in both integer and fractional orders, laying the foundation for fractional calculus in discrete systems. Additionally, the operator ∇^e is iteratively defined as $\nabla^e = \nabla(\nabla^{e-1})$, where ∇ denotes the backward difference operator.

Definition 6.14 ([57]) A set $\mathcal{L}$ in Banach space is said to be *uniformly Cauchy* if, for any $\mu > 0$, there exists an integer N such that for any two indices t_1, t_2 greater than N, the distance between any two elements $h(t_1)$ and $h(t_2)$ in $\mathcal{L}$ satisfies:

$$|h(t_1) - h(t_2)| < \mu.$$

This definition is crucial for analyzing convergence properties in sequence sets within Banach spaces.

Definition 6.15 ([58]) A system solution $h(t, t_0, \varphi)$ is said to be *finite-time stable* with respect to the parameters $(t_0, \delta, \epsilon, \Theta)$, where $\Theta = [t_0, t_0 + T]$, if there exist constants δ, ϵ, and a time $T > 0$ such that for every initial condition function φ, a second initial condition ψ with $\|\varphi - \psi\|_C < \delta$ implies that:

$$\|h(t, t_0, \varphi) - u(t, t_0, \psi)\| < \epsilon, \forall t \in \Theta.$$

This concept is used to determine if systems reach stability within a specific time frame.

Definition 6.16 ([59]) For $\alpha > 0$ and a real number a, let $n = \lfloor \alpha \rfloor + 1$, and consider a function $h : \mathbb{N}_a \to \mathbb{C}$, where $\rho(s) = s - 1$. The following operators are defined:

1. The *backward difference operator* for $h(t)$ is given by:

$$\nabla h(t) = h(t) - h(t-1), \quad t \in \mathbb{N}_1.$$

2. The *nabla α-order sum operator* is defined as:

$$\nabla_a^{-\alpha} h(t) = \frac{1}{\Gamma(\alpha)} \sum_{s=a+1}^{t} (t - \rho(s))^{\overline{\alpha - 1}} h(s), \quad t \in \mathbb{N}_{a+1}.$$

3. The *nabla α-order Caputo fractional difference operator* is defined as:

$$c\nabla_a^{\alpha} h(t) = \nabla_a^{-(n-\alpha)} \nabla^n h(t) = \frac{1}{\Gamma(n-\alpha)} \sum_{s=a+1}^{t} (t-\rho(s))^{\overline{n-\alpha-1}} \nabla^n h(s), \quad t \in \mathbb{N}_{a+1}.$$

4. Let $\mu > -1$. Then for the nabla sum operator, we have:

$$\nabla_a^{-\alpha}(t-a)^{\overline{\mu}} = \frac{\Gamma(\mu+1)}{\Gamma(\alpha+\mu+1)}(t-a)^{\alpha+\mu}, \quad t \in \mathbb{N}_a.$$

In particular, when $0 < \alpha < 1$ and $a = 0$, then $\mathbb{N}_0 = \{0, 1, 2, \dots\}$, and $n = 1$. According to this definition, we have:

$$\nabla_0^{-\alpha} h(t) = \frac{1}{\Gamma(\alpha)} \sum_{s=1}^{t} (t-\rho(s))^{\alpha-1} h(s), \quad t \in \mathbb{N}_1,$$

and

$${}^c\nabla_0^{\alpha} h(t) = \nabla_0^{-(1-\alpha)} \nabla h(t) = \frac{1}{\Gamma(1-\alpha)} \sum_{s=1}^{t} (t-\rho(s))^{\overline{-\alpha}} \nabla h(s), \quad t \in \mathbb{N}_1.$$

Also, for $\mu > -1$, the nabla sum operator is:

$$\nabla_0^{-\alpha} t^{\overline{\mu}} = \frac{\Gamma(\mu+1)}{\Gamma(\alpha+\mu+1)} t^{\overline{\alpha+\mu}}, \quad t \in \mathbb{N}_0.$$

Remark 6.7 ([59]) Based on Definition 6.16, the following property holds for the nabla fractional sum operator:

$$\nabla_a^{-\alpha}(uf(t) + vg(t)) = u\nabla_a^{-\alpha} f(t) + v\nabla_a^{-\alpha} g(t),$$

for any constants $u, v \in \mathbb{R}$ and functions $f(t), g(t) \in \mathbb{C}$. This linearity property simplifies the manipulation of fractional sums in complex systems.

Lemma 6.6 ([60]) *Assume that $\alpha > 0$ and $y : \mathbb{N}_a \to \mathbb{R}$. The following identity holds for the fractional difference operator:*

$$\nabla_a^{-\alpha}\, {}^c\nabla_a^{\alpha} y(t) = y(t) - \sum_{k=0}^{n-1} \frac{(t-a)^{\overline{k}}}{k!} \nabla^k y(a).$$

If $0 < \alpha \le 1$, this simplifies to:

$$\nabla_a^{-\alpha}\, {}^c\nabla_a^{\alpha} y(t) = y(t) - y(a).$$

This lemma establishes a key relationship between the fractional difference operator and the original function.

Lemma 6.7 ([59]) *For a function $h : \mathbb{N}_a \to \mathbb{C}$, we have the following extension of Lemma 6.6 into the complex domain:*

$$\nabla_a^{-\alpha}\, {}^c\nabla_a^{\alpha} h(t) = h(t) - \sum_{k=0}^{n-1} \frac{(t-a)^{\overline{k}}}{k!} \nabla^k h(a).$$

For $0 < \alpha \leq 1$, this simplifies to:

$$\nabla_a^{-\alpha}\, {}^c\nabla_a^{\alpha} h(t) = h(t) - h(a).$$

This result shows how fractional difference operators work with complex-valued functions.

Lemma 6.8 ([59]) *Consider the fractional difference system with a delay:*

$${}^c\nabla_a^{\alpha} \omega(t) = f(t, \omega(t), \omega(t-\tau)), \quad \omega(s) = \varphi(s), \; s \in [a-\tau, a].$$

The solution to this system can be expressed as:

$$\omega(t) = \varphi(a) + \frac{1}{\Gamma(\alpha)} \sum_{s=a+1}^{t} (t-\rho(s))^{\overline{\alpha-1}} f(s, \omega(s), \omega(s-\tau)), \quad t \in \mathbb{N}_a.$$

This lemma provides an integral representation of the solution for fractional systems with time delays.

In 2020, the authors of [59] explored the existence and finite-time stability of discrete fractional-order complex-valued neural networks with time delays, without decomposing them into real-valued systems. To establish their key results, they introduced a new discrete Caputo fractional difference equation in the complex domain, extending the theory of FoNNs from the real field to the complex field based on discrete fractional calculus. By applying Arzelà-Ascoli's theorem, inequality scaling techniques, and Krasnosel'skiĭ's fixed point theorem, they derived sufficient delay-dependent conditions that ensured the existence and finite-time stability of solutions for the proposed networks. Their findings were validated through two numerical examples with simulations, which demonstrated the feasibility of the theoretical results. Moreover, they observed that as the fractional order decreased, the discrete fractional-order complex-valued neural networks more readily achieved finite-time stability.

In [59], considered a class of discrete FoCvNNs with time delays:

$$c\nabla_a^\alpha h_j(t) = -c_j h_j(t) + \sum_{k=1}^{n} a_{jk} f_k(h_k(t)) + \sum_{k=1}^{n} b_{jk} g_k(h_k(t-\tau)) + W_j, \quad j = 1, 2, \ldots, n, \ t \in \mathbb{N}_a. \tag{6.26}$$

In vector form, the above equation can be written as:

$$c\nabla_a^\alpha h(t) = -Ch(t) + Af(h(t)) + Bg(h(t-\tau)) + W, \quad t \in \mathbb{N}_a. \tag{6.27}$$

The initial condition of system (6.26) is given by:

$$h_j(s) = \varphi_j(s), \quad s \in [-\tau + a, -\tau + a + 1, \ldots, a]. \tag{6.28}$$

However, they considered following assumptions in order to prove main results of [59]:

Assumption 7: There exist positive constants l_k and λ_k such that the following conditions hold for all $k = 1, 2, \ldots, n$, and for any $\alpha_1, \alpha_2 \in \mathbb{C}$:

- The difference between the activation functions f_k at two points α_1 and α_2 is bounded by:
$$|f_k(\alpha_1) - f_k(\alpha_2)| \le l_k |\alpha_1 - \alpha_2|.$$
- Similarly, the difference between the activation functions g_k at two points α_1 and α_2 is bounded by:
$$|g_k(\alpha_1) - g_k(\alpha_2)| \le \lambda_k |\alpha_1 - \alpha_2|.$$

These conditions ensure that both activation functions are Lipschitz continuous with constants l_k and λ_k, respectively, providing control over the growth of the functions as they vary across the complex plane.

Assumption 8: There exist two positive constants M and N such that for all $k = 1, 2, \ldots, n$, and for any $\alpha \in \mathbb{C}$, the activation functions f_k and g_k are bounded by:

$$|f_k(\alpha)| \le M \quad \text{and} \quad |g_k(\alpha)| \le N.$$

This assumption guarantees that both activation functions are bounded, providing an upper limit on their magnitude, which is crucial for ensuring the stability of the system's behavior.

Theorem 6.18 ([59]) *Under the framework of Assumptions 7 and 8, if the following two conditions are satisfied:*

$$\mathcal{I} < 1, \quad \frac{\mathcal{P}\|\varphi\|_C + \mathcal{R}}{1 - \mathcal{I}} \le \eta,$$

then the system 6.26 has a solution within the bounded set

$$\Omega = \{h : h = (h_1, h_2, \ldots, h_n)^T, \|h\| \le \eta, h_j \in L([a, T], C)\},$$

where:

$$\mathcal{I} = \frac{\max_{1\leq j\leq n}\{c_j\}}{\alpha B(T-a,\alpha)} + \frac{\sum_{j=1}^{n}\max_{1\leq k\leq n}\{|a_{jk}|l_k\}}{\alpha B(T-a,\alpha)} + \frac{\sum_{j=1}^{n}\max_{1\leq k\leq n}\{|b_{jk}|\lambda_k\}}{\alpha B(T-a-\tau,\alpha)},$$

$$\mathcal{P} = 1 + \frac{\sum_{j=1}^{n}\max_{1\leq k\leq n}\{|b_{jk}|\lambda_k\}}{\alpha B(T-a+1,\alpha)},$$

$$\mathcal{R} = \frac{1}{\alpha B(T-a,\alpha)}\|W\| + \frac{1}{\alpha B(T-a,\alpha)}\sum_{j=1}^{n}\sum_{k=1}^{n}|a_{jk}|M + \frac{1}{\alpha B(T-a,\alpha)}\sum_{j=1}^{n}\sum_{k=1}^{n}|b_{jk}|N.$$

This theorem ensures the existence of a solution for the discrete FoCvNNs system under specified conditions for boundedness and stability.

Theorem 6.19 ([59]) *Under Assumption 7, the system 6.26 is finite-time stable with respect to the parameters* $(t_0 = a, \delta, \epsilon, \Theta = [a, T])$*, if the following conditions are met:*

1. *There exist three positive constants* T*,* δ*, and* ϵ *such that* $\delta < \epsilon$*.*
2. *The following inequalities hold:*

$$\mathcal{I} < 1, \quad \mathcal{P}\cdot\frac{1}{1-\mathcal{I}} < \frac{\epsilon}{\delta},$$

where:

$$\mathcal{I} = \frac{\max_{1\leq j\leq n}\{c_j\}}{\alpha B(T-a,\alpha)} + \frac{\sum_{j=1}^{n}\max_{1\leq k\leq n}\{|a_{jk}|l_k\}}{\alpha B(T-a,\alpha)} + \frac{\sum_{j=1}^{n}\max_{1\leq k\leq n}\{|b_{jk}|\lambda_k\}}{\alpha B(T-a-\tau,\alpha)},$$

$$\mathcal{P} = 1 + \frac{\sum_{j=1}^{n}\max_{1\leq k\leq n}\{|b_{jk}|\lambda_k\}}{\alpha B(T-a+1,\alpha)}.$$

The theorem guarantees that under the above conditions the system (6.26) is stable in a finite time T in the sense that for any initial conditions $h_j(s) = \varphi_j(s)$, where $s \in [a-\tau, a]$ the system will be stable in the time interval $[a, T]$. This conclusion in the study of discrete FoCvNNs highlights the importance of order in the finite-time stability. Namely, it has been noted that lower order can enable such stability, enabling the system to approach the desired performance indicators much faster and over a shorter period. When the conditions outlined in the respective theorems are satisfied, it becomes evident from the properties of the Beta function that the relationship $T - a - \tau > 0$ holds, further implying $T - a > 1$ due to the positive integer nature of τ. This relationship is vital as it underpins the system's operational parameters.

For given parameters such as $c_j, a_{jk}, b_{jk}, l_k, \lambda_k, T, \tau, \eta, \epsilon$, and δ, the values of $\mathcal{I}$ and $\mathcal{P}$ are noted to decrease with a reduction in the order α. This observation suggests that the stability conditions outlined in the theorems are met more readily at lower orders. Thus, this property of lower-order systems reinforces the practical applicability of discrete FoCvNNs in various scenarios, simplifying the process of ensuring system stability.

In broader discussions of stability criteria for FoCvNNs, various authors[see, for example, [44, 51, 58, 61–64] have examined aspects like uniform stability and global asymptotic stability in the context of fractional-order networks with time delays. While traditional approaches often involve separating the activation functions into real and imaginary components–leading to the transformation of CvNNs into equivalent RvNNs–this technique tends to complicate the analysis by increasing system dimensions. The unique approach taken in this paper diverges from this trend, maintaining the validity of existence and finite-time stability criteria without necessitating such transformations. This not only simplifies the analysis but also enhances the robustness of the findings in the realm of discrete FoCvNNs.

6.3.2 L_p-Sense

By setting $a = 0$ in Lemmas 6.7 and 6.8, the analysis leads to the establishment of several new lemmas that extend the foundational properties of the nabla fractional difference operator to a broader context [65]. Specifically, this adjustment allows for the derivation of results applicable to discrete fractional-order systems starting from the origin. The resulting lemmas clarify the relationships between the operators and the complex-valued functions, enabling a more nuanced understanding of the dynamics inherent in FoNNs. Moreover, the authors of [65] followed the same pattern as followed in [59], but they extend the results to L^p-space.

In [59], the norm for the complex-valued functions in the context of discrete FoCvNNs is expressed as:

$$\|h\|_C = \sup_{a<t\leq T} \sum_{j=1}^{n} |h_j(t)|.$$

This norm indicates the supremum of the sum of the absolute values of the components of the complex state vector $h(t)$ over the interval $(a, T]$, ensuring boundedness of the solution within the specified range.

In contrast, the provided norm in [65] for the space of bounded functions is defined as:

$$||h||_p = \sup_{0\leq t\leq T} \left[\sum_{i=1}^{n} |h_i(t)|^p \right]^{\frac{1}{p}}, \quad \text{for } h \in \mathscr{L}([-\tau, 0], \mathbb{C}^n).$$

This norm incorporates a p-norm formulation, emphasizing the p-th power of the sum of the absolute values of the components, allowing for different levels of emphasis on the magnitude of the components depending on the choice of p.

Here, considered system is,

$$c\nabla_0^{\delta} h_j(t) = -c_j h_j(t) + \sum_{k=1}^{n} a_{jk} f_k(h_k(t)) + \sum_{k=1}^{n} b_{jk} g_k(h_k(t-\tau)) + W_j, \quad j = 1, 2, \ldots, n, \ t \in \mathbb{N}_a. \tag{6.29}$$

Theorem 6.20 ([65]) *The system described by Eq. 6.29 has at least one solution within the interval* $[0, T]$ *if the following criteria are satisfied:*

($\mathscr{C}_1$). *The value of* ζ *must be less than 1.*
($\mathscr{C}_2$).

$$\frac{||\psi|| + \frac{n^{\frac{1}{p}}\Gamma(T+\delta)}{\Gamma(\delta+1)\Gamma(T)}\left[||\Theta|| + \mathscr{A}\left(\sum_{j=1}^{n}\alpha_j^p\right)^{\frac{1}{p}} + \mathscr{B}\left(\sum_{j=1}^{n}\sigma_j^p\right)^{\frac{1}{p}}\right]}{1 - \frac{n^{\frac{1}{p}}\Gamma(T+\delta)c_0}{\Gamma(\delta+1)\Gamma(T)}} \le \theta,$$

where ζ *is defined as* $\frac{c_0 n^{\frac{1}{p}}\Gamma(T+\delta)}{\Gamma(\delta+1)\Gamma(T)}$, *and* c_0 *is the maximum value among all* c_i.

Theorem 6.21 ([65]) *Under Assumption* $\mathcal{A}$, *the system* (6.29) *is finite-time stable in view of* $(\ell_0 = 0, \omega, \varrho, [0, T])$, *whenever three non-negative constants* T, ω, ϱ *are such that*

$$\Im < 1, \quad \frac{\wp}{1-\Im} < \frac{\varrho}{\omega},$$

where

$$\wp = 1 + \frac{\Gamma(T+1+\delta)}{\Gamma(\delta+1)\Gamma(T+1)}\eta_k\left(\sum_{j=1}^{n}\sigma_j^p\right)^{\frac{1}{p}},$$

and

$$\Im = \left(c_0\frac{\Gamma(T+\delta)}{\Gamma(\delta+1)\Gamma(T)} + \frac{\Gamma(T+\delta)}{\Gamma(\delta+1)\Gamma(T)}\mu_k\left(\sum_{j=1}^{n}\alpha_j^p\right)^{\frac{1}{p}} + \frac{\Gamma(T-\tau+\delta)}{\Gamma(\delta+1)\Gamma(T-\tau)}\eta_k\left(\sum_{j=1}^{n}\sigma_j^p\right)^{\frac{1}{p}}\right).$$

In [65], the two theorems discussed stability properties of complex-valued systems governed by fractional calculus, especially within the setting of L^p-spaces. The first theorem inequates that on certain conditions, including the values of constants and some inequalities, the system, which is described by Eq. 6.29 ensures the existence of at least one solution in the interval $[0, T]$ in the L^p-sense. This establishes a framework basis of the effect of parameters on the behavior of a system. The second theorem, in its turn, builds on this idea by stating that the same system can exhibit finite-time stability when specific non-negative constants are used, and the inequalities with these constants hold. All these theorems give a thorough analysis of how the interactions between the system parameters and bounds on them can result in trustworthy solutions and stable dynamics in complex-valued neural networks and, thus, extend the theory of discrete fractional-order systems in the L^p setting.

6.3.3 Synchronization

In the work of [66], the authors explore the following:

$$^{c}\nabla_0^{\alpha} x_j(t) = -c_j x_j(t) + \sum_{k=1}^{n} a_{jk} f_k(x_k(t)) + \sum_{k=1}^{n} b_{jk} g_k(x_k(t-\tau)) + W_j, \quad j \in s,\ t \in \mathbb{N}_0. \tag{6.30}$$

This is a dynamic of a discrete-time FoCvNNs with time delays. $x_j(t)$ is used to represent the state of the j-th neuron at time t and c_j are the self-feedback weights of the individual neurons. The summations incorporate the influences of the activation functions f_k and g_k of the connected neurons k, adjusted for the respective weights a_{jk} and b_{jk}. Additionally, W_j represents external inputs to the j-th neuron. The equation holds for indices j within the set s and for time t in the set of non-negative integers $\mathbb{N}_0$.

Assumption 9: There exist two positive constants, $\mathcal{F}_k$ and $\mathcal{G}_k$, such that for all $k \in \mathfrak{s}$ and for any $u, v \in \mathbb{C}$, the following inequalities hold:

$$|f_k(u) - f_k(v)| \le \mathcal{F}_k |u - v|,$$

$$|g_k(u) - g_k(v)| \le \mathcal{G}_k |u - v|.$$

Definition 6.17 The *nabla discrete Mittag-Leffler function*, as defined by [67, 68], is characterized for parameters $\lambda \in \mathbb{R}$ with $|\lambda| < 1$ and complex numbers $\alpha, \beta, \gamma \in \mathbb{C}$ such that $\Re(\alpha) > 0$. This function with two parameters is expressed as:

$$E_{\overline{\alpha,\beta}}(\lambda, \gamma) = \sum_{k=0}^{\infty} \lambda^k \frac{\gamma^{\overline{k\alpha+\beta-1}}}{\Gamma(\alpha k + \beta)}.$$

Specifically, when $\beta = 1$, the function simplifies to:

$$E_{\bar{\alpha}}(\lambda, \gamma) \triangleq E_{\overline{\alpha,1}}(\lambda, \gamma) = \sum_{k=0}^{\infty} \lambda^k \frac{\gamma^{\overline{k\alpha}}}{\Gamma(\alpha k + 1)}.$$

Definition 6.18 ([66]) An equilibrium point $x^* = \left(x_1^*, x_2^*, \ldots, x_n^*\right)^T$ of the system described by Eq. 6.30 is considered *globally Mittag-Leffler stable* if there are two positive constants $\mathcal{L}$ and d, with $0 < \mathcal{L} < 1$, such that for any solution $x(t)$ stemming from an initial value ϕ_0, the following inequality holds:

$$\left\|x(t) - x^*\right\| \le \left[m\left(\phi_0 - x^*\right) E_{\bar{\alpha}}(-\mathcal{L}, t)\right]^d, \quad t \in \mathbb{N}_0,$$

where $\alpha \in (0, 1)$, $m(0) = 0$, and $m(x) \ge 0$ is locally Lipschitz in terms of $x \in \mathbb{C}^n$.

Definition 6.19 The system defined by Eq. 6.30 and its corresponding response system represented by Eq. 6.32 are said to achieve *global Mittag-Leffler synchronization* if the synchronization error system outlined in Eq. 6.33 exhibits global Mittag-Leffler stability for any solution $\varrho(t)$ initialized at ϱ_0. This is expressed by the following condition:

$$\|\varrho(t)\| \leq \left[m\left(\varrho_0\right) E_{\bar{\alpha}}(-\mathcal{L}, t)\right]^d, \quad t \in \mathbb{N}_0,$$

where $\mathcal{L}$ and d are positive constants, $0 < \mathcal{L} < 1, \alpha \in (0, 1)$, and $m(0) = 0$, with $m(\varrho) \geq 0$ being locally Lipschitz concerning $\varrho \in \mathbb{C}^n$.

Lemma 6.9 ([66]) *For the range* $-1 < \lambda < 0$, *the function* $E_{\bar{\alpha}}(\lambda, t)$ *is strictly decreasing and satisfies the inequality* $0 < E_{\bar{\alpha}}(\lambda, t) \leq 1$ *for all* $t \in \mathbb{N}_0$.

Lemma 6.10 ([67]) *The discrete Laplace transform, denoted by* $\mathcal{N}\{\cdot\}$, *for a function* $f(t) : \mathbb{N}_0 \to \mathbb{R}$, *as well as for the nabla fractional sum* $\nabla_a^{-\alpha} f(t)$, *the Caputo left fractional difference* ${}^c\nabla_a^{\alpha} f(t)$, *and the nabla discrete convolution of functions* $f(t)$ *and* $g(t)$, *can be expressed as follows:*

$$\mathcal{N}_a\{f(t)\}(z) = \sum_{z=a+1}^{\infty} (1-t)^{z-1} f(z),$$

$$\begin{aligned}
\mathcal{N}_a\left\{\nabla_a^{-\alpha} f(t)\right\}(z) &= z^{-\alpha} \mathcal{N}_a\{f(z)\}(z), \\
\mathcal{N}_a\left\{{}^c\nabla_a^{\alpha} f(t)\right\}(z) &= z^{\alpha} \mathcal{N}_a\{f(z)\}(z) - (1-z)^a z^{\alpha-1} f(a), \\
\mathcal{N}_a\{f(t) * g(t)\}(z) &= \mathcal{N}_a\{f(z)\}(z) \mathcal{N}\{g(z)\}(z).
\end{aligned}$$

It is important to note that $\mathcal{N} = \mathcal{N}_0$.

Lemma 6.11 ([67]) *The discrete Laplace transforms for Mittag-Leffler functions are represented as follows:*

$$\mathcal{N}\left\{E_{\bar{\alpha}}(\lambda, t)\right\}(z) = \frac{z^{\alpha-1}}{z^{\alpha} - \lambda},$$

$$\mathcal{N}\left\{E_{\overline{\alpha,\alpha}}(\lambda, t)\right\}(z) = \frac{1}{z^{\alpha} - \lambda}.$$

Lemma 6.12 ([66]) *For a function* $\ell(t) \in \mathbb{R}$, *the following inequality holds:*

$$\frac{1}{2}\, {}^c\nabla_0^{\alpha} \ell^2(t) \leq \ell(t)\, {}^c\nabla_0^{\alpha} \ell(t), \quad \text{for } \alpha \in (0, 1),\ t \in \mathbb{N}_0.$$

Lemma 6.13 ([66]) *For a function* $\ell(t) \in \mathbb{C}$, *the following relationship is satisfied:*

$${}^c\nabla_0^{\alpha} \ell(t)\overline{\ell(t)} \leq \ell(t)\, {}^c\nabla_0^{\alpha} \overline{\ell(t)} + \overline{\ell(t)}\, {}^c\nabla_0^{\alpha} \ell(t), \quad \text{for } \alpha \in (0, 1),\ t \in \mathbb{N}_0.$$

Lemma 6.14 ([66]) *Let $\mathcal{V}(t, x(t))$ be a positive definite and decreasing scalar function defined on $\mathbb{N}_0 \times \mathbb{C}^n \to \mathbb{R}$. There exists a constant $0 < \mathcal{L} < 1$ such that the following inequality holds:*

$$^{c}\nabla_0^{\alpha}\mathcal{V}(t, x(t)) \leq -\mathcal{L}\mathcal{V}(t, x(t)), \quad 0 < \alpha < 1.$$

Consequently, it follows that:

$$\mathcal{V}(t, x(t)) \leq \mathcal{V}(0, x(0))E_{\bar{\alpha}}(-\mathcal{L}, t), \quad t \in \mathbb{N}_0.$$

Lemma 6.15 ([69]) *For any two complex numbers $\omega, \nu \in \mathbb{C}$ and a real number $\rho > 0$, the following inequality holds:*

$$\omega\bar{\nu} + \bar{\omega}\nu \leq \rho\omega\bar{\omega} + \frac{1}{\rho}\bar{\nu}\nu.$$

Lemma 6.16 ([37]) *Given two sequences of complex numbers $a = (a_1, a_2, \ldots, a_n)$ and $b = (b_1, b_2, \ldots, b_n)$, the following inequality holds:*

$$\left|\sum_{k=1}^{n} a_k b_k\right|^2 \leq \left(\sum_{k=1}^{n} |a_k|^2\right)\left(\sum_{k=1}^{n} |b_k|^2\right),$$

with equality occurring if and only if the sequences a and b are proportional.

Theorem 6.22 ([66]) *Considering that Assumption 9 is valid, if the following condition is met:*

$$0 < \mathcal{K} = \left(\sum_{j=1}^{n}\sum_{k=1}^{n}\left(\frac{|a_{jk}|\mathcal{F}_k + |b_{jk}|\mathcal{G}_k}{c_k}\right)^2\right)^{\frac{1}{2}} < 1,$$

then the system defined by Eq. 6.30 possesses a unique equilibrium point.

According to Theorem 6.22, define $\breve{x}_j(t) = x_j(t) - x_j^*$. The system can then be expressed as follows:

$$^{c}\nabla_0^{\alpha}\breve{x}_j(t) = -c_j\breve{x}_j(t) + \sum_{k=1}^{n} a_{jk}\left(f_k\left(\breve{x}_k(t) + x_k^*\right) - f_k\left(x_k^*\right)\right) + \sum_{k=1}^{n} b_{jk}\left(g_k\left(\breve{x}_k(t-\tau) + x_k^*\right) - g_k\left(x_k^*\right)\right), \tag{6.31}$$

$$j \in \mathfrak{s}, \; t \in \mathbb{N}_0.$$

The initial condition for the system is given by:

$$\breve{x}_j(s) = \phi_{0j}(s) - x_j^*, \quad s \in [-\tau, -\tau + 1, \ldots, 0].$$

Next, they derived the sufficient conditions for the global Mittag-Leffler stability of the equilibrium point for the system defined above, with the specified initial condition.

Theorem 6.23 ([66]) *Under Assumption 9, if there are positive constants* $\rho_1, \rho_2, \mathcal{F}_j, \mathcal{G}_j$ *such that:*

$$\mathcal{L}_1 = \min_{1\leq j\leq n}\left(c_j + \overline{c_j} - \rho_1 - \rho_2 - \mathcal{F}_j^2 \sum_{k=1}^{n} \frac{1}{\rho_1}\overline{a_{kj}}a_{kj}\right) > 0,$$

$$\mathcal{L}_2 = \max_{1\leq j\leq n}\left(\mathcal{G}_j^2 \sum_{k=1}^{n} \frac{1}{\rho_2}\overline{b_{kj}}b_{kj}\right) > 0,$$

and

$$0 < \mathcal{L} = \mathcal{L}_1 - \mathcal{L}_2 < 1,$$

for $j \in \mathfrak{s}$, *then the equilibrium point* x^* *of the system is globally Mittag-Leffler stable.*

The authors in [66] examined the global synchronization of the system (6.30) alongside the corresponding response system (6.32) using a linear feedback controller. The response system, represented by the discrete-time fractional-order complex-valued neural networks (FCVNNs), is described by the equation:

$$^{c}\nabla_0^{\alpha} y_j(t) = -c_j y_j(t) + \sum_{k=1}^{n} a_{jk} f_k(y_k(t)) + \sum_{k=1}^{n} b_{jk} g_k(y_k(t-\tau)) + W_j + u_j(t), \quad j \in \mathfrak{s},\ t \in \mathbb{N}_0, \tag{6.32}$$

where $y_j(t)$ denotes the state of the corresponding response system, and $u_j(t)$ represents an appropriate control input. The control input is formulated as:

$$u_j(t) = -\zeta_j(y_j(t) - x_j(t)),$$

with $\zeta_j > 0$ indicating the control gain. The initial condition for the system (6.32) is given by:

$$y_j(s) = \varphi_{0j}(s), \quad j \in \mathfrak{s},\ s \in [-\tau, -\tau+1, \ldots, 0],$$

where $\varphi_0 = (\varphi_{01}, \varphi_{02}, \ldots, \varphi_{0n})^T$ and $\varphi_{0j} \in (C[-\tau, 0], \mathbb{C})$.

Let the synchronization error be defined as $\varrho_j(t) = y_j(t) - x_j(t)$. Based on the dynamics of systems (6.30) and (6.32), the synchronization error system can be formulated as:

$$^{c}\nabla_0^{\alpha} \varrho_j(t) = -c_j \varrho_j(t) - \zeta_j \varrho_j(t) + \sum_{k=1}^{n} a_{jk} f_k(\varrho_k(t)) + \sum_{k=1}^{n} b_{jk} g_k(\varrho_k(t-\tau)), \quad j \in \mathfrak{s},\ t \in \mathbb{N}_0. \tag{6.33}$$

Here, $f_k(\varrho_k(t)) = f_k(y_k(t)) - f_k(x_k(t))$ and $g_k(\varrho_k(t-\tau)) = g_k(y_k(t-\tau)) - g_k(x_k(t-\tau))$. The initial conditions for the error system (6.33) are specified as:

$$\varrho_{0j}(s) = \varphi_{0j}(s) - \phi_{0j}(s), \quad j \in \mathfrak{s},\ s \in [-\tau, -\tau+1, \ldots, 0],$$

where $\varrho_0 = (\varrho_{01}, \varrho_{02}, \dots, \varrho_{0n})^T$ and $\varrho_{0j} \in (C[-\tau, 0], \mathbb{C})$. The subsequent focus is on establishing a sufficient criterion for global Mittag-Leffler synchronization in the synchronization error system (6.33).

Theorem 6.24 ([66]) *Under Assumption 9, if there are positive constants $\rho_3, \rho_4, \mathcal{F}_j, \mathcal{G}_j > 0$ such that:*

$$\mathcal{L}_3 = \min_{1 \le j \le n} \left(\overline{c_j} + c_j + \overline{\zeta_j} + \zeta_j - \rho_3 - \rho_4 - \mathcal{F}_j^2 \sum_{k=1}^{n} \frac{1}{\rho_3} \overline{a_{kj}} a_{kj} \right) > 0,$$

$$\mathcal{L}_4 = \max_{1 \le j \le n} \left(\mathcal{G}_j^2 \sum_{k=1}^{n} \frac{1}{\rho_4} \overline{b_{kj}} b_{kj} \right) > 0,$$

and $0 < \mathcal{L}' = \mathcal{L}_3 - \mathcal{L}_4 < 1$ for $j \in \mathfrak{s}$, then the drive system (6.30) *and the corresponding response system* (6.32) *achieve global Mittag-Leffler synchronization.*

As a conclusion, the authors of [66] drew upon *contraction mapping theory*, Cauchy's inequality, and the fractional derivative inequality, along with the principles of complex-valued functions. By developing appropriate Lyapunov functions, they proposed various sufficient conditions that ensured the existence and uniqueness of equilibrium points, as well as their global Mittag-Leffler stability. Furthermore, the authors established criteria for achieving global Mittag-Leffler synchronization in discrete-time fractional-order complex-valued neural networks (FCVNNs) with time delays.

Example 6.7 ([66]) Consider a discrete-time fractional-order complex-valued neural network consisting of three neurons. The system is described by the equation:

$$^{c}\nabla_0^{\alpha} x_j(t) = -c_j x_j(t) + \sum_{k=1}^{3} a_{jk} f_k(x_k(t)) + \sum_{k=1}^{3} b_{jk} g_k(x_k(t-\tau)) + W_j + u_j(t), \quad j = 1, 2, 3, \ t \in \mathbb{N}_0. \tag{6.34}$$

In this scenario, we set $\alpha = 0.945$ and define $x_j(t)$ as a complex number expressed in terms of its real and imaginary components: $x_j(t) = x_j^R(t) + x_j^I(t)\mathbf{i}$, where $x_j^R(t)$ and $x_j^I(t)$ are real values. The activation functions are defined as follows:

$$f_k(x_k) = 0.2 \tanh(x_k^R) + 0.2 \tanh(x_k^I)\mathbf{i}, \quad g_k(x_k) = 0.25 \tanh(x_k^R) + 0.25 \tanh(x_k^I)\mathbf{i}.$$

For this example, we take $\tau = 2$, and the matrices C, A, and B are specified as:

$$C = \begin{pmatrix} 1.4 & 0 & 0 \\ 0 & 1.336 & 0 \\ 0 & 0 & 1.44 \end{pmatrix}, \quad A = \begin{pmatrix} 0.8 - 0.6\mathbf{i} & 0.78 - 0.5\mathbf{i} & 0.45 + 0.7\mathbf{i} \\ -0.6 - 0.5\mathbf{i} & 0.45 + 0.7\mathbf{i} & 0.58 + 0.6\mathbf{i} \\ -0.72 + 0.54\mathbf{i} & 0.46 + 0.48\mathbf{i} & 0.12 - 0.41\mathbf{i} \end{pmatrix},$$

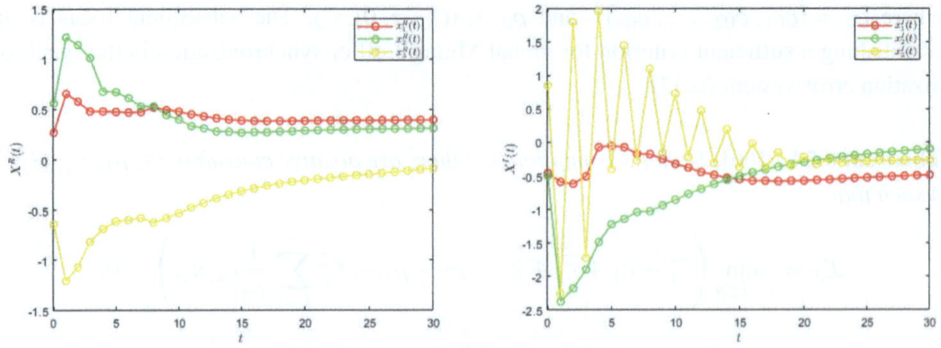

Fig. 6.8 The state trajectories of the real and imaginary parts of system (6.34)

$$B = \begin{pmatrix} 0.42 + 0.53\mathbf{i} & -0.52 + 0.64\mathbf{i} & 0.45 + 0.6\mathbf{i} \\ -0.84 + 0.56\mathbf{i} & -0.47 - 0.32\mathbf{i} & 0.48 - 0.37\mathbf{i} \\ 0.48 - 0.47\mathbf{i} & 0.45 + 0.52\mathbf{i} & 0.51 - 0.62\mathbf{i} \end{pmatrix}, \quad W = \begin{pmatrix} -0.0031 - 0.0179\mathbf{i} \\ -0.0029 - 0.0048\mathbf{i} \\ -0.0003 + 0.0087\mathbf{i} \end{pmatrix}.$$

For any x_j and $\widetilde{x}_j$ in $\mathbb{C}$, we can derive the following relationship:

$$\begin{aligned} \left|f_j(x_j) - f_j(\widetilde{x}_j)\right| &= \left|0.2\tanh(x_j^R) + 0.2\tanh(x_j^I)\mathbf{i} - 0.2\tanh(\widetilde{x_j^R}) - 0.2\tanh(\widetilde{x_j^I})\mathbf{i}\right| \\ &= 0.2\sqrt{\left(\tanh(x_j^R) - \tanh(\widetilde{x_j^R})\right)^2 + \left(\tanh(x_j^I) - \tanh(\widetilde{x_j^I})\right)^2}. \end{aligned}$$

This can be further simplified to:

$$= 0.2\left|\tanh(x_j) - \tanh(\widetilde{x_j})\right| \le 0.2\left|x_j - \widetilde{x_j}\right|.$$

It can be easily verified that Assumption 9 holds true with $\mathcal{F}_1 = \mathcal{F}_2 = \mathcal{F}_3 = 0.2$ and $\mathcal{G}_1 = \mathcal{G}_2 = \mathcal{G}_3 = 0.25$. Consequently, the calculation reveals that $0 < \mathcal{K} = 0.746 < 1$, indicating that system 6.30 has a unique equilibrium point according to Theorem.. Setting the constants $\rho_1 = 0.8$ and $\rho_2 = 1.0526$, we compute $\mathcal{L}_1 = \min\{0.8264, 0.7198, 0.9488\} = 0.7198 > 0$ and $\mathcal{L}_2 = \max\{0.1145, 0.0877, 0.0935\} = 0.1145 > 0$. This gives us $0 < \mathcal{L} = 0.6053 < 1$. Therefore, the conditions are satisfied, and according to Theorem 2, the equilibrium point x^* of the system is globally Mittag-Leffler stable. Figure 6.8 illustrates the state trajectories of the real and imaginary parts of the system, with initial conditions $\phi_1(t) = 0.2712 - 0.456\mathbf{i}$, $\phi_2(t) = 0.5601 - 0.5057\mathbf{i}$, and $\phi_3(t) = -0.6416 + 0.8481\mathbf{i}$, confirming the global Mittag-Leffler stability of the discrete-time FoCvNNs.

Example 6.8 ([66]) Consider a FoCvNNs with two neurons, represented as the drive system. The dynamics are defined by:

$$ {}^{c}\nabla_0^{\alpha} x_j(t) = -c_j x_j(t) + \sum_{k=1}^{2} a_{jk} f_k(x_k(t)) + \sum_{k=1}^{2} b_{jk} g_k(x_k(t-\tau)) + W_j, \quad j = 1, 2, \ t \in \mathbb{N}_0. \tag{6.35} $$

Here, set $\alpha = 0.875$ and define $x_j(t)$ as a complex number expressed in terms of its real and imaginary components: $x_j(t) = x_j^R(t) + x_j^I(t)\mathbf{i}$, where both $x_j^R(t)$ and $x_j^I(t)$ are real components. The activation functions are given by:

$$ f_j(x_j) = 0.6 \tanh(x_j^R) + 0.6 \tanh(x_j^I)\mathbf{i}, \quad g_j(x_j) = 0.4 \tanh(x_j^R) + 0.4 \tanh(x_j^I)\mathbf{i}, $$

with a delay $\tau = 1$. The matrices C, A, and B are defined as follows:

$$ C = \begin{pmatrix} 2.0968 & 0 \\ 0 & 2.1935 \end{pmatrix}, \quad A = \begin{pmatrix} 0.78 - 0.42\mathbf{i} & -0.32 + 0.54\mathbf{i} \\ -0.64 + 0.53\mathbf{i} & -0.35 - 0.7\mathbf{i} \end{pmatrix}, $$

$$ B = \begin{pmatrix} -0.64 - 0.53\mathbf{i} & 0.52 - 0.74\mathbf{i} \\ -0.27 + 0.43\mathbf{i} & -0.47 + 0.52\mathbf{i} \end{pmatrix}, \quad W = \begin{pmatrix} 0.8 - 0.52\mathbf{i} \\ -0.56 + 0.2\mathbf{i} \end{pmatrix}. $$

The corresponding response system is described by:

$$ {}^{c}\nabla_0^{\alpha} y_j(t) = -c_j y_j(t) + \sum_{k=1}^{2} a_{jk} f_k(y_k(t)) + \sum_{k=1}^{2} b_{jk} g_k(y_k(t-\tau)) + W_j + u(t), \quad j = 1, 2, \ t \in \mathbb{N}_0, \tag{6.36} $$

where the control input is defined as $u(t) = -\zeta_j(y_j(t) - x_j(t))$. The error between the drive and response systems is represented by $\varrho_j(t) = y_j(t) - x_j(t)$. The error dynamics can be expressed as:

$$ {}^{c}\nabla_0^{\alpha} \varrho_j(t) = -c_j \varrho_j(t) - \zeta_j \varrho_j(t) + \sum_{k=1}^{2} a_{jk} f_k(\varrho_k(t)) + \sum_{k=1}^{2} b_{jk} g_k(\varrho_k(t-\tau)), \quad j = 1, 2, \ t \in \mathbb{N}_0. \tag{6.37} $$

The initial conditions for the error system indicate that if there is no control input ($\zeta_1 = \zeta_2 = 0$), the phase trajectories of $x_j(t)$ and $y_j(t)$ will not achieve global Mittag-Leffler synchronization. By selecting a control gain matrix $\zeta = \mathrm{diag}(0.7, 1.2)$, the dynamics of the drive-response systems demonstrate improved synchronization. Verifying the parameters shows that Assumption 9 is satisfied with $\mathcal{F}_1 = \mathcal{F}_2 = 0.6$ and $\mathcal{G}_1 = \mathcal{G}_2 = 0.4$. Setting $\rho_3 = 2.21$ and $\rho_4 = 0.08$, we can compute that $\mathcal{L}_3 = \min\{3.0632, 4.3331\} = 3.0632 > 0$ and $\mathcal{L}_4 = \max\{1.8966, 2.6186\} = 2.6186 > 0$. This leads to $0 < \mathcal{L}' = 0.4446 < 1$, indicating that the conditions for stability are met. Therefore, according to the established theorems, the error system is globally Mittag-Leffler stable, confirming that the drive system and the corresponding response system can achieve global Mittag-Leffler synchronization (Figs. 6.9, 6.10, 6.11, 6.12, 6.13 and 6.14).

In the study of discrete-time fractional-order complex-valued neural networks (FCVNNs), two examples illustrate the intricate dynamics and synchronization properties of such systems. The first example features a three-neuron FCVNN defined by the equation

$$^{c}\nabla_{0}^{\alpha} x_j(t) = -c_j x_j(t) + \sum_{k=1}^{3} a_{jk} f_k(x_k(t)) + \sum_{k=1}^{3} b_{jk} g_k(x_k(t-\tau)) + W_j, \quad j = 1, 2, 3, \ t \in \mathbb{N}_0.$$

In this scenario, the authors set $\alpha = 0.945$ and define $x_j(t)$ as a complex number expressed in terms of its real and imaginary components: $x_j(t) = x_j^R(t) + x_j^I(t)\mathbf{i}$, where $x_j^R(t)$ and $x_j^I(t)$ are real values. The activation functions are defined as follows:

$$f_k(x_k) = 0.2\tanh(x_k^R) + 0.2\tanh(x_k^I)\mathbf{i}, \quad g_k(x_k) = 0.25\tanh(x_k^R) + 0.25\tanh(x_k^I)\mathbf{i}.$$

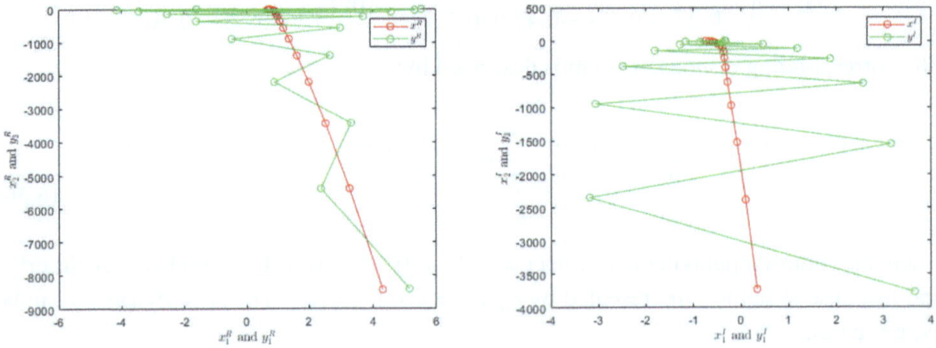

Fig. 6.9 Phase trajectories of system (6.35) and (6.36) without control

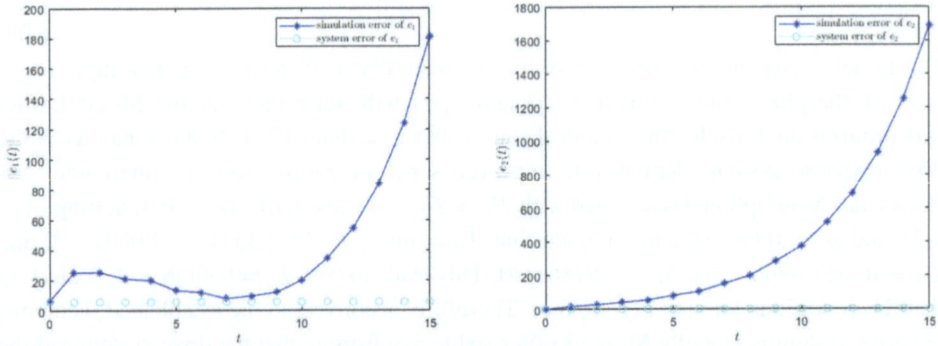

Fig. 6.10 Synchronization errors of error system (6.37) without control

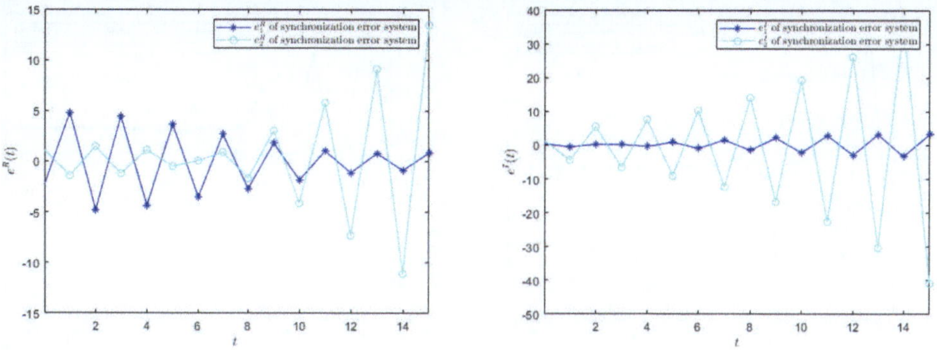

Fig. 6.11 State trajectories of error system (6.37) without control

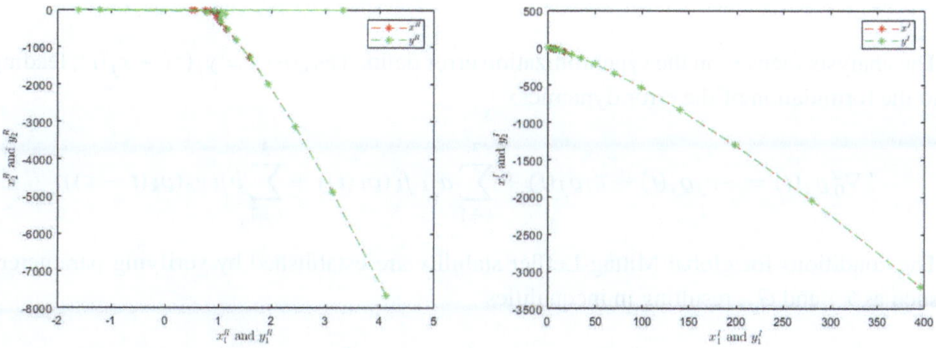

Fig. 6.12 Phase trajectories of response system (6.35) and (6.36) with control

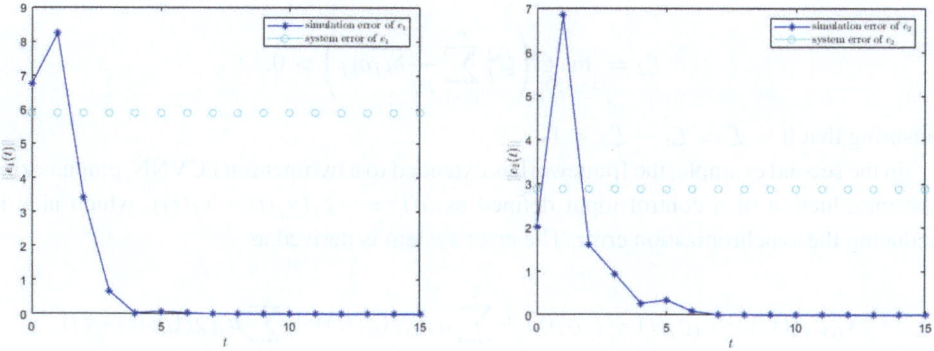

Fig. 6.13 Synchronization errors of error system (6.37) with control

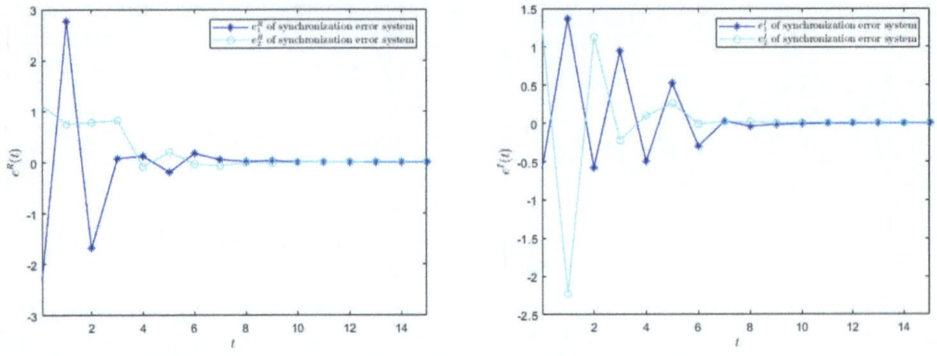

Fig. 6.14 State trajectories of error system (6.37) with control

The analysis focuses on the synchronization error defined as $\varrho_j(t) = y_j(t) - x_j(t)$, leading to the formulation of the error dynamics:

$$^c\nabla_0^\alpha \varrho_j(t) = -c_j\varrho_j(t) - \zeta_j\varrho_j(t) + \sum_{k=1}^{3} a_{jk} f_k(\varrho_k(t)) + \sum_{k=1}^{3} b_{jk} g_k(\varrho_k(t-\tau)).$$

The conditions for global Mittag-Leffler stability are established by verifying parameters such as $\mathcal{F}_j$ and $\mathcal{G}_j$, resulting in inequalities

$$\mathcal{L}_1 = \min_{1\le j\le n}\left(c_j + \overline{c_j} - \rho_1 - \rho_2 - \mathcal{F}_j^2 \sum_{k=1}^{n} \frac{1}{\rho_1}\overline{a_{kj}} a_{kj}\right) > 0,$$

$$\mathcal{L}_2 = \max_{1\le j\le n}\left(\mathcal{G}_j^2 \sum_{k=1}^{n} \frac{1}{\rho_2}\overline{b_{kj}} b_{kj}\right) > 0,$$

ensuring that $0 < \mathcal{L} = \mathcal{L}_1 - \mathcal{L}_2 < 1$.

In the second example, the framework is extended to a two-neuron FCVNN, emphasizing the introduction of a control input defined as $u(t) = -\zeta_j(y_j(t) - x_j(t))$, which aids in reducing the synchronization error. The error system is derived as

$$^c\nabla_0^\alpha \varrho_j(t) = -c_j\varrho_j(t) - \zeta_j\varrho_j(t) + \sum_{k=1}^{2} a_{jk} f_k(\varrho_k(t)) + \sum_{k=1}^{2} b_{jk} g_k(\varrho_k(t-\tau)).$$

The successful tuning of control gains is critical for achieving effective synchronization. The conditions for Assumption 9 are verified with parameters $\mathcal{F}_1 = \mathcal{F}_2 = 0.6$ and $\mathcal{G}_1 = \mathcal{G}_2 = 0.4$, while the constants ρ_3 and ρ_4 are set to ensure positive $\mathcal{L}_3$ and $\mathcal{L}_4$, leading to $0 < \mathcal{L}' < 1$. Therefore, the derived criteria affirm that the error system exhibits global Mittag-Leffler stability, indicating successful global synchronization between the drive and response systems. This comprehensive analysis highlights the complex interplay between system parameters and their effects on stability and synchronization in discrete-time FCVNNs.

Open Questions

Prove or disprove the following:

6.1 α-stability and α-synchronization for Caputo–Katugampola fractional-order time-delayed neural networks in view of [22].

6.2 Finite-time stability of FoNNs with Atangana-Baleanu fractional derivative with complex order in view of [70].

6.3 Existence and uniform stability of complex-valued neural networks with time delays in the view of tempered Ψ-Caputo fractional derivatives [71].

References

1. McCulloch. W.S., Pitts. W.: A logical calculus of the ideas imminent in nervous activity. Bull. Math. Biophys. 115–133 (1943)
2. Tanner, M.A., Jacobs, R.A.: Neural Networks and Related Statistical Latent Variable Models. In: Smelser, N.J., Baltes, P.B. (eds.) International Encyclopedia of the Social & Behavioral Sciences, Pergamon, pp. 10526–10534 (2001)
3. Bahrami, A., Lynch, M., Dagli, C.H.: Intelligent design retrieval and packaging system: application of neural networks in design and manufacturing. Int. J. Prod. Res. **33**(2), 405–426 (1995)
4. Cheng, B., Titterington, D.M.: Neural networks: a review from the statistical perspective. Stat. Sci. **9**, 2–54 (1994)
5. Reinschmidt, K.F.: Neural networks: next step for simulation and control. Power Eng. **95**(11), 41–45 (1991)
6. Sorsa, T., Koivo, H.N., Koivisto, H.: Neural networks in process fault diagnosis. IEEE Trans. Syst., Man, Cybern. **21**(4), 815–825 (1991)
7. Smith, A.E., Dagli, C.H.: Controlling industrial processes through supervised, feedforward neural networks. Comput. Ind. Eng. **21**(1–4), 247–251 (1991)
8. Chowdhury, P.: Neural networks: basics of architecture- a quick guide to basic neural networks concept. https://medium.com/@praneshchowdhury/neural-networks-understanding-the-basics-of-architecture-31d2cf2afd1a
9. Boroomand, A., Menhaj, M.B.: Fractionalorder hopfield neural networks. In: Proceedings of International Conference on Neural Information Processing, pp. 883–890. Springer, Berlin (2008)
10. Kaslik, E., Sivasundaram, S.: Dynamics of fractional-order neural networks. In: The 2011 international joint conference on neural networks. IEEE, pp. 611–618 (2011)
11. Kaslik, E., Sivasundaram, S.: Nonlinear dynamics and chaos in fractional-order neural networks. Neural Netw. **32**, 245–256 (2012)
12. Sierociuk, D., Petráš, I.: Modeling of heat transfer process by using discrete fractional-order neural networks. In: 2011 16th international conference on methods & models in automation & robotics, pp. 146–150. IEEE (2011)

13. Chen, L., Chai, Y., Wu, R., Ma, T., Zhai, H.: Dynamic analysis of a class of fractional-order neural networks with delay. Neurocomputing **111**, 190–194 (2013)
14. Ran-Chao, W., Xin-Dong, H., Li-Ping, C.: Finitetime stability of fractional-order neural networks with delay. Commun. Theor. Phys. **60**(2), 189 (2013)
15. Chen, L., Liu, C., Wu, R., He, Y., Chai, Y.: Finite time stability criteria for a class of fractional-order neural networks with delay. Neural Comput. Appl. **27**(3), 549–556 (2016)
16. Song, C., Cao, J.: Dynamics in fractional-order neural networks. Neurocomputing **142**, 494–498 (2014)
17. Chen, L., Qu, J., Chai, Y., Wu, R., Qi, G.: Synchronization of a class of fractional-order chaotic neural networks. Entropy **15**(8), 3265–3276 (2013)
18. Podlubny, I.: Fractional Differential Equation. AcademicPress, SanDiego (1999)
19. Li, C.P., Deng, W.H.: Remarks on fractional derivatives. Appl. Math. Comput. **187**(2), 777–784 (2007)
20. Zhang, S.: Positive solutions for boundary value problems of nonlinear fractional differential equations. Electron. J. Differ. Equ. **2006**, 1–12 (2006)
21. Krasnosel'skii, M.: Topological Methods in the Theory of Nonlinear Integral Equations. Pergamon, Elmsford (1964)
22. Yu, J., Hu, C., Jiang, H.: α-Stability and α-synchronization for fractional-order neural networks. Neural Netw. **35**, 82–87 (2012)
23. Zou, T., et al.: Stability analysis of a class of fractional-order neural networks. TELKOMNIKA Indones. J. Electr. Eng. **12**(2), 1086–1093 (2014)
24. Panda, S.K., Vijayakumar, V., Agarwal, R.P.: On the uniform stability of fractional-order neural networks and solutions to nonlinear mixed integral equations via nonlinear F-contractions. Complex Anal. Oper. Theory **19**, 191 (2025). https://doi.org/10.1007/s11785-025-01815-2
25. Wardowski, D.: Fixed points of a new type of contractive mappings in complete metric spaces. Fixed Point Theory Appl. **94**, 6 (2012). https://doi.org/10.1186/1687-1812-2012-94
26. Turinici, M.: Wardowski implicit contractions in metric spaces (2013). arxiv:1212.3164v2 [Math.GN]
27. Wardowski, D.: Solving existence problems via F-contractions. Proc. Am. Math. Soc. **146**(4), 1585–1598 (2018)
28. Cao, Y., Bai, C.: Existence and stability analysis of fractional order BAM neural networks with a time delay. Appl. Math. **6**, 2057–2068 (2015). https://doi.org/10.4236/am.2015.612181
29. Aguilar, C.J.Z., G6mez-Aguilar, J.F., Alvarado-MartíÂnez, V.M., Romero-Ugalde, H.M.: Fractional order neural networks for system identification. Chaos, Solitons Fractals **130**, 109444 (2020). https://doi.org/10.1016/j.chaos.2019.109444
30. Liu, X., Chen, L., Zhao, Y., Song, X.: Dynamic stability of a class of fractional-order nonlinear systems via fixed point theory. Math. Meth. Appl. Sci. **45**, 77–92 (2021)
31. Ran-Chao, W., Xin-Dong, H., Li-Ping, C.: Finite-time stability of fractional-order neural networks with delay. Commun. Theor. Phys. **60**(2), 189 (2013)
32. Xu, C., Li, P.: On finite-time stability for fractional-order neural networks with proportional delays. Neural Process Lett. (2018). https://doi.org/10.1007/s11063-018-9917-2
33. Jmal, A., Ben, M.A., Nagy, A.M., Naifar, O.: Finite-time stability for Caputo-Katugampola fractional-order time-delayed neural networks. Neural Process. Lett. **50**, 607–621 (2019)
34. Katugampola, U.N.: New approach to a generalized fractional integral. Appl. Math. Comput. **218**(3), 860–865 (2011)
35. Katugampola, U.N.: A new approach to generalized fractional derivatives. Bull. Math. Anal. Appl. **6**(4), 1–15 (2014)
36. Kuczma, M.: An introduction to the theory of functional equations and inequalities: Cauchy's equation and Jensen's inequality. Springer Science & Business Media, Berlin (2009)

37. Mitrinovic, N.D.: Analytic inequalities. Springer, New York (1970)
38. Ben Makhlouf, A., Nagy, A.M.: Finite-time stability of linear Caputo–Katugampola fractional-order time delay systems. Asian J. Control (2018). https://doi.org/10.1002/asjc.1880
39. Ke, Y., Miao, C.: Stability analysis of fractional-order Cohen-Grossberg neural networks with time delay. Int. J. Comput. Math. **92**(6), 1102–1113 (2015)
40. Yang, X., Song, Q., Liu, Y., Zhao, Z.: Finite-time stability analysis of fractional-order neural networks with delay. Neurocomputing **152**, 19–26 (2015)
41. Dai, M., Jiang, Y., Du, J., et al.: Finite time stability of Caputo–Katugampola fractional order time delay projection neural networks. Neural Process Lett. **54**, 4851–4867 (2022). https://doi.org/10.1007/s11063-022-10838-1
42. Xiao, S., Li, J.: New Result on Finite-Time Stability for Caputo–Katugampola Fractional-Order Neural Networks with Time Delay. Neural Process Lett 55, 7951–7966 (2023). https://doi.org/10.1007/s11063-023-11291-4
43. Yang, Z., Zhang, J., Hu, J., Mei, J.: New results on finite-time stability for fractional-order neural networks with proportional delay. Neurocomputing **442**, 327–336 (2021). https://doi.org/10.1016/j.neucom.2021.02.082
44. Rakkiyappan, R., Cao, J., Velmurugan, G.: Existence and uniform stability analysis of fractional-order complex-valued neural networks with time delays. IEEE Trans. Neural Netw. Learn. Syst. **26**(1), 84–97 (2015). https://doi.org/10.1109/TNNLS.2014.2311099
45. Panda, S.K., Kalla, K.S., Nagy, A.M.: Priyanka L, Numerical simulations and complex valued fractional order neural networks via $(\varepsilon - \mu)$-uniformly contractive mappings. Chaos, Solitons Fractals **173**, 113738 (2023)
46. Wang, L., Song, Q., Liu, Y., Zhao, Z., Alsaadi, F.E.: Global asymptotic stability of impulsive fractional-order complex-valued neural networks with time delay. Neurocomputing **243**, 49–59 (2017). https://doi.org/10.1016/j.neucom.2017.02.086
47. Kaslik, E., Rădulescu, I.R.: Dynamics of complex-valued fractional-order neural networks. Neural Netw. **89**, 39–49 (2017). https://doi.org/10.1016/j.neunet.2017.02.011
48. Li, H., Kao, Y., Bao, H., Chen, Y.: Uniform stability of complex-valued neural networks of fractional order with linear impulses and fixed time delays. IEEE Trans. Neural Netw. Learn. Syst. **33**(10), 5321–5331 (2022). https://doi.org/10.1109/TNNLS.2021.3070136
49. Wang, Y., Zheng, C.D., Lin, M.: Robust stability of complex-valued fractional-order neural networks with uncertain parameters based on new integral inequalities. Int. J. Mach. Learn. Cyber. **14**, 4377–4391 (2023)
50. Liu, X., Chen, L., Zhao, Y.: Uniform stability of a class of fractional-order fuzzy complex-valued neural networks in infinite dimensions. Fractal Fract. **6**, 281 (2022). https://doi.org/10.3390/fractalfract6050281
51. Rakkiyappan, R., Udhayakumar, K., Velmurugan, G., et al.: Stability and Hopf bifurcation analysis of fractional-order complex-valued neural networks with time delays. Adv. Differ. Equ. **2017**, 225 (2017). https://doi.org/10.1186/s13662-017-1266-3
52. Panda, S.K., et al.: Stability analysis for complex-valued neural networks with fractional order. Chaos, Solitons Fractals **175**, 114045 (2023)
53. Anbalagan, P., Ramachandran, R., Cao, J., et al.: Global robust synchronization of fractional order complex valued neural networks with mixed time varying delays and impulses. Int. J. Control Autom. Syst. **17**, 509–520 (2019). https://doi.org/10.1007/s12555-017-0563-7
54. Panda, S.K., Abdeljawad, T., Nagy, A.M.: On uniform stability and numerical simulations of complex valued neural networks involving generalized Caputo fractional order. Sci. Rep. **14**, 4073 (2024). https://doi.org/10.1038/s41598-024-53670-4
55. Goodrich, C., Peterson, A.C.: Discrete Fractional Calculus. Springer, Berlin (2015)
56. Ganji, M., Gharari, F.: The discrete delta and Nabla Mittag- Leffler distributions. Commun. Stat. - Theory Methods **47**, 4568–4589 (2018)

57. Alzabut, J., Abdeljawad, T., Baleanu, D.: Nonlinear delay fractional difference equations with applications on discrete fractional Lotka-Volterra competition model. J. Comput. Anal. Appl. **25**, 889–898 (2018)
58. Wang, L., Song, Q., Liu, Y., Zhao, Z., Alsaadi, F.E.: Finite-time stability analysis of fractional-order complex-valued memristor-based neural networks with both leakage and time-varying delays. Neurocomputing **245**, 86–101 (2017)
59. You, X., Song, Q., Zhao, Z.: Existence and finite-time stability of discrete fractional-order complex-valued neural networks with time delays. Neural Netw. **123**, 248–260 (2020). https://doi.org/10.1016/j.neunet.2019.12.012
60. Abdeljawad, T.: On delta and Nabla Caputo fractional differences and dual identities. Discret. Dyn. Nat. Soc. (2013)
61. Ding, X., Cao, J., Zhao, X., Alsaadi, F.E.: Finite-time stability of fractional-order complex-valued neural networks with time delays. Neural Process. Lett. **46**, 561–580 (2017)
62. Wang, L., Song, Q., Liu, Y., Zhao, Z., Alsaadi, F.E.: Global asymptotic stability of impulsive fractional-order complex-valued neural networks with time delay. Neurocomputing **243**, 49–59 (2017)
63. Zhang, L., Song, Q., Zhao, Z.: Stability analysis of fractional-order complex-valued neural networks with both leakage and discrete delays. Appl. Math. Comput. **298**, 296–309 (2017)
64. Song, Q., Yang, X., Li, C., Huang, T., Chen, X.: Stability analysis of nonlinear fractional-order systems with variable-time impulses. J. Franklin Inst. **354**, 2959–2978 (2017)
65. Panda, S.K., Vijayakumar, V., Nagy, A.M.: Complex-valued neural networks with time delays in the LP sense: numerical simulations and finite time stability. Chaos, Solitons Fractals **177**, 114263 (2023). https://doi.org/10.1016/j.chaos.2023.114263
66. You, X., Song, Q., Zhao, Z.: Global Mittag-Leffler stability and synchronization of discrete-time fractional-order complex-valued neural networks with time delay. Neural Netw. **122**, 382–394 (2020). https://doi.org/10.1016/j.neunet.2019.11.004
67. Abdeljawad, T., Baleanu, D.: Discrete fractional differences with nonsingular discrete Mittag-Leffler kernels. Adv. Differ. Equ. **2016**(232) (2016)
68. Abdeljawad, T., Baleanu, D.: Monotonicity analysis of a Nabla discrete fractional operator with discrete Mittag-Leffler kernel. Chaos, Solitons Fractals **102**, 106–110 (2017)
69. Yang, S., Yu, J., Hu, C., Jiang, H.: Quasi-projective synchronization of fractional-order complex-valued recurrent neural networks. Neural Netw. **104**, 104–113 (2018)
70. Fernandez, A.: A complex analysis approach to Atangana–Baleanu fractional calculus. Math. Methods Appl. Sci. 1–18 (2019)
71. Medved, M., Brestovanská, E.: Differential equations with tempered Ψ-Caputo fractional derivative. Math. Model. Anal. **26**(4), 631–650 (2021)

Modeling and Simulation 7

Mathematical modeling refers to the dire action of describing practical systems, phenomena, or issues with mathematical abstractions and frameworks. Modeling aims at reducing the complexity of systems, whilst maintaining their key properties, so that they can be analyzed, predicted, and decisions made. The chapter is devoted to the modeling of so-called computer viruses and infectious diseases where the main tools of the analysis are the fractional calculus and the fixed point theory. Computer virus spread can be effectively modeled mathematically as a way of gaining insight into this phenomenon and as a means of making predictions about it. They infect, spread and disrupt their host systems in much the same way as biological viruses do and once again this means that mathematical techniques, especially those of epidemiology can be used to analyze their behaviour. Computer viruses are transmitted over the networks via the vulnerability in the systems. To analyze such process, it is common to use models that divide the devices in three states: Susceptible (S), which are vulnerable devices; Infected (I), which are device already infected and Recovered (R), which are devices that have been cleaned or secured against the virus. These classes assist in simplifying and dissecting the manner in which a virus spreads and is finally contained. A popular model is the SIR model which is used to trace how the devices are flowing in and out of the three categories. The model takes parameters like the transmission rate (β), where the rate determines how fast the virus spreads, and the recovery rate (γ), where the rate determines how well infected devices are recuperated. There is also another simpler model that does not consider recovery and is only concerned with the spread of the virus; the SI model.

An important parameter in such models is the **basic reproduction number** (R_0) which is a measure of the average number of devices that an infected device can infect. When R_0 is exceeding 1, then the virus is expected to spread substantially; otherwise, when R_0 is smaller than 1, the virus will eventually become extinct. The spread can be combated by

S. K. Panda et al., *Fixed Point Theory from Early Foundations to Contemporary Challenges*,
Synthesis Lectures on Mathematics & Statistics,
https://doi.org/10.1007/978-3-032-03585-1_7

modification of factors such as the transmission rate (by use of firewalls) or the recovery rate (by use of antivirus software). Threshold effects are also uncovered by mathematical models, whereby minute shifts in security measures can drastically change the spread of the virus. As an illustration, outbreaks can be averted by escalating the antivirus updates or enhancing the user understanding. Such observations are used to formulate efficient mitigation measures, such as the use of firewalls or enhanced software patching procedures. Nevertheless, the computer viruses are not that easy to model. Digital networks are intricate, time-varying and usually non-homogenous. Further, devices and users are both heterogeneous in terms of behavior and vulnerability. Predictions are further confounded by human variables, such as user neglect, or slow software updates.

In spite of this, mathematical models are important in cybersecurity. They assist in optimization of resources, policy design and the creation of tools to decrease risks. These models can enhance the resilience and security of digital systems face dynamic threats by borrowing concepts on biological virus containment.

Infectious diseases modeling on the other hand, is based on the mathematical descriptions and predictions of infectious diseases dynamics in populations. Such classical models as the SIR and SEIR (Susceptible-Exposed-Infected-Recovered) models describe the transition between compartments using systems of ordinary differential equations (ODEs). For example, the rate of change of the susceptible population, $\frac{dS}{dt}$, often depends on the basic reproduction number R_0, which is defined as:

$$R_0 = \frac{\beta}{\gamma},$$

where β is the transmission rate and γ is the recovery rate.

The extension to fractional calculus brings in fractional-order derivatives, which can include memory effects and anomalous diffusion. This makes it a closer model of real-world phenomena, including later reaction to interventions or non-homogenous interaction of populations. The theory of fixed point may as well be used to prove the existence and the stability of equilibrium states and this will provide a better understanding of the disease control and eradication interventions.

7.1 Computer Virus Modeling

Mathematical modeling is a systematic approach to learning and forecasting the dynamics of real-life systems, such as the propagation of computer viruses in computer systems. Computer viruses may be compared to biological ones- they reproduce, are transmitted and produce harm to their host systems. On the basis of this similarity the tools of mathematics and especially those motivated by epidemiology can be used to learn about and manage the spread of computer viruses. Modeling of computer viruses is the mathematical modeling and study of the distribution of malicious software over digital networks. Computer viruses, just like biological ones, reproduce and spread using the weaknesses in devices and communications Systems. The models of this field frequently use ideas of epidemiology, like the Susceptible-Infected-Recovered (SIR) model, to define the infection phases on a network. These models allow predicting the speed of the virus spreading, the influence of various types of network structures, and the efficiency of antivirus measures with the help of equations and algorithms. The lessons learned are used to develop effective cybersecurity protocols to reduce the chance of a mass destruction. The history of computer viruses dates back to the 70 s when the Creeper program was created which was aimed at disrupting and changing the operations of computer systems [1]. After that some sophisticate virus programs such as Rabbit, Univac, Michelangelo and CIH virus (also referred as Chernobyl) were created which showed how the malicious software was becoming so advanced [2]. These developments did not only reveal the weak points of digital systems but also demonstrated the importance of computational tools in solving sophisticated problems in different disciplines.

As computational technologies have developed, researchers have been able to use sophisticated software packages to construct and examine complex network behaviours including virus spread and system vulnerability. As a few examples, network security problems, linear ordering, halting problems, and network anomaly detection have been studied [3–8]. These studies demonstrate the strengths of computational modeling in shedding light on a variety of real-world problems. To address the urgent demand of efficient methods to curb the spread of the computer viruses, mathematical models which were initially used in epidemiology have been borrowed and applied in understanding the dynamics of the viruses in network systems. A prominent example is the SIR (Susceptible-Infected-Removed) model, which has been further extended to include an antidotal population, forming the SIRA model. This adaptation allows for the analysis of virus behavior in networked systems using nonlinear differential equations [9]. The SIRA model is defined by the following system of equations:

$$
\begin{aligned}
\frac{dS(t)}{dt} &= -\beta_{SA} S(t)A(t) - \gamma S(t)I(t) + \delta R(t),\\
\frac{dI(t)}{dt} &= \gamma S(t)I(t) - \beta_{IA} A(t)I(t) - \epsilon I(t),\\
\frac{dR(t)}{dt} &= \epsilon I(t) - \delta R(t),\\
\frac{dA(t)}{dt} &= \beta_{SA} S(t)A(t) + \beta_{IA} A(t)I(t),
\end{aligned} \tag{7.1}
$$

where S, I, R, and A represent the susceptible, infected, removed, and antidotal populations, respectively. The parameters β_{SA}, β_{IA}, γ, δ, and ϵ are positive constants representing the rates of infection, removal, and recovery.

To better capture memory effects and time-dependent dynamics, the model is extended to a fractional-order system using the Caputo derivative, as follows:

$$
\begin{aligned}
D_t^{\alpha} S(t) &= -\beta_{SA} S(t)A(t) - \gamma S(t)I(t) + \delta R(t),\\
D_t^{\alpha} I(t) &= \gamma S(t)I(t) - \beta_{IA} A(t)I(t) - \epsilon I(t),\\
D_t^{\alpha} R(t) &= \epsilon I(t) - \delta R(t),\\
D_t^{\alpha} A(t) &= \beta_{SA} S(t)A(t) + \beta_{IA} A(t)I(t),
\end{aligned} \tag{7.2}
$$

where D_t^{α} denotes the Caputo fractional derivative of order $\alpha \in (0, 1]$ [10, 11].

In [12], the solution of this fractional model was obtained by the authors through Multi-Step Generalized Differential Transform Method (MSGDTM). The method provides very accurate approximations even in long interval and it agrees with classical Runge-Kutta in the integer order derivative situation. The authors of [12] conducted graphical analysis to depict how the compartments $S(t)$, $I(t)$, $R(t)$, and $A(t)$ vary with time by changing parameters setting, which confirmed the significance of the fractional-order derivatives in predicting long-term behavior of the system.

Just this year, 2022, a new epidemiological model with modifications was proposed in the authors of [13] to analyze the computer viruses dynamical behavior. The model which will be denoted as the SIRA model, adds another antidotal population to represent the systems having antivirus programs. By using the Caputo fractional derivative, the paper has given a detailed study of the model behavior, which includes the wave-like nature of computer virus spread in network systems. In order to obtain the solutions of this model, the authors used the Fractional Natural Decomposition Method (FNDM) and the Variational Iteration Method (VIM) to make sure that they have obtained the reliable and accurate solutions. This study presented the preferred values of the parameters to reduce the propagation of the viruses, with the theoretical information on the uniqueness of the solutions with the Lipschitz condition and Banach space. Further, numerical simulations depicted important information of the distributions of waves and the dynamics of the infected, susceptible, and removed population and systems with antivirus-equipped nodes. The proposed study provides a powerful framework of computer virus dynamic mitigation and highlights the promise of fractional-order mathematical model-based approach to dealing with nonlinear phenomena in computational and network systems.

In shorter words, authors of [13] considered a modified SIRA model to investigate computer virus dynamics on network systems. In this model the network is compartmentalized into four sets: $S(t)$, $I(t)$, $R(t)$, and $A(t)$ denoting susceptible computers, infected computers, removed computers (by infection or other causes) and non-infected computers with antivirus software, respectively. The system is initially modeled using a set of ordinary differential equations (ODEs), expressed as:

$$\frac{dS(t)}{dt} = -\beta_{SA}S(t)A(t) - \gamma S(t)I(t) + \delta R(t),$$
$$\frac{dI(t)}{dt} = \gamma S(t)I(t) - \beta_{IA}A(t)I(t) - \epsilon I(t),$$
$$\frac{dR(t)}{dt} = \epsilon I(t) - \delta R(t),$$
$$\frac{dA(t)}{dt} = \beta_{SA}S(t)A(t) + \beta_{IA}A(t)I(t).$$

Here, β_{SA} and β_{IA} denote the infection rates involving interactions with antivirus-equipped computers, γ represents the infection rate between susceptible and infected computers, δ is the removal rate, and ϵ corresponds to the recovery rate of infected computers.

To better represent real-world scenarios where historical interactions play a significant role, the authors of [13] extended the SIRA model by incorporating the Caputo fractional derivative D_t^α, where $\alpha \in (0, 1]$. This extension allows the model to capture memory effects and time-dependent dynamics more accurately. The fractional SIRA model is given by:

$$D_t^\alpha S(t) = -\beta_{SA}S(t)A(t) - \gamma S(t)I(t) + \delta R(t),$$
$$D_t^\alpha I(t) = \gamma S(t)I(t) - \beta_{IA}A(t)I(t) - \epsilon I(t),$$
$$D_t^\alpha R(t) = \epsilon I(t) - \delta R(t),$$
$$D_t^\alpha A(t) = \beta_{SA}S(t)A(t) + \beta_{IA}A(t)I(t).$$

The outcome of the study in [13] showed that the considered fractional derivative plays a significant role in enhancing the capabilities of the model in considering the aspects of memory and delay that are maintained by network systems. They applied numerical methods, which included Fractional Natural Decomposition Method (FNDM) and Variational Iteration Method (VIM) in the investigation of the fractional model. To take an example, the VIM-based correction of $S(t)$ will be:

$$S_{k+1}(t) = S_k(t) + \int_0^t \lambda(\tau)\left[D_t^\alpha S_k(\tau) + \beta_{SA}S_k(\tau)A_k(\tau) + \gamma S_k(\tau)I_k(\tau) - \delta R_k(\tau)\right]d\tau,$$

where $\lambda(\tau) = \frac{(-1)^\alpha (t-\tau)^{\alpha-1}}{\Gamma(\alpha)}$ is the Lagrange multiplier.

The writers of [13] committed to the reality that the Lipschitz condition ensures the uniqueness of the solutions:

$$\|F(t, x_1) - F(t, x_2)\| \leq L\|x_1 - x_2\|,$$

where F denotes the vector field of the equations, x_1 and x_2 represent solution states, and L is the Lipschitz constant.

To verify the model, the authors of [13] conducted numerical simulations so as to obtain optimal values of the parameters such as β_{SA}, β_{IA}, and γ which minimize the transmission of the virus. Their results enabled the superiority of the fractional-order model to capture the wave-like characteristic of infections and accurately predict the dynamics of systems. With the help of the study, it was proven beyond doubt that the fractional SIRA model is more accurate than its integer-order corresponding model, presenting an influential outline of computer virus transmission in network systems and the means of restraining the impact.

In [14], the authors introduced a fractional-order SIRA model of computer viruses in terms of the Atangana-Baleanu fractional derivative (AB-fractional derivative). This derivative brings in the memory effects and non-singular kernel in the analysis of computer virus propagation. The model divides the total population T into four compartments:

- $S(t)$: Susceptible computers,
- $I(t)$: Infected computers,
- $R(t)$: Removed computers,
- $A(t)$: Antivirus-protected computers.

The mathematical model incorporating the AB-fractional derivative is given as:

$$\begin{aligned}
{}^{ABC}D^{\rho}S(t) &= -\alpha_1 S(t)A(t) - \beta S(t)I(t) + \sigma R(t),\\
{}^{ABC}D^{\rho}I(t) &= \beta S(t)I(t) - \alpha_2 A(t)I(t) - \delta I(t),\\
{}^{ABC}D^{\rho}R(t) &= \delta I(t) - \sigma R(t),\\
{}^{ABC}D^{\rho}A(t) &= \alpha_1 S(t)A(t) + \alpha_2 A(t)I(t),
\end{aligned}$$

where:

- ${}^{ABC}D^{\rho}$: Atangana-Baleanu Caputo fractional derivative with order $\rho \in (0, 1]$,
- α_1, α_2: Transition rates to antivirus-protected states,
- β: Infection rate,
- δ: Recovery rate,
- σ: Restoration rate to susceptibility.

Meanwhile, we should notice that the Atangana-Baleanu fractional derivative in the Caputo sense is defined as:

$$ {}^{ABC}D^{\rho}f(t) = \frac{B(\rho)}{1-\rho}\int_0^t f'(\tau)E_{\rho}\left(-\frac{\rho(t-\tau)^{\rho}}{1-\rho}\right)d\tau, $$

where E_{ρ} is the Mittag-Leffler function, and $B(\rho)$ is a normalization constant. Here are some assumptions in their study:

- The total population $T = S + I + R + A$ is constant.

- No influx ($N = 0$) and no mortality effects ($\mu = 0$) are considered.
- Initial conditions: $S(0) = S_0$, $I(0) = I_0$, $R(0) = R_0$, $A(0) = A_0$.

Using **fixed-point theory**, the authors established the existence and uniqueness of the solution. The conditions for uniqueness rely on satisfying the Lipschitz criterion for the model's kernel functions K_1, K_2, K_3, and K_4, defined as:

$$\begin{aligned} K_1(S, A) &= -\alpha_1 SA - \beta SI + \sigma R, \\ K_2(S, I) &= \beta SI - \alpha_2 AI - \delta I, \\ K_3(I, R) &= \delta I - \sigma R, \\ K_4(S, I, A) &= \alpha_1 SA + \alpha_2 AI. \end{aligned}$$

Moreover, the authors of [14] also investigated an optimal control approach by introducing a control variable $u(t)$, representing antivirus treatment. The control-augmented equations are:

$$\begin{aligned} {}^{ABC}D^{\rho} I(t) &= \beta S(t)I(t) - \alpha_2 A(t)I(t) - \delta I(t) - u(t)I(t), \\ {}^{ABC}D^{\rho} R(t) &= \delta I(t) - \sigma R(t) + u(t)I(t). \end{aligned}$$

The objective is to minimize the cost functional:

$$J = \frac{1}{2}\int_0^T \left(qI^2(t) + ru^2(t)\right) dt,$$

where q and r are weighting parameters. Figure 7.1 depict the numerical simulations of the model of the fractional-order computer virus that demonstrate the dynamics of susceptible, infected, removed, and antivirus-protected states.

The authors of [15] introduced a comprehensive mathematical framework to simulate the propagation of computer viruses within networks. The model categorizes the population into five compartments:

- $S(t)$: Susceptible computers that are vulnerable to infection,
- $P(t)$: Non-infected computers protected by antivirus software,
- $E(t)$: Exposed computers, recently infected or at risk of infection,
- $I(t)$: Infected computers, serving as hosts for the virus,
- $R(t)$: Removed computers, no longer part of the active network due to infection or malfunction.

The evolution of these compartments is governed by the following system of ordinary differential equations (ODEs):

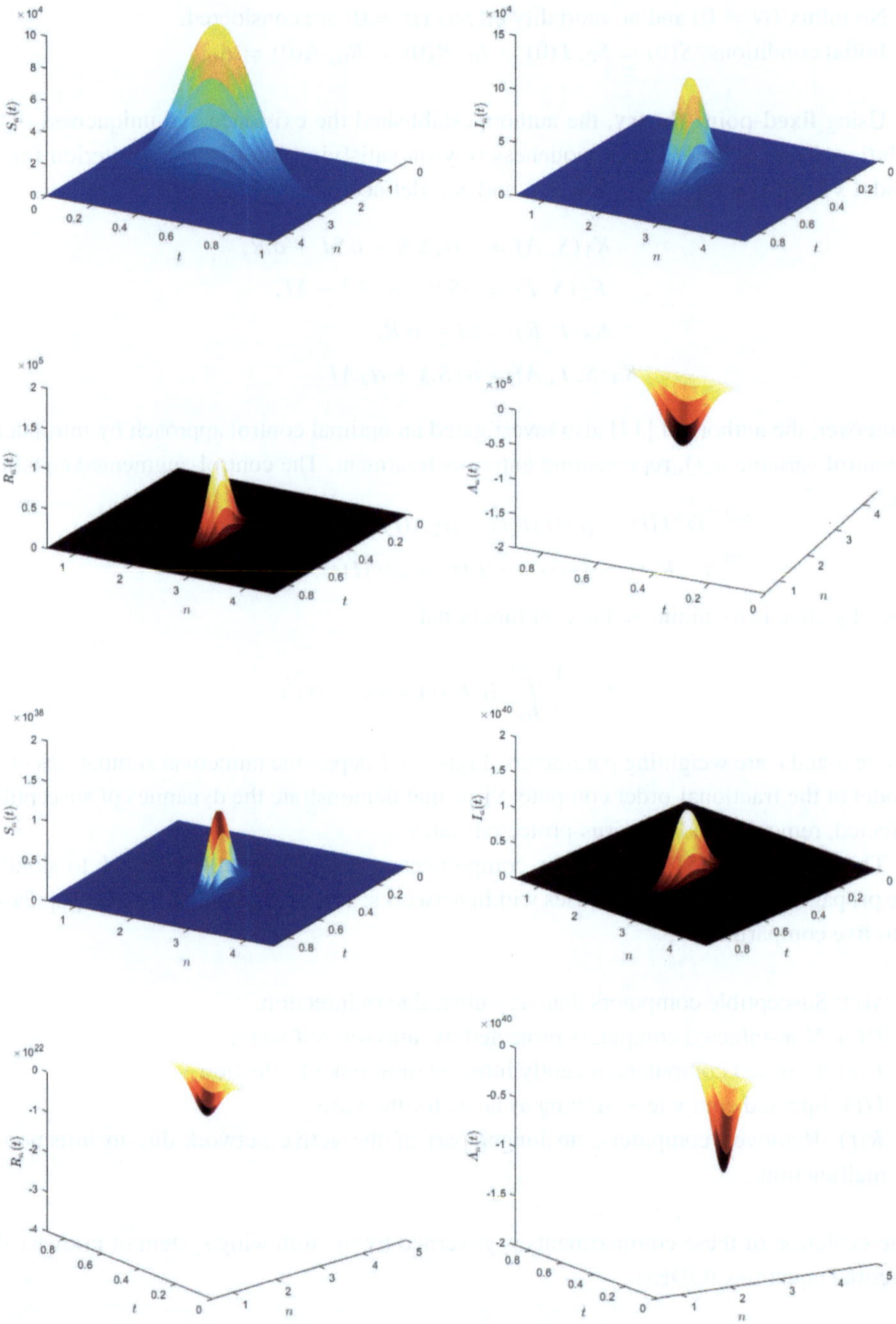

Fig. 7.1 SIRA model without control and with control

$$\frac{dS}{dt} = \Pi - \beta SI - pS + \alpha P + \omega R - \mu S,$$
$$\frac{dP}{dt} = pS - \alpha P - \mu P,$$
$$\frac{dE}{dt} = \beta SI - \delta E - \mu E,$$
$$\frac{dI}{dt} = \delta E - \gamma I - \mu I,$$
$$\frac{dR}{dt} = \gamma I - \mu R - \omega R,$$

where the parameters are defined as:

- β: Infection rate, representing the likelihood of susceptible computers being infected through interaction with infected computers,
- p: Rate of antivirus adoption in susceptible computers,
- α: Decay rate of antivirus effectiveness, transitioning protected computers to the susceptible class,
- ω: Rate at which removed computers rejoin the susceptible class,
- δ: Rate at which exposed computers progress to the infected class,
- γ: Rate of recovery for infected computers,
- μ: Rate of removal due to hardware or software failures,
- Π: Arrival rate of new susceptible computers into the system.

To account for memory effects and non-local dynamics in computer virus propagation, the authors reformulated the model using the Liouville-Caputo fractional derivative ${}^{LC}D_t^{\varphi}$, where $0 < \varphi \leq 1$. The fractional-order model is expressed as:

$$\begin{aligned}
{}^{LC}D_t^{\varphi} S(t) &= \Pi - \beta S(t)I(t) - pS(t) + \alpha P(t) + \omega R(t) - \mu S(t),\\
{}^{LC}D_t^{\varphi} P(t) &= pS(t) - \alpha P(t) - \mu P(t),\\
{}^{LC}D_t^{\varphi} E(t) &= \beta S(t)I(t) - \delta E(t) - \mu E(t),\\
{}^{LC}D_t^{\varphi} I(t) &= \delta E(t) - \gamma I(t) - \mu I(t),\\
{}^{LC}D_t^{\varphi} R(t) &= \gamma I(t) - \mu R(t) - \omega R(t).
\end{aligned}$$

The initial conditions for the model are:

$$S(0) \geq 0, \quad P(0) \geq 0, \quad E(0) \geq 0, \quad I(0) \geq 0, \quad R(0) \geq 0.$$

Additionally, the total population at any time t is:

$$N = S(t) + P(t) + E(t) + I(t) + R(t).$$

The fractional-order derivative introduces long-term memory effects, enhancing the model's capability to capture real-world virus dynamics in heterogeneous networks. The parameters $\beta, p, \alpha, \delta, \omega, \gamma$, and μ significantly influence the trajectory of the system. The authors showed how the fractional orders and the bifurcation conditions influence the viruses spread, which can help to develop more stable cybersecurity measures and improve the antivirus practices. In [15] though the authors used both **Banach Fixed Point Theorem** and **Schaefer Fixed Point Theorem** in proving that the solution to the fractional-order computer virus model is unique and exists. Researchers have relied heavily on fixed point theory to study mathematical models of many phenomena including the spread of computer viruses. This is a theoretical framework to assure the well-posedness of models, that is, the existence and uniqueness of solutions. Researchers used a fixed-point theory in the fractional-order computer virus model to prove the reliability of the system. After transforming the fractional differential equations into equations of the form of integral equations, the fixed point theory was applied to prove the existence of at least one solution of the system and the uniqueness of this solution with the given conditions. The mathematical technique used was the establishment of required mathematical properties of the operator corresponding to the system, namely, continuity, compactness, boundedness in a certain functional space. This rigorous application of fixed point theory not only confirms the existence of solutions but also reinforces the model's effectiveness in representing the dynamic behavior of computer virus propagation. For further insights and detailed applications of fixed point theory in computer virus models and/or computer virus models with associated simulations, readers are encouraged to refer to [16–30].

7.2 Infectious Disease Modeling

Infectious disease modeling often leverages **fixed point theory** and **fractional derivatives** to capture the complexity and memory-dependent nature of disease dynamics. See for example, consider a fractional-order SIR (Susceptible-Infectious-Recovered) model represented by:

$$
\begin{aligned}
D^{\alpha} S(t) &= -\beta S(t) I(t), \text{ for susceptible individuals,} \\
D^{\alpha} I(t) &= \beta S(t) I(t) - \gamma I(t), \text{ for infectious individuals,} \\
D^{\alpha} R(t) &= \gamma I(t), \text{ for recovered individuals,}
\end{aligned}
$$

where D^{α} is the Caputo fractional derivative of order $\alpha \in (0, 1]$, and β and γ are transmission and recovery rates, respectively. The fixed point of the system, typically determined in the reduced phase space, corresponds to the equilibrium state of the disease, where $\dot{S} = \dot{I} = \dot{R} = 0$. By employing Banach's fixed point theorem in appropriate Banach spaces, the existence and uniqueness of these equilibria can be rigorously established. Fractional derivatives, with their nonlocal properties, provide a refined framework to account for historical effects and long-term dependencies in disease progression, which are often crucial in accurately predicting outbreak behavior and devising control strategies.

However, infectious diseases emerge from complex interactions between microbial agents (bacteria, fungi, parasites, viruses excluding prions) and their hosts, influenced by environmental and population factors [31]. Effective management of such diseases relies on understanding these dynamics to develop interventions like antimicrobial treatments, vaccines, and preventive strategies [32]. However, the unpredictability of pathogens, driven by drug resistance, globalization, and urbanization, continues to pose significant challenges [33–35]. Historical examples, such as the spread of smallpox, TB, and diphtheria during colonialism and trade expansions, highlight the critical impact of infectious diseases on human populations [36]. Despite advancements like rapid vaccine development for SARS-CoV-2, neglected diseases such as malaria, HIV, and TB still contribute to high morbidity and mortality, especially in low-income regions [37–39].

Over the past several decades, the study of infectious disease dynamics has evolved into a multidisciplinary field encompassing epidemiology, public health, mathematics, sociology, biology, and even philosophy [40–43]. This integration allows for the exploration of disease transmission using mathematical models to quantify and predict the spread and impact of infections across populations [44]. Mathematical modeling serves as a crucial tool to test hypotheses, analyze the sensitivity of disease dynamics to parameter variations, and predict key outcomes [45–47]. Such models not only enhance understanding of disease behavior but also provide valuable insights for designing and evaluating control strategies. Theoretical advancements address challenges arising from the nonlinear and multilayered nature of infection networks, while practical applications focus on managing large datasets and implementing effective data collection systems. By incorporating both theoretical and experimental insights, mathematical models offer a robust platform to simulate the progression of infectious diseases, assess potential interventions, and develop public health policies. These models have become indispensable for understanding transmission dynamics at local, regional, and global scales, particularly in the context of emerging diseases and pandemic scenarios [48, 49].

Infectious diseases remain a significant global health challenge despite remarkable advancements in understanding their transmission dynamics. The World Health Organization (WHO) reports that millions of individuals contract diseases such as malaria, tuberculosis (TB), HIV/AIDS, and cholera annually, with developing nations bearing the brunt of these burdens [50]. For instance, malaria affects 300–500 million people every year, while sexually transmitted diseases (STDs) impact over 333 million individuals globally. Furthermore, approximately 33 million people live with HIV/AIDS, 14 million are infected with tuberculosis, and 3–5 million experience cholera outbreaks annually. These illnesses disproportionately impact underdeveloped regions, where inadequate healthcare systems exacerbate their severity. However, industrialized nations are not immune; for example, diarrheal diseases affect 48 million individuals annually in the United States, causing thousands of hospitalizations and fatalities [51]. The influenza virus alone infects 10–40% of the global population each year, with pandemics like H1N1 demonstrating the devastating potential of emerging pathogens [52]. Notably, serological studies reveal the widespread presence of viral antibodies in various populations, indicating significant exposure to pathogens such as herpes simplex virus (HSV) and human herpes virus (HHV) [53]. This demonstrates that infectious diseases, while more prevalent in developing countries, also persist as a critical concern in industrialized regions. Addressing these challenges requires targeted public health interventions, global collaborations, and comprehensive mathematical models to predict, manage, and mitigate disease spread [54].

Mathematical modeling of infectious diseases has played a pivotal role in epidemiology for over a century, offering insights into disease spread and control strategies. The advent of computational tools, rapid diagnostic testing, genetic sequencing, and internet-based data sharing has revolutionized the ability to track and predict infectious disease outbreaks [55, 56]. Modern mathematical models provide a foundation for designing public health policies, enabling researchers to test hypotheses, define optimal data collection methods, and estimate sample sizes for precise analyses [57, 58]. Such models have played a critical role in the study of diseases such as COVID-19, dengue, and malaria, providing a system to forecast the impacts of such interventions as vaccination and quarantine policies. Among the first models, the compartmental SIR model by Kermack and McKendrick, provided a basis of interpreting the dynamics of susceptible, infected, and removed populations [59]. This model has later been generalized to include other compartments such as exposed (SEIR) and vaccinated (SVEIR) to enable more detailed representations of the dynamics of disease. These models can explain the biological mechanisms of spread of diseases but also consider behavioral and environmental traits which affect infection spread [60]. Recent advancements in modeling have focused on addressing challenges such as nonlinearities, stochasticity, and multi-layered infection networks, which are critical for understanding complex outbreaks. By integrating theoretical and practical perspectives, these models have become indispensable tools for researchers, public health officials, and policymakers in combating infectious diseases worldwide [61].

Infectious diseases and epidemics represent one of the most significant global health challenges, particularly in low-income nations, where they remain a leading cause of mortality and disability. Diseases such as HIV, tuberculosis (TB), and malaria collectively account for roughly 10% of annual fatalities worldwide, underscoring their devastating impact [37, 38]. Over recent decades, the frequency and intensity of infectious disease outbreaks have surged, leading to severe economic consequences, widespread disabilities, and countless deaths [39]. Advances in modern transportation have inadvertently facilitated the rapid dissemination of communicable diseases, enabling microbial pathogens and their hosts to spread globally through international travel and trade networks [45]. For instance, the Zika virus has demonstrated explosive transmission across continents, exacerbated by the lack of background immunity and the absence of an effective vaccine [46]. Disease outbreaks, while devastating, offer opportunities to gather crucial data for analyzing transmission dynamics and refining control measures. However, in certain instances, data collection remains a challenge due to inherent pathological limitations. Historical perspectives reveal that mathematical modeling in epidemiology has a rich legacy, beginning with Daniel Bernoulli's pioneering work in the 18th century [62, 63], which demonstrated the potential of immunity to reduce smallpox mortality and laid the foundation for eradication strategies [51, 52]. In modern times, mathematical models, such as [41, 42] have become indispensable for evaluating the potential impact of control and eradication efforts on morbidity and mortality rates. These models bridge the gap between clinical observations of specific subpopulations and broader population-level dynamics, providing a comprehensive framework for understanding infectious diseases [44]. Moreover, they contribute to enhancing the understanding of herd immunity and the indirect protection offered by immunization programs, further solidifying their role in global health planning [64].

Panda et al. [65] provided valuable insights into the application of fractional derivatives in conjunction with fuzzy mappings to model the dynamics of the novel coronavirus. Their exploration spanned fractional operators, including Atangana-Baleanu, Caputo-Fabrizio, and Caputo classes. In parallel, Atangana and Goufo critically evaluated models addressing long-term economic growth, emphasizing the role of capital accumulation, labor expansion, and technological progress [66]. Further, Atangana and Araz introduced innovative mathematical concepts that have been pivotal in understanding the transmission dynamics of infectious diseases. They highlighted the significance of the reproductive number in assessing the stability of disease spread, proposing frameworks for predicting stability and instability under varying conditions [67]. To tackle core challenges in fractional calculus, Atangana and Baleanu developed a derivative characterized by nonlocal and nonsingular kernels, capable of modeling chaotic behavior absent in local derivatives [68]. These derivatives also proved effective in solving fractional heat transfer problems, showcasing unique properties and practical applications.

Expanding this work, Atangana applied iterative techniques to analyze nonlinear equations, such as the Fisher's reaction–diffusion equation, demonstrating the stability of solutions through fixed-point theory [69]. Atangana and Badr [70] followed up by applying the fixed-point theorem to analyze linked solutions existence and investigating their singularity. They computed coupled solutions of modified systems using iterative techniques, and this broadened the theoretical basis of fractional calculus. Atangana and Secer added a presentation in great details of the different definitions of fractional calculus adapted to real-world problems, with numerical approximations and simulations to solve practical problems [71, 72]. At the same time, Tchoundjeu et al. [73] concentrated on the problem of agriculture in the African Humid Tropics, suggesting the way to solve the issues of malnutrition and improve food security via the consumption of local vegetables, which was also based on the idea of the connection between the environment and health. Aguilar and Atangana [74] studied the possibility of various mathematical laws, such as the power-law, exponential decay, and Mittag-Leffler laws describing physical phenomena. They presented a Mittag-Leffler kernel of three fractional orders, which interpolates between exponential and power-law behavior, and resolved a number of numerical examples, showing the flexibility of the kernel.

Recent developments A new fractional model of human liver dynamics based on the exponential kernel and Caputo-Fabrizio derivative [75], and applications of the Hilfer fractional derivative in epidemic models [76]. The increasing trend in study of fractional-order systems emphasizes the versatility of the systems as opposed to the conventional fixed-order systems. Fractional calculus overcomes these shortcomings of the previous methods by introducing hybrid operators such as exponential decay and Mittag-Leffler kernels, which makes the fractional calculus a key tool in research and engineering. A relatively recent edition to the fractional calculus is the fractal-fractional operator which is an extension of the integer-order derivatives, providing new parameters to tune simulations. Such developments have led to improved disease modeling by incorporating memory effects, multistage processes and improved data fitting capacities, fractional Epidemic models are present in [77].

7.2.1 COVID-19 Mathematical Models

The onset of the coronavirus pandemic in 2020 proved to be one of the most serious problems in modern history as the majority of people faced the impact on their health and stability of the world order. The severe respiratory infections causing virus had high contagion rates and spread fast through human contact. The coronavirus was first identified officially in Wuhan, China, on December 31, 2019, but soon spread to other areas, such as Europe, North America, and Asia-Pacific, to become a global crisis. The symptoms would come very quickly, usually between 2 and 10 d, which further complicates the containment measures. Effective vaccines and medications were unavailable at the period to reduce the effects of the virus, which caused massive deaths all over the world. The dynamic developments of the pandemic required the researchers to look towards fractional-order dynamical models, which had sophisticated capabilities in terms of modeling the dynamics of a virus spread. They were fractional order isolated class models that were used to depict the dynamics of the COVID-19 and were solved through numerical methods including the fourth-order Runge-Kutta (RK4) method and adaptive predictor-corrector methodology [78]. Yadav et al. [79] employed the q-homotopy analysis technique, a powerful analytical tool, and incorporated the Liouville-Caputo approach to analyze fractional operators. Furthermore, the generalized Adams-Bashforth-Moulton method was utilized to quantitatively test proposed models, showcasing the flexibility of fractional calculus.

A significant addition to these studies involved fixed-point theory, which was leveraged to establish the existence and uniqueness of solutions for fractional COVID-19 models. By applying fixed-point theorems, researchers validated the mathematical soundness of these models, ensuring their ability to describe epidemic dynamics accurately. For example, the authors of [80] developed a piecewise fractional COVID-19 model incorporating quarantine and vaccination strategies, employing the Newton polynomial approach to numerically solve the equations. Fixed-point theory was instrumental in verifying the stability and feasibility of their solutions.

A mathematical model that considered a number of parameters was recently proposed by Altaf and Atangana [81] to depict the spread of the coronavirus-19. To learn more about the variables in the mathematical model below, readers might refer [81].

$$\begin{aligned}
\frac{dS_b}{dt} &= \Pi_b - \lambda_b S_b - \frac{\eta_b S_b I_b}{N_b}, \\
\frac{dE_b}{dt} &= \frac{\eta_b S_b I_b}{N_b} - (\mu_b + \theta_b) E_b, \\
\frac{dI_b}{dt} &= \theta_b E_b - (\tau_b + \mu_b) I_b, \\
\frac{dR_b}{dt} &= \tau_b I_b - \mu_b R_b, \\
\frac{dS_h}{dt} &= \Pi_h - \mu_h S_h - \frac{\eta_{bh} S_h I_b}{N_h} - \frac{\eta_h S_h I_h}{N_h}, \\
\frac{dE_h}{dt} &= \frac{\eta_{bh} S_h I_b}{N_h} + \frac{\eta_h S_h I_h}{N_h} - (\mu_h + \theta_h) E_h, \\
\frac{dI_h}{dt} &= \theta_h E_h - (\mu_h + \tau_h) I_h, \\
\frac{dR_h}{dt} &= \tau_h I_h - \mu_h R_h.
\end{aligned} \tag{7.3}$$

The equations describe the dynamics of disease transmission within and between bat and host populations, dividing each into four compartments: susceptible (S), exposed (E), infected (I), and recovered or removed (R). The total populations of bats (N_b) and hosts (N_h) are represented as the sum of their respective compartments. Susceptible populations (S_b, S_h) decrease due to interactions with infected individuals, modeled through transmission terms like $\frac{\eta_b S_b I_b}{N_b}$ for bats and $\frac{\eta_{bh} S_h I_b}{N_h} + \frac{\eta_h S_h I_h}{N_h}$ for hosts, as well as due to natural mortality (μ_b, μ_h). Exposed compartments (E_b, E_h) increase as susceptible individuals are exposed to infection, transitioning to the infected compartments (I_b, I_h) at rates η_b and η_h, respectively. Infected populations grow as exposed individuals become infectious but decline due to recovery or removal at rates τ_b and τ_h. The recovered compartments (R_b, R_h) represent individuals who have recovered from the disease or have been otherwise removed from the infectious cycle. Birth rates (Π_b, Π_h) replenish the susceptible populations, while transmission coefficients (η_b, η_{bh}, η_h) and recovery rates (τ_b, τ_h) influence the overall dynamics. This compartmental framework effectively models the spread of infection through bats and hosts while accounting for natural demographic changes and disease progression.

This focuses on modeling the interaction between the coronavirus reservoir (seafood market) and the human population. The total human population is divided into five compartments:

- $S_p(t)$: Susceptible individuals;
- $E_p(t)$: Exposed individuals;

- $I_p(t)$: Infected individuals (symptomatic);
- $A_p(t)$: Asymptomatic infected individuals;
- $R_p(t)$: Recovered or removed individuals.

Additionally, the virus concentration in the seafood market is represented by $M(t)$. The dynamics of these compartments are described using a system of nonlinear differential equations:

$$
\begin{aligned}
\frac{dS_p}{dt} &= \Pi_p - \mu_p S_p - \eta_p \frac{S_p(I_p + \psi A_p)}{N_p} - \eta_w S_p M,\\
\frac{dE_p}{dt} &= \eta_p \frac{S_p(I_p + \psi A_p)}{N_p} + \eta_w S_p M - (1-\theta_p)\omega_p E_p - \theta_p \rho_p E_p - \mu_p E_p,\\
\frac{dI_p}{dt} &= (1-\theta_p)\omega_p E_p - (\tau_p + \mu_p) I_p,\\
\frac{dA_p}{dt} &= \theta_p \rho_p E_p - (\tau_{ap} + \mu_p) A_p,\\
\frac{dR_p}{dt} &= \tau_p I_p + \tau_{ap} A_p - \mu_p R_p,\\
\frac{dM}{dt} &= b \frac{M I_h}{N_h} + \varrho_p I_p + \varpi_p A_p - \pi M.
\end{aligned}
$$

- Π_p: Birth rate of the human population.
- μ_p: Natural mortality rate.
- ρ_p: Disease transmission coefficient for interactions between susceptible and infected humans.
- w: Transmissibility multiplier for asymptomatic individuals relative to symptomatic ones.
- η_w: Transmission coefficient from the seafood market to the susceptible population.
- θ_p: Proportion of exposed individuals developing asymptomatic infections.
- ω_p, ρ_p: Transition rates from exposed to symptomatic and asymptomatic infected classes.
- τ_p, τ_{ap}: Recovery or removal rates for symptomatic and asymptomatic infected individuals.
- M: Contribution of the virus by infected hosts to the seafood market.
- ϱ_p, ϖ_p: Contribution of symptomatic and asymptomatic infected individuals to the virus reservoir.
- π: Removal rate of the virus from the seafood market.
- N_p: Total human population $(S_p + E_p + I_p + A_p + R_p)$.

Key Assumptions of Altaf and Atangana [81]

- The seafood market serves as a reservoir of the virus and can infect susceptible individuals.
- Symptomatic and asymptomatic individuals contribute differently to the virus concentration in the seafood market.

- The model excludes interactions between bats and hosts after the virus reaches the seafood market.

Above model describes the dynamics of COVID-19 transmission, including environmental contamination, and provides a framework for analyzing disease spread and intervention strategies. Figure 7.2 depicts the full flow transfer between the bat and host populations. Figure 7.3 shows the whole transmission between humans and the seafood reserve. From the above work of Altaf and Atangana [81], very recently, Panda et al. [65] offered significant insights into the application of fractional derivatives combined with fuzzy mappings to model the dynamics of the novel coronavirus. Their study encompassed various fractional operators, including the Atangana-Baleanu, Caputo-Fabrizio, and Caputo classes. They then used the acquired numerical solutions to show numerical simulations for various values of Atangana-Baleanu fractional order. One can see the numerical solutions in below Fig. 7.4.

Farman et al. [82] introduced innovative fractional derivatives based on Atangana-Baleanu concepts to enhance the modeling of COVID-19 transmission. They also utilized the Mittag-Leffler kernel, which accounts for memory and hereditary properties, to provide more realistic simulations of epidemic behavior. Additionally, a four-compartment COVID-19 model explored in [83] employed fuzzy and random approaches, with numerical solutions obtained via the Adams-Bashforth method. Numerical simulations under the ABC derivative framework further demonstrated the robustness of these models.

By combining a fixed-point theory and enhanced methods of fractional calculus, scientists have been able to seize the nonlocal dynamics and intricacy of the COVID-19 pandemic. Such methods have helped gain useful insights into the dynamics of the spread of the virus and have given clues to the design of public health interventions and have opened the door to more complex epidemic modeling. The COVID-19 pandemic which is suggested

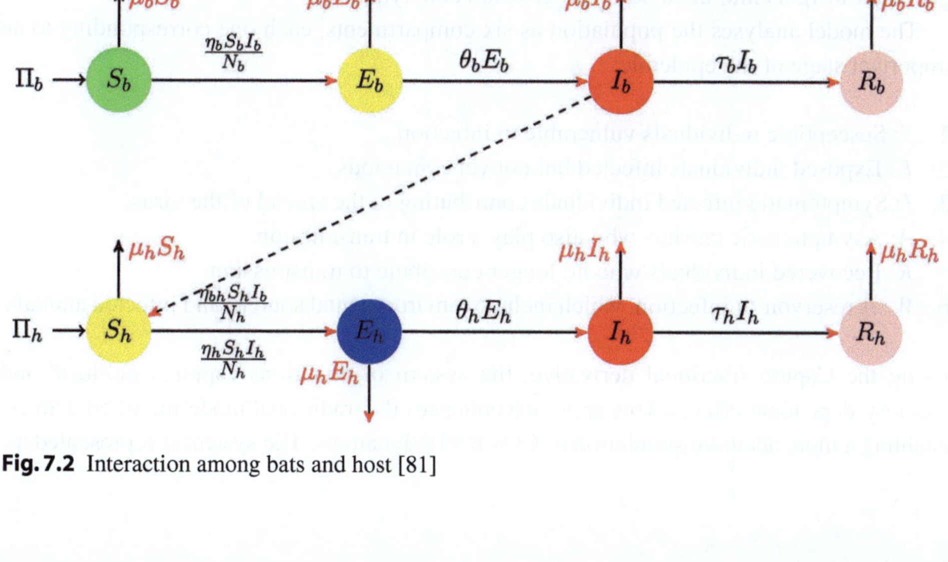

Fig. 7.2 Interaction among bats and host [81]

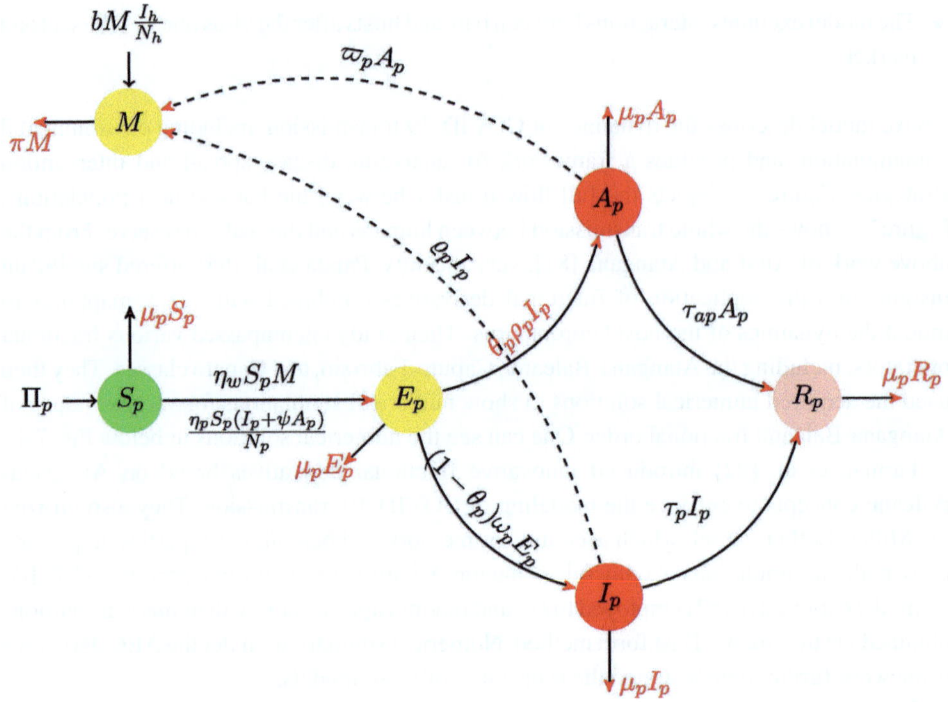

Fig. 7.3 Interaction among the people and the reservoir (seafood) market [81]

to be of zoonotic origin highlights the importance of having well-developed mathematical backgrounds to study its dynamics. A power-law kernel fractional-order model can serve as an effective initiative to consider the memory effects and hereditary nature of the disease to get better insights into the disease progression and dynamics.

The model analyses the population as six compartments, each one corresponding to an important stage of the epidemic:

1. S: Susceptible individuals vulnerable to infection.
2. E: Exposed individuals infected but not yet contagious.
3. I: Symptomatic infected individuals contributing to the spread of the virus.
4. A: Asymptomatic carriers who also play a role in transmission.
5. R: Recovered individuals who no longer contribute to transmission.
6. W: A reservoir of infection, which includes environmental sources and infected animals.

Using the Caputo fractional derivative, the system of equations captures nonlocal and memory-dependent effects. This approach enhances the traditional modeling of epidemics, enabling a more accurate simulation of COVID-19 dynamics. The system is represented as:

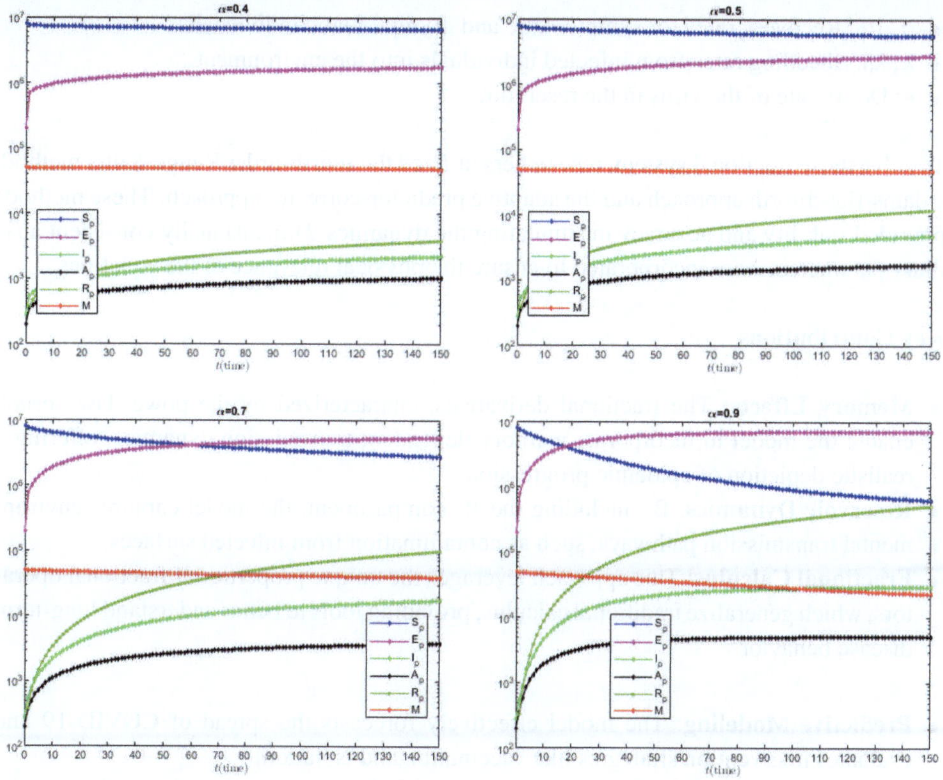

Fig. 7.4 Numerical simulation for various fractional orders

$$
\begin{aligned}
f^{(m-1)}\,{}^{c}\mathcal{D}_t^m S(t) &= -b_p S(I + kA) - b_w SW,\\
f^{(m-1)}\,{}^{c}\mathcal{D}_t^m E(t) &= b_p S(I + kA) + b_w SW - (1-d)\gamma E - d\gamma_0 E - qE,\\
f^{(m-1)}\,{}^{c}\mathcal{D}_t^m I(t) &= (1-d)\gamma E - (\mu + q)I,\\
f^{(m-1)}\,{}^{c}\mathcal{D}_t^m A(t) &= d\gamma_0 E - (\mu_0 + q)A,\\
f^{(m-1)}\,{}^{c}\mathcal{D}_t^m R(t) &= \mu I + \mu_0 A - qR,\\
f^{(m-1)}\,{}^{c}\mathcal{D}_t^m W(t) &= \lambda I + \lambda_0 A - \epsilon W.
\end{aligned}
$$

Here,

- b_p: Transmission rate between humans.
- b_w: Environmental transmission rate from the reservoir.
- k: Relative impact of asymptomatic individuals compared to symptomatic ones.
- γ, γ_0: Transition rates from exposed to symptomatic and asymptomatic states.
- q: Behavioral adjustment rate.

- μ, μ_0: Recovery rates for symptomatic and asymptomatic individuals.
- λ, λ_0: Shedding rates from infected individuals into the environment.
- ϵ: Decay rate of the virus in the reservoir.

To solve those fractional system, researchers utilized the fourth-order Runge-Kutta method, Adams-Baschforth approach and the adaptive predictor-corrector approach. These methods provided stability and accuracy in simulating the dynamics. Dimensionally consistent auxiliary parameters were incorporated to ensure the physical relevance of the equations.

Key Contributions

- **Memory Effects**: The fractional derivatives, characterized by the power-law kernel, enable the model to incorporate memory-dependent transmission dynamics, offering a realistic depiction of epidemic progression.
- **Reservoir Dynamics**: By including the W compartment, the model captures environmental transmission pathways, such as contamination from infected surfaces.
- **Fractional Calculus**: This approach leverages the unique properties of fractional operators, which generalize traditional calculus, providing tools to better understand long-term disease behavior.

- **Predictive Modeling**: The model effectively forecasts the spread of COVID-19 and evaluates intervention strategies like vaccination and isolation.
- **Policy Formulation**: By incorporating memory effects and environmental factors, the model aids policymakers in designing more effective mitigation measures.

Through the use of fractional calculus and advanced numerical techniques, this model provides a comprehensive framework for analyzing COVID-19 dynamics, particularly when traditional models fall short [84–86].

7.2.2 HIV/AIDS Disease Modeling

Acquired Immunodeficiency Syndrome (AIDS) is a chronic, potentially life-threatening condition caused by the human immunodeficiency virus (HIV). This virus targets the immune system, impairing its ability to combat infections and diseases. Although there is no definitive cure for HIV/AIDS, advancements in antiretroviral therapy (ART) have significantly slowed disease progression and reduced mortality rates globally. These treatments have also increased the accessibility of preventive and therapeutic measures in resource-limited countries, spearheaded by international organizations.

Khan et al. [87] investigated the application of Liouville-Caputo and Atangana-Baleanu-Caputo derivatives in a fractional-order HIV/AIDS model, employing Laplace and Sumudu transforms to derive iterative solutions. Recognizing the role of fractional-order dynamics, another study [88] developed a model to evaluate the impact of viral screening on individuals unaware of their HIV status. This model incorporated predictor–corrector methods to simulate interactions between infected and non-infected populations. Additionally, [89] reformulated a Caputo-Fabrizio fractional-order model to analyze HIV/AIDS transmission, demonstrating through a three-step fractional Adams-Bashforth method that implementing multiple control strategies simultaneously yields better outcomes than employing them individually.

Further, a compartmental fractional model incorporating antiretroviral therapy was analyzed in [90], utilizing Caputo-Fabrizio and fractal-fractional derivatives. Solutions were obtained using the Sumudu transform. Real-world data from Botswana inspired Omame et al. [91] to propose a dual variability model for co-infection of COVID-19 and HIV, analyzing the interplay between the two diseases. Simulations in [92] also explored how modifications in COVID-19 dynamics influenced HIV prevalence. Arafa et al. [93] generalized these concepts, while Huo et al. [94] reported notable advancements in fractional modeling for HIV. Moore et al. [95] applied the Caputo-Fabrizio derivative to replace first-order derivatives in a fractional-order HIV/AIDS model. Their work incorporated non-singular, exponentially decaying kernels to better capture the dynamics described by the Caputo-Fabrizio framework.

The HIV/AIDS epidemic has prompted the development of advanced mathematical models to better understand its dynamics and implement effective control measures. One such approach integrates fractional differential operators with the Mittag–Leffler kernel to account for memory effects and nonlocal influences in the disease progression.

This fractional-order model divides the human population into five distinct classes:

1. $S(t)$: The population of individuals who are susceptible and unaware of HIV.
2. $A(t)$: The susceptible population that is aware of HIV risks.
3. $I(t)$: Individuals infected with HIV.
4. $T(t)$: Those undergoing antiretroviral therapy (ART).
5. $Q(t)$: The population suffering from AIDS.

The governing equations for this system utilize the Mittag–Leffler kernel to capture the hereditary and memory-dependent behaviors of disease spread:

$$
\begin{aligned}
{}^{ABC}\mathcal{D}_t^r S(t) &= \Pi - \frac{\gamma SI}{N} - (d+\kappa)S,\\
{}^{ABC}\mathcal{D}_t^r A(t) &= dS - (1-\phi)\frac{\gamma AI}{N} - \kappa A,\\
{}^{ABC}\mathcal{D}_t^r I(t) &= \frac{\gamma}{N}\big(S + (1-\phi)A\big)I + \beta T + \omega Q - (\sigma+\eta+\kappa)I,\\
{}^{ABC}\mathcal{D}_t^r T(t) &= \sigma I - (\beta+\kappa)T,\\
{}^{ABC}\mathcal{D}_t^r Q(t) &= \eta I - (\omega+\chi+\kappa)Q,
\end{aligned}
$$

where ϕ, β, σ, and ω are parameters representing awareness, treatment, and progression rates.

An iterative method of obtaining approximate solutions to this system was done by taking the Laplace transform of the system. Compared with integer-order derivatives, the fractional Atangana-Baleanu-Caputo (ABC) operator using the Mittag- Leffler kernel enhances the versatility and accuracy of the model. These simulations were carried out numerically to demonstrate the fact that such an approach can be employed in the process of investigating the dynamics of the HIV/AIDS spreading and intervention strategies [96].

The model can be used in the assessment of the effects of awareness-raising activities, treatment programs and prevention interventions. The model allows a more realistic form of HIV/AIDS dynamics because it can capture the memory-dependent dynamics of disease spread. It can also be used to simulate the situation where different intervention strategies are employed and this can be very useful in public health planning [97].

7.2.3 Tuberculosis (TB) Disease Modeling

Tuberculosis (TB) is one of the international health concerns, especially in resource-constrained facilities where the lack of appropriate treatment and action worsens the situation. The fractional-order TB model with the Atangana-Baleanu-Caputo (ABC) derivative is a new method to research the TB dynamics with the memory and hereditary effects. This model categorizes the population into five compartments: susceptible (S), exposed (E), infected (I), treated (T), and recovered (R) individuals.
The fractional-order model [98, 99] is represented as:

$$
\begin{aligned}
{}^{ABC}\mathcal{D}_{t0}^{\lambda} S(t) &= W - \phi S(1+(1-T)) - \psi S,\\
{}^{ABC}\mathcal{D}_{t0}^{\lambda} E(t) &= \phi S(1+(1-T)) - (\psi+\tau)E + (1-\xi)T,\\
{}^{ABC}\mathcal{D}_{t0}^{\lambda} I(t) &= \tau E - (\psi+\gamma_1+\chi)I,\\
{}^{ABC}\mathcal{D}_{t0}^{\lambda} T(t) &= \chi I - (\psi+\gamma_2+\zeta)T,\\
{}^{ABC}\mathcal{D}_{t0}^{\lambda} R(t) &= \xi\zeta T - \psi R.
\end{aligned}
$$

The system is initialized with the following conditions:

$$S(0) = S_0, \quad E(0) = E_0, \quad I(0) = I_0, \quad T(0) = T_0, \quad R(0) = R_0.$$

Here,

- W: Recruitment rate of new individuals into the population.
- ϕ: Effective contact rate leading to exposure.
- ψ: Natural death rate applicable to all compartments.
- τ: Progression rate from exposed to infected.
- χ: Rate at which infected individuals begin treatment.
- ξ: Fraction of treated individuals who recover successfully.
- γ_1, γ_2: TB-induced death rates for infected and treated individuals, respectively.
- ζ: Transition rate from treatment to recovery.

To solve the system, numerical methods such as the fractional Adams-Bashforth approach were employed. These methods utilize fractional calculus to address memory-dependent dynamics. Fixed-point theory and nonlinear analysis were also applied to verify the existence and uniqueness of solutions. Additionally, graphical simulations highlighted the model's applicability and its capability to capture the long-term effects of interventions. There are vast findings in this pursuit, see for example [100–102].

Key Findings of [98, 99]

- **Memory Effects**: The inclusion of the ABC derivative reflects the long-term impacts of disease progression and intervention strategies. This memory effect offers a more realistic representation compared to traditional integer-order models.
- **Simulation Results**: Simulations revealed that fractional-order models provide deeper insights into the dynamics of TB, especially under scenarios of incomplete treatment and varying vaccination strategies.
- **Policy Implications**: The model underscores the importance of improving treatment adherence and coverage while optimizing resource allocation for TB interventions.

More briefly, by incorporating fractional-order dynamics and leveraging numerical techniques, the model demonstrates the critical role of memory effects in understanding and addressing TB transmission dynamics. The lessons learnt through the approach help in developing improved public health TB control strategies. The theory of fixed points has had a central role in the construction and study of fractional-order models in different areas, but mostly in epidemiology and dynamic systems. In [71], Atangana and Secer employed fixed point theory in the study of existence and uniqueness of solutions of coupled fractional differential equations. They used iterative methods to guarantee the uniqueness of solutions so that real epidemiological problems in the world could be modeled. Elsewhere, Aman Ullah et al. [103] used the fixed point theory to prove the existence and uniqueness of solutions of a fractional-order model of childhood diseases. Analytical solutions were obtained by using the natural transform method and the Adomian decomposition method to show that fixed point theory is effective in solving complex nonlinear systems. Also, bifurcation and stability properties of fractional-order epidemic models with the effect of fear were investigated based on the fixed point theory [104]. These models provided significant insights into how fear influences disease transmission dynamics and impacts outbreak control strategies.

Fixed point theory has also been applied to cancer modeling. For instance, fractional diffusion equations were analyzed to understand tumor dynamics, using fixed point theory to validate the solutions' convergence and stability [105–110]. Additionally, readers can examine how well fixed point theory works with a variety of models, including those for malaria, TB, and typhoid disease. (see, [111–121]).

7.3 Existence of Solution for the Fractal-Fractional Model of the Computer Virus Problem Using (φ, F)-Contraction

Nevertheless, in this section, we investigate the fractal-fractional model of the computer virus problem using the Atangana–Baleanu operator in conjunction with a nonlinear F-contraction framework.

We consider a fractal-fractional model of computer viruses as in [122]. The model is formulated using the Atangana–Baleanu operator in the Caputo sense with fractal-fractional derivative of order α and fractal dimension β. The system consists of four compartments: susceptible computers $S(t)$, infected computers $I(t)$, recovered or removed systems $R(t)$, and protected systems with antivirus $A(t)$. The model is governed by the following system of equations:

$$
\begin{aligned}
{}_0^{FFM}D_t^{\alpha,\beta} S(t) &= -\alpha_{SA} S(t)A(t) - \beta S(t)I(t) + \sigma R(t),\\
{}_0^{FFM}D_t^{\alpha,\beta} I(t) &= \beta_1 S(t)I(t) - \alpha_{IA} A(t)I(t) - \delta I(t),\\
{}_0^{FFM}D_t^{\alpha,\beta} R(t) &= \delta I(t) - \sigma R(t),\\
{}_0^{FFM}D_t^{\alpha,\beta} A(t) &= \alpha_{SA} S(t)A(t) + \alpha_{IA} A(t)I(t),
\end{aligned}
\tag{7.4}
$$

with initial conditions $S(0) = S_0, I(0) = I_0, R(0) = R_0,$ and $A(0) = A_0$. The model parameters represent various transition dynamics: α_{SA} denotes the rate at which susceptible computers receive antivirus protection, β is the infection rate due to contact between susceptible and infected nodes, β_1 corresponds to the contribution of infected-susceptible interaction, α_{IA} is the antivirus application on infected systems, δ is the recovery rate, and σ accounts for systems returning to the susceptible class.

We adopt the Atangana–Baleanu operator defined in the Caputo sense [123]. The fractal-fractional derivative of a function $u(t)$ with order α and dimension β is given by:

$$
{}^{FFM}D_t^{\alpha,\beta}(u(t)) = \frac{AB(\alpha)}{1-\alpha}\frac{d}{dt^{\beta}}\int_a^t u(z)E_{\alpha}\left[-\frac{\alpha}{1-\alpha}(t-z)^{\alpha}\right]dz, \quad 0<\alpha,\beta\le 1, \tag{7.5}
$$

where $AB(\alpha)$ is a normalization function satisfying $AB(0) = AB(1) = 1$, and $E_{\alpha}(\cdot)$ is the Mittag-Leffler function.

Correspondingly, the fractal-fractional integral of the function $u(t)$ is presented in [123] as:

$$
{}^{FFM}I_t^{\alpha,\beta}(u(t)) = \frac{\beta\alpha}{AB(\alpha)\Gamma(\alpha)}\int_0^t z^{\beta-1}u(z)(t-z)^{\alpha-1}dz + \frac{\beta(1-\alpha)t^{\beta-1}}{AB(\alpha)}u(t), \quad 0<\alpha,\beta\le 1. \tag{7.6}
$$

To establish the existence of solutions for the system, we reformulate it using the Atangana–Baleanu–Riemann (ABR) operator:

$$
\begin{aligned}
{}_0^{ABR}D_t^{\alpha}S(t) &= \beta t^{\beta-1}Q(t,S,I,R,A),\\
{}_0^{ABR}D_t^{\alpha}I(t) &= \beta t^{\beta-1}W(t,S,I,R,A),\\
{}_0^{ABR}D_t^{\alpha}R(t) &= \beta t^{\beta-1}M(t,S,I,R,A),\\
{}_0^{ABR}D_t^{\alpha}A(t) &= \beta t^{\beta-1}Z(t,S,I,R,A),
\end{aligned} \tag{7.7}
$$

where the nonlinear terms are specified as:

$$
\begin{aligned}
Q(t,S,I,R,A) &= -\alpha_{SA}S(t)A(t) - \beta S(t)I(t) + \sigma R(t),\\
W(t,S,I,R,A) &= \beta_1 S(t)I(t) - \alpha_{IA}A(t)I(t) - \delta I(t),\\
M(t,S,I,R,A) &= \delta I(t) - \sigma R(t),\\
Z(t,S,I,R,A) &= \alpha_{SA}S(t)A(t) + \alpha_{IA}A(t)I(t).
\end{aligned} \tag{7.8}
$$

To proceed with the analysis, we express the system in abstract vector form:

$$
{}_0^{ABR}D_t^{\alpha}\chi(t) = \beta t^{\beta-1}\Lambda(t,\chi(t)), \quad \chi(0)=\chi_0, \tag{7.9}
$$

where $\chi(t) = [S(t), I(t), R(t), A(t)]^{\top}$, and $\Lambda(t,\chi(t)) = [Q, W, M, Z]^{\top}$. Applying the inverse AB fractional operator, the equivalent integral form becomes:

$$
\chi(t) = \chi(0) + \frac{\beta t^{\beta-1}(1-\alpha)}{AB(\alpha)}\Lambda(t,\chi(t)) + \frac{\alpha\beta}{AB(\alpha)\Gamma(\alpha)}\int_0^t \lambda^{\beta-1}(t-\lambda)^{\beta-1}\Lambda(\lambda,\chi(\lambda))d\lambda. \tag{7.10}
$$

Let us denote the components as follows:

$$
\chi(t) = \begin{bmatrix} S(t)\\ I(t)\\ R(t)\\ A(t)\end{bmatrix}, \quad \Lambda(t,\chi(t)) = \begin{bmatrix} Q(t,S,I,R,A)\\ W(t,S,I,R,A)\\ M(t,S,I,R,A)\\ Z(t,S,I,R,A)\end{bmatrix}. \tag{7.11}
$$

On the other hand, Wardowski in [124] introduced an alternative contraction framework, now known as the F-*contraction*, which generalizes classical notions in a nontrivial way. A mapping $\mathbb{T} : X \to X$, where (X, d) is a metric space, is said to be an F-contraction if there exists a function $F : (0, \infty) \to \mathbb{R}$ and a constant $\tau > 0$ such that the inequality

$$\tau + F(d(\mathbb{T}x, \mathbb{T}y)) \leq F(d(x, y)), \tag{7.12}$$

holds for all $x, y \in X$ with $\mathbb{T}x \neq \mathbb{T}y$. The function F is assumed to satisfy the following conditions:

(F1) F is strictly increasing; i.e., $t_1 > t_2$ implies $F(t_1) > F(t_2)$ for all $t_1, t_2 > 0$;
(F2) For any sequence $(t_n) \subset (0, \infty)$, we have $t_n \to 0$ if and only if $F(t_n) \to -\infty$;
(F3) There exists $k \in (0, 1)$ such that $\lim_{t \to 0^+} t^k F(t) = 0$.

Wardowski [124] demonstrated that such F-contractions on complete metric spaces possess a unique fixed point. By choosing specific functions for F, many classical contraction principles can be retrieved as special cases. For instance, selecting $F(t) = \ln(t)$ for $t > 0$ leads us back to the Banach contraction principle.

Subsequent literature has explored various extensions and generalizations of this concept (see, for example, [125–132]). In [133], Turinici proposed a relaxation of condition (F2), replacing it with:

(F2′) $\lim_{t \to 0^+} F(t) = -\infty$,

and argued that the implication

(F2″) $F(t_n) \to -\infty \Rightarrow t_n \to 0$,

follows from the monotonicity condition (F1).

Inspired by these developments, the author of [134] extended the above framework further by allowing the relaxation parameter τ to be governed by a function. Specifically, the author of [134] defined the notion of a *nonlinear* F-*contraction*, or more formally, a (φ, F)-*contraction*, where the map $T : X \to X$ satisfies

$$\varphi(d(x, y)) + F(d(\mathbb{T}x, \mathbb{T}y)) \leq F(d(x, y)), \tag{7.13}$$

for all $x, y \in X$ with $\mathbb{T}x \neq \mathbb{T}y$. The involved functions $\varphi : (0, \infty) \to (0, \infty)$ and $F : (0, \infty) \to \mathbb{R}$ are required to meet the following conditions:

(H1) The function F satisfies (F1) and the relaxed (F2′);
(H2) For all $t \geq 0$, $\liminf_{s \to t^+} \varphi(s) > 0$;
(H3) The contraction inequality above holds for all $x, y \in X$ with $\mathbb{T}x \neq \mathbb{T}y$.

Theorem 7.1 (Fixed Point Theorem for Nonlinear F-Contractions [134]) *Let (X, d) be a complete metric space and let $\mathbb{T} : X \to X$ be a (φ, F)-contraction. Then $\mathbb{T}$ has a unique fixed point.*

In-order to obtain our claims, we will need the following assumptions.

To apply fixed-point results, we must need the following hypotheses:

(C_1) There exist positive constants G_Λ and M_Λ such that:

$$|\Lambda(t, \chi(t))| \leq G_\Lambda |\chi(t)| + M_\Lambda. \tag{7.14}$$

(C_2) For all $\chi, \bar{\chi} \in U$, there exists a constant $L_\Lambda > 0$ satisfying:

$$|\Lambda(t, \chi(t)) - \Lambda(t, \bar{\chi}(t))| \leq L_\Lambda |\chi(t) - \bar{\chi}(t)|. \tag{7.15}$$

(C_3)

$$\rho = \left(\frac{\beta T^{\beta-1}(1-\alpha)}{AB(\alpha)} + \frac{\alpha\beta}{AB(\alpha)\Gamma(\alpha)} T^{\alpha+\beta-1} H(\alpha, \beta) \right) G_\Lambda < 1,$$

where $H(\alpha, \beta)$ is a function dependent on the fractional and fractal orders.

(C_4)

$$\left(\frac{\beta T^{\beta-1}(1-\alpha)}{AB(\alpha)} + \frac{\alpha\beta}{AB(\alpha)\Gamma(\alpha)} T^{\alpha+\beta-1} H(\alpha, \beta) \right) G_\Lambda \leq \frac{e^{2\|\chi-\bar{\chi}\|-n}}{e^{\left(\frac{\beta T^{\beta-1}(1-\alpha)}{AB(\alpha)} + \frac{\alpha\beta}{AB(\alpha)\Gamma(\alpha)} T^{\alpha+\beta-1} H(\alpha,\beta) \right) G_\Lambda \|\chi-\bar{\chi}\|}}.$$

These conditions ensure the continuity of the operator L and support the applications of fixed-point theory to guarantee the existence and uniqueness of solutions to the fractal-fractional computer virus model.

To establish the existence of solutions, we consider the Banach space $U = V \times V \times V \times V \times V$, where $V = C[0, T]$ is the space of continuous real-valued functions on the interval $[0, T]$. The norm on U is defined as

$$\|\chi\| = \max_{t \in [0,T]} |S(t) + I(t) + R(t) + A(t)|.$$

Theorem 7.2 *If $(C_1) - (C_4)$ are satisfied, then the system (7.7) has a unique solution in U.*

Proof Now we define the operator $L : U \to U$ by,

$$L(\chi)(t) = \chi(0) + \frac{\beta t^{\beta-1}(1-\alpha)}{AB(\alpha)} \Lambda(t, \chi(t)) + \frac{\alpha\beta}{AB(\alpha)\Gamma(\alpha)} \int_0^t (t-\lambda)^{\beta-1} \Lambda(\lambda, \chi(\lambda)) d\lambda. \tag{7.16}$$

Define $\varphi : (0, \infty) \to (0, \infty)$ piecewise as follows:

$$\varphi(t) = \begin{cases} -t+\alpha_1, & 0<t<\alpha_1, \\ -t+\alpha_n, & \alpha_{n-1}\le t<\alpha_n, \quad n\ge 2, \end{cases}$$

where $(\alpha_n)_{n\in\mathbb{N}\cup\{0\}}$ is the strictly increasing sequence.
Obviously $\liminf_{s\to t^+}\varphi(s)>0$ for any $t\ge 0$ and $F(\alpha)=log\alpha+\alpha,\ \alpha>0$.
Let us consider the operator $L:U\to U$ defined by

$$L(\chi)(t)=\chi(0)+\frac{\beta t^{\beta-1}(1-\alpha)}{AB(\alpha)}\Lambda(t,\chi(t))+\frac{\alpha\beta}{AB(\alpha)\Gamma(\alpha)}\int_0^t (t-\lambda)^{\beta-1}\Lambda(\lambda,\chi(\lambda))d\lambda.$$

Let $\chi,\bar{\chi}\in U$. We estimate the norm of the difference $L(\chi)-L(\bar{\chi})$ as follows:

$$\begin{aligned}
\|L(\chi)-L(\bar{\chi})\| &= \max_{t\in[0,T]}\Bigg|\frac{\beta t^{\beta-1}(1-\alpha)}{AB(\alpha)}[\Lambda(t,\chi(t))-\Lambda(t,\bar{\chi}(t))] \\
&\quad +\frac{\alpha\beta}{AB(\alpha)\Gamma(\alpha)}\int_0^t (t-\lambda)^{\beta-1}[\Lambda(\lambda,\chi(\lambda))-\Lambda(\lambda,\bar{\chi}(\lambda))]\,d\lambda\Bigg| \\
&\le G_\Lambda\|\chi-\bar{\chi}\|\left(\frac{\beta T^{\beta-1}(1-\alpha)}{AB(\alpha)}+\frac{\alpha\beta}{AB(\alpha)\Gamma(\alpha)}\int_0^T (T-\lambda)^{\beta-1}d\lambda\right) \\
&\le \left(\frac{\beta T^{\beta-1}(1-\alpha)}{AB(\alpha)}+\frac{\alpha\beta T^{\alpha+\beta-1}}{AB(\alpha)\Gamma(\alpha)}H(\alpha,\beta)\right)G_\Lambda\|\chi-\bar{\chi}\| \\
&\le \left(\frac{e^{2\|\chi-\bar{\chi}\|-n}}{e^{\left(\frac{\beta T^{\beta-1}(1-\alpha)}{AB(\alpha)}+\frac{\alpha\beta}{AB(\alpha)\Gamma(\alpha)}T^{\alpha+\beta-1}H(\alpha,\beta)\right)G_\Lambda\|L(\chi)-L(\bar{\chi})\|}}\right)\|\chi-\bar{\chi}\|,\ for\ n\ge 2, \\
&\le e^{\left(2\|\chi-\bar{\chi}\|-n-\left(\frac{\beta T^{\beta-1}(1-\alpha)}{AB(\alpha)}+\frac{\alpha\beta}{AB(\alpha)\Gamma(\alpha)}T^{\alpha+\beta-1}H(\alpha,\beta)\right)G_\Lambda\|L(\chi)-L(\bar{\chi})\|\right)}\|\chi-\bar{\chi}\| \\
&\le \left(e^{2\|\chi-\bar{\chi}\|-n-\|L(\chi)-L(\bar{\chi})\|}\right)\|\chi-\bar{\chi}\|,
\end{aligned}$$

which yields that,

$$\begin{aligned}
&\frac{\|L(\chi)-L(\bar{\chi})\|}{\|\chi-\bar{\chi}\|}\le e^{2\|\chi-\bar{\chi}\|-n-\|L(\chi)-L(\bar{\chi})\|} \\
&\Rightarrow \log\left(\frac{\|L(\chi)-L(\bar{\chi})\|}{\|\chi-\bar{\chi}\|}\right)\le 2\|\chi-\bar{\chi}\|-n-\|L(\chi)-L(\bar{\chi})\| \\
&\Rightarrow -\|\chi-\bar{\chi}\|+n+\log(\|L(\chi)-L(\bar{\chi})\|)+\|L(\chi)-L(\bar{\chi})\|\le\log(\|\chi-\bar{\chi}\|)+\|\chi-\bar{\chi}\| \\
&\Rightarrow \varphi(\|\chi-\bar{\chi}\|)+F(\|L(\chi)-L(\bar{\chi})\|)\le F(\|\chi-\bar{\chi}\|).
\end{aligned}$$

Similarly, it can be proved for $n=1$ as well.

Thus all the conditions of Theorem 7.2 satisfied. Hence the system (7.7) has a unique solution.

Overall, fixed point theory offers a robust mathematical foundation for solving fractional-order models, ensuring the validity and reliability of solutions. Its application spans a range of disciplines, offering theoretical rigor and practical relevance for complex dynamic systems.

Open Questions

7.1 In [13], Is it possible to derive a rigorous analysis of the sensitivity of the fractional-order SIRA system to small disturbances in the order α and in what way does this sensitivity affect control design?

7.2 What are the sufficient and necessary conditions, on which the SIRA model with Caputo derivatives exists multiple endemic equilibria, and how can the bifurcation theory be employed to describe that behavior?

7.3 **In** [122],
(a) What is the qualitative and quantitative Atangana-Baleanu operator compared to other kernels (e.g., Mittag-Leffler vs. exponential) in modelling cyber-epidemics?
(b) What is the order of rigorous error and convergence study of Atangana-Toufik method in the present model? Is it amenable to improvement or to hybridization with spectral or meshless techniques?
(c) Can one build high-order adaptive numerical methods, like adaptive mesh refinement or predictor-corrector methods, which are more accurate and efficient than the Atangana-Toufik method when acting on the proposed fractal-fractional virus model?

References

1. Uhde, A.: A short history of computer viruses (2017). https://content.sentrian.com.au/blog/a-short-history-of-computer-viruses
2. Denning, J.: Computers Under Attack. Addison-Wesley, Reading, Mass (1990)
3. Köhler, S., Schindelhauer, C., Ziegler, M.: On approximating real-world halting problems. In: International Symposium Fundamentals of Computation Theory FCT 2005: Fundamentals of Computation Theory, pp. 454–466 (2005)
4. You, S., Cho, B.H., Yook, S., Kima, J.Y., Shon, Y.M., Seo, D.W., Kim, I.Y.: Unsupervised automatic seizure detection for focal-onset seizures recorded with behind-the-ear EEG using an anomaly-detecting generative adversarial network. Comput. Methods Prog. Biomed. **105472**, 1–20 (2020)
5. Cattani, C.: A review on harmonic wavelets and their fractional extension. J. Adv. Eng. Comput. **2**(4), 224–38 (2018)
6. Gao, W., Yel, G., Baskonus, H.M., Cattani, C.: Complex solitons in the conformable (2+1)-dimensional ablowitz-kaup-Newell-segur equation. Aims. Math. **5**(1), 507–521 (2020)
7. Singh, J., Kumar, D., Hammouch, Z., Atangana, A.: A fractional epidemiological model for computer viruses pertaining to a new fractional derivative. Appl. Math. Comput. **2018**(316), 504–15 (2018)
8. Owolabi, K.M., Hammouch, Z.: Mathematical modeling and analysis of two variable system with noninteger-order derivative. Chaos **29**(1), 013145 (2019)
9. Piqueira, J.R.C., Araujo, V.O.: A modified epidemiological model for computer viruses. Appl. Math. Comput. **213**, 355–60 (2009)

10. Mittag-Leffler, G.M.: Sur la nouvelle fonction Ea x. C R Acad. Sci. Paris **137**, 554–8 (1903)
11. Kilbas, A., Srivastava, H.M., Trujillo, J.J.: Theory and Applications of Fractional Differential Equations. Elsevier, New York (2006)
12. Handam, A.H., Freihat, A.A.: A new analytic numeric method solution for fractional modified epidemiological model for computer viruses. Appl. Appl. Math. **10**(2), 919–36 (2015)
13. Gao, W., Baskonus, H.M.: Deeper investigation of modified epidemiological computer virus model containing the Caputo operator. Chaos, Solitons Fractals **158**, 112050 (2022)
14. Ravichandran, C., Logeswari, K., Khan, A., Abdeljawad, T., Gómez-Aguilar, J.: An epidemiological model for computer virus with Atangana-Baleanu fractional derivative. Res. Phys. Article 106601 (2023)
15. Ahmad, I., Bakar, A.A., Jan, R., Yussof, S.: Dynamic behaviors of a modified computer virus model: insights into parameters and network attributes Alex. Eng. J. **103**, 266–277 (2024)
16. Boulaaras, S., Rehman, Z.U., Abdullah, F.A., Jan, R., Abdalla, M., Jan, A.: Coronavirus dynamics, infections and preventive interventions using fractional-calculus analysis. AIMS Math. **8**(4), 8680–8701 (2023)
17. Mishra, B.K., Saini, D.: Mathematical models on computer viruses. Appl. Math. Comput. **187**(2), 929–936 (2007)
18. Piqueira, J.R., Vasconcelos, A.A., Gabriel, C.E., Araujo, V.O.: Dynamic models for computer viruses. Comput. Secur. **27**(7–8), 355–359 (2008)
19. Yang, L.-X., Yang, X., Wen, L., Liu, J.: A novel computer virus propagation model and its dynamics. Int. J. Comput. Math. **89**(17), 2307–2314 (2012)
20. Ansari, M.A., Arora, D., Ansari, S.P.: Chaos control and synchronization of fractional order delay-varying computer virus propagation model. Math. Methods Appl. Sci. **39**(5), 1197–1205 (2016)
21. Bonyah, E., Atangana, A., Khan, M.A.: Modeling the spread of computer virus via Caputo fractional derivative and the beta-derivative. Asia Pac. J. Comput. Eng. **4**, 1–15 (2017)
22. Raza, A., Fatima, U., Rafiq, M., Ahmed, N., Khan, I., Nisar, K.S., Iqbal, Z.: Mathematical analysis and design of the nonstandard computational method for an epidemic model of computer virus with delay effect: application of mathematical biology in computer science. Res. Phys. **21**, Article 103750 (2021)
23. Alhebshi, R.M., Ahmed, N., Baleanu, D., Fatima, U., Dayan, F., Rafiq, M., Raza, A., Ahmad, M.O., Mahmoud, E.E.: Modeling of computer virus propagation with fuzzy parameters. Comput. Mater. Continua **74**(3) (2023)
24. Hoang, M.T.: Dynamics of a fractional-order epidemiological model for computer viruses São Paulo. J. Math. Sci. 1–22 (2023)
25. Iqbal, Z., Rehman, M.A.-U., Imran, M., Ahmed, N., Fatima, U., Akgül, A., Rafiq, M., Raza, A., Djuraev, A.A., Jarad, F.: A finite difference scheme to solve a fractional order epidemic model of computer virus. AIMS Math. **8**, 2337–2359 (2023)
26. Zarin, R., Khaliq, H., Khan, A., Ahmed, I., Humphries, U.W.: A numerical study based on Haar wavelet collocation methods of fractional-order antidotal computer virus model. Symmetry **15**(3), 621 (2023)
27. Ehsan, M.S., Rafiq, M., Raza, A.: A competitive numerical model for virus transmission in a computer network. In: 2023 4th International Conference on Computing, Mathematics and Engineering Technologies, pp. 1–6. ICoMET, IEEE (2023)
28. Coronel, A., Huancas, F., Hess, I., Lozada, E., Novoa-Muñoz, F.: Analysis of a SEIR-KS mathematical model for computer virus propagation in a periodic environment. Mathematics **8**(5), 761 (2020)
29. Al-Jarrah, A., Alquran, M., Freihat, A., Al-Omari, S., Nonlaopon, K.: A fractional-order model for computer viruses and some solution associated with residual power series method. Appl. Math. Sci. Eng. **31**(1), Article 2214301 (2023)

30. Ebenezer, B., Farai, N., Kwesi, A.-A.S.: Fractional dynamics of computer virus propagation. Computer **10**, 12 (2015)
31. da Costa, A.L.P., Neto, O.A.R., Silva-Junior, A.C.S.: Conditioners of the infectious diseases dynamics. Estaçnao Cientıfica (UNIFAP) **8**(3), 09–23 (2019)
32. Wang, Z., Andrews, M.A., Wu, Z.X., Wang, L., Bauch, C.T.: Coupled disease-behavior dynamics on complex networks: a review. Phys. Life Rev. **15**, 1–29 (2015)
33. Fauci, A.S.: Infectious diseases: considerations for the 21st century. Clin. Infect. Dis. **32**(5), 675–685 (2001)
34. Puri, N., Coomes, E.A., Haghbayan, H., Gunaratne, K.: Social media and vaccine hesitancy: new updates for the era of COVID-19 and globalized infectious diseases. Human Vacc. Immunother. **16**(11), 2586–2593 (2020)
35. Wu, T., Perrings, C., Kinzig, A., Collins, J.P., Minteer, B.A., Daszak, P.: Economic growth, urbanization, globalization, and the risks of emerging infectious diseases in China: a review. Ambio **46**, 18–29 (2017)
36. Martin, D.L., Gustafson, T.L., Pelosi, J.W., Suarez, L., Pierce, G.V.: Contaminated produce–a common source for two outbreaks of Shigella gastroenteritis. Am. J. Epidemiol. **124**(2), 299–305 (1986)
37. Misra, A.K., Sharma, A., Shukla, J.B.: Modeling and analysis of effects of awareness programs by media on the spread of infectious diseases. Math. Comput. Model. **53**(5–6), 1221–1228 (2011)
38. Huo, J., Zhao, H., Zhu, L.: The effect of vaccines on backward bifurcation in a fractional order HIV model. Nonlinear Anal.: Real World Appl. **26**, 289–305 (2015)
39. Abdelouahab, M.S., Hamri, N.E.: The Grunwald-Letnikov fractional-order derivative with fixed memory length. Mediterr. J. Math. **13**(2), 557–572 (2016)
40. Hethcote, H.W.: The mathematics of infectious diseases. SIAM Rev. **42**(4), 599–653 (2000)
41. Du, M., Wang, Z., Hu, H.: Measuring memory with the order of fractional derivative. Sci. Rep. **3**(1), 3431 (2013)
42. Rida, S.Z., Rady, A.A., Arafa, A.A.M., Khalil, M.: Approximate solution of a fractional order model of HCV infection with drug therapy effect. Int. J. Appl. **1**(2), 108–116 (2012)
43. Taubenberger, J.K., Morens, D.M.: 1918 Influenza: the mother of all pandemics. Revista Biomed. **17**(1), 69–79 (2006)
44. Jit, M., Brisson, M.: Modelling the epidemiology of infectious diseases for decision analysis: a primer. Pharmacoeconomics **29**, 371–386 (2011)
45. Mack, A., Choffnes, E.R., Relman, D.A. (Eds.): Infectious Disease Movement in a Borderless World: Workshop Summary. National Academies Press (2010)
46. World Health Organization.: Zika virus: Fact sheet (2016)
47. Swaddiwudhipong, W., Karintraratana, S., Kavinum, S.: A common-source outbreak of shigellosis involving a piped public water supply in northern Thai communities. J. Trop. Med. Hygiene **98**, 145–145 (1995)
48. Hethcote, H.W.: Three basic epidemiological models. Appl. Math. Ecol. 119–144 (1989)
49. Chen, Y., Liu, F., Yu, Q., Li, T.: Review of fractional epidemic models. Appl. Math. Model. **97**, 281–307 (2021)
50. Straif-Bourgeois S., Ratard R.: Infectious disease epidemiology. In: Handbook of Epidemiology, pp. 1327–1362 (2005)
51. Mounts, A.W., Holman, R.C., Clarke, M.J., Bresee, J.S., Glass, R.I.: Trends in hospitalizations associated with gastroenteritis among adults in the United States, 1979–1995. Epidemiol. Infect. **123**(1), 1–8 (1999)
52. Mandell, G.L., Bennett, J.E., Dolin, R.: Plasmodium species. Mandell, Douglas and Bennet's principles and practise of infectious diseases, 5th ed., Churchill Livingstone, an Imprint of Elsevier, Philadelphia, pp. 2817–2830 (2000)

53. Georgopoulos, A.P., James, L.M.: Immunogenetic profiles of 9 human herpes virus envelope glycoproteins. Sci. Rep. **14**, 20924 (2024). https://doi.org/10.1038/s41598-024-71558-1
54. Grassly, N., Fraser, C.: Mathematical models of infectious disease transmission. Nat. Rev. Microbiol. **6**, 477–487 (2008). https://doi.org/10.1038/nrmicro1845
55. Elsevier. (n.d.): Ecological paradigms lost (2005). Retrieved February 11, 2023, from https://www.elsevier.com/books/ecological-paradigms-lost/beisner/978-0-12-088459-9
56. Dietz, K., Heesterbeek, J.A.P.: Daniel Bernoulli's epidemiological model revisited. Math. Biosci. **180**(1–2), 1–21 (2002)
57. Anderson, R.M., May, R.M.: Infectious Diseases of Humans: Dynamics and Control. Oxford University Press (1992)
58. Glasser, J., Meltzer, M., Levin, B.: Mathematical modeling and public policy: responding to health crises. Emerg. Infect. Dis. **10**(11), 2050–2051 (2004)
59. Kermack, W.O., McKendrick, A.G.: A contribution to the mathematical theory of epidemics. Proc. Royal Soc. London. Ser. A, Contain. Papers Math. Phys. Char. **115**(772), 700–721 (1927)
60. Kermack, W.O., McKendrick, A.G.: Contributions to the mathematical theory of epidemics—I. Bltn. Mathcal. Biol. **53**, 33–55 (1991). https://doi.org/10.1007/BF02464423
61. Okabe, Y.; Shudo, A.: A mathematical model of epidemics—a tutorial for students. Mathematics **8**, 1174 (2020). https://doi.org/10.3390/math8071174
62. Glomski, M., Ohanian, E.: Eradicating a disease: lessons from mathematical epidemiology. College Math. J. **43**(2), 123–132 (2012)
63. Scherer, A., McLean, A.: Mathematical models of vaccination. Brit. Med. Bull. **62**(1), 187–199 (2002)
64. Huppert, A., Katriel, G.: Mathematical modelling and prediction in infectious disease epidemiology. Clin. Microbiol. Infect. **19**(11), 999–1005 (2013)
65. Panda, S.K., Atangana, A., Abdeljawad, T.: Existence results and numerical study on novel coronavirus 2019-Ncov/Sars- Cov-2 model using differential operators based on the generalized Mittag-Leffler kernel and fixed points. Fractals **30**(08), 2240214 (2022)
66. Atangana, A., Goufo, E.F.D.: Modern and generalized analysis of exogenous growth models. Chaos, Solitons Fractals **163**, 112605 (2022)
67. Atangana, A., Igret Araz, S.: Advanced analysis in epidemiological modeling: detection of wave. medRxiv, 2021–09 (2021)
68. Atangana, A., Baleanu, D.: Caputo-Fabrizio derivative applied to groundwater flow within confined aquifer. J. Eng. Mech. D4016005 (2016)
69. Atangana, A.: On the new fractional derivative and application to nonlinear Fisher's reaction-diffusion equation. Appl. Math. Comput. **273**, 948–956 (2016)
70. Atangana, A., Alkahtani, B.S.T.: Analysis of the Keller–Segel model with a fractional derivative without singular kernel. Entropy **17**, 4439–4453 (2015). https://doi.org/10.3390/e17064439
71. Atangana, A., Secer, A.: A note on fractional order derivatives and table of fractional derivatives of some special functions. In: Abstract and Applied Analysis, vol. 2013. Hindawi
72. Atangana, A., Nieto, J.J.: Numerical solution for the model of RLC circuit via the fractional derivative without singular kernel. Adv. Mech. Eng. **7**(10), 1687814015613758 (2015)
73. Tchoundjeu, Z., Asaah, E.K., Anegbeh, P., Degrande, A., Mbile, P., Facheux, C., Simons, A.J.: Putting participatory domestication into practice in West and Central Africa. For., Trees Livelihoods **16**(1), 53–69 (2006)
74. Gomez-Aguilar, J.F., Atangana, A.: New insight in fractional differentiation: power, exponential decay and Mittag-Leffler laws and applications. Eur. Phys. J. Plus **132**, 1–21 (2017)
75. Baleanu, D., Jajarmi, A., Mohammadi, H., Rezapour, S.: A new study on the mathematical modelling of human liver with Caputo-Fabrizio fractional derivative. Chaos, Solitons Fractals 134,1097052020

76. Jajarmi, A., Baleanu, D., Sajjadi, S.S., Nieto, J.J.: Analysis and some applications of a regularized Psi-Hilfer fractional derivative. J. Comput. Appl. Math. **415**, 114476 (2022)
77. Nisar, K.S., Farman, M., Abdel-Aty, M., Cao, J.: A review on epidemic models in sight of fractional calculus. Alex. Eng. J. **75**, 81–113 (2023)
78. Zhang, Z., Zeb, A., Egbelowo, O.F., Erturk, V.S.: Dynamics of a fractional order mathematical model for COVID-19 epidemic. Adv. Differ. Equ. **2020**(1), 1–16 (2020)
79. Yadav, S., Kumar, D., Singh, J., Baleanu, D.: Analysis and dynamics of fractional order Covid-19 model with memory effect. Results Phys. **24**, 104017 (2021)
80. Zeb, A., Atangana, A., Khan, Z.A., Djillali, S.: A robust study of a piecewise fractional order COVID-19 mathematical model. Alex. Eng. J. **61**(7), 5649–5665 (2022)
81. Atangana, A., Muhammad Altaf Khan: Modeling the dynamics of novel coronavirus (2019-nCov) with fractional derivative. Alex. Eng. J. **59**(4), 2379–2389 (2020)
82. Farman, M., Akgul, A., Nisar, K.S., Ahmad, D., Ahmad, A., Kamangar, S., Saleel, C.A.: Epidemiological analysis of fractional order COVID-19 model with Mittag-Leffler kernel. AIMS Math. **7**(1), 756–783 (2022)
83. ur Rahman, M. , Arfan, M., Shah, K., Gomez-Aguilar, J.F.: Investigating a nonlinear dynamical model of COVID-19 disease under fuzzy Caputo, random and ABC fractional order derivative. Chaos, Solitons Fractals **140**, 110232 (2020)
84. Ullah, M.Z., Alzahrani, A.K., Baleanu, D.: An efficient numerical technique for a new fractional tuberculosis model with nonsingular derivative operator. J. Taibah Univ. Sci. **13**(1), 1147–1157 (2019)
85. Chen, T.M., Rui, J., Wang, Q.P., Zhao, Z.Y., Cui, J.A., Yin, L.: A mathematical model for simulating the phase-based transmissibility of a novel coronavirus. Infect. Dis. Poverty **9**(1), 1–8 (2020)
86. Tuan, N.H., Mohammadi, H., Rezapour, S.: A mathematical model for COVID-19 transmission by using the Caputo fractional derivative. Chaos, Solitons Fractals **140**, 110107 (2020)
87. Khan, A., Gomez-Aguilar, J.F., Khan, T.S., Khan, H.: Stability analysis and numerical solutions of fractional order HIV/AIDS model. Chaos, Solitons Fractals **122**, 119–128 (2019)
88. Babaei, A., Jafari, H., Ahmadi, M.: A fractional order HIV/ AIDS model based on the effect of screening of unaware infectives. Math. Methods Appl. Sci. **42**(7), 2334–2343 (2019)
89. Wang, H., Jahanshahi, H., Wang, M.K., Bekiros, S., Liu, J., Aly, A.A.: A Caputo-Fabrizio fractional-order model of HIV/AIDS with a treatment compartment: sensitivity analysis and optimal control strategies. Entropy **23**(5), 610 (2021)
90. Farman, M., Akgul, A., Tekin, M.T., Akram, M.M., Ahmad, A., Mahmoud, E.E., Yahia, I.S.: Fractal fractional-order derivative for HIV/AIDS model with Mittag-Leffler kernel. Alex. Eng. J. **61**(12), 10965–10980 (2022)
91. Omame, A., Isah, M.E., Abbas, M., Abdel-Aty, A.H., Onyenegecha, C.P.: A fractional order model for Dual Variants of COVID-19 and HIV co-infection via Atangana-Baleanu derivative. Alex. Eng. J. **61**(12), 9715–9731 (2022)
92. Srivastava, P.K., Banerjee, M., Chandra, P.: Modeling the drug therapy for HIV infection. J. Biol. Syst. **17**(02), 213–223 (2009)
93. Arafa, A.A.M., Rida, S.Z., Khalil, M.: A fractional-order model of HIV infection with drug therapy effect. J. Egypt. Math. Soc. **22**(3), 538–543 (2014)
94. Huo, H.F., Chen, R., Wang, X.Y.: Modelling and stability of HIV/AIDS epidemic model with treatment. Appl. Math. Model. **40**(13–14), 6550–6559 (2016)
95. Moore, E.J., Sirisubtawee, S., Koonprasert, S.: A Caputo- Fabrizio fractional differential equation model for HIV/AIDS with treatment compartment. Adv. Differ. Equ. **2019**(1), 1–20 (2019)
96. Shaikh, A., Nisar, K.S., Jadhav, V., Elagan, S.K., Zakarya, M.: Dynamical behaviour of HIV/AIDS model using fractional derivative with Mittag-Leffler kernel. Alex. Eng. J. **61**(4), 2601–2610 (2022)

97. Khan, M.A., Odinsyah, H.P.: Fractional model of HIV transmission with awareness effect. Chaos, Solitons Fractals **138**, 109967 (2020)
98. Ullah, I., Ahmad, S., Al-Mdallal, Q., Khan, Z.A., Khan, H., Khan, A.: Stability analysis of a dynamical model of tuberculosis with incomplete treatment. Adv. Differ. Equ. **2020**(1), 1–14 (2020)
99. Rahman, M.U., Arfan, M., Shah, Z., Kumam, P., Shutaywi, M.: Nonlinear fractional mathematical model of tuberculosis (TB) disease with incomplete treatment under Atangana-Baleanu derivative. Alex. Eng. J. **60**(3), 2845–2856 (2021)
100. Wojtak, W., Silva, C.J., Torres, D.F.: Uniform asymptotic stability of a fractional tuberculosis model. Math. Model. Nat. Phenom. **13**(1), 9 (2018)
101. Yang, Y., Li, J., Ma, Z., Liu, L.: Global stability of two models with incomplete treatment for tuberculosis. Chaos, Solitons Fractals **43**(1–12), 79–85 (2010)
102. Ahmad, S., Ullah, R., Baleanu, D.: Mathematical analysis of tuberculosis control model using nonsingular kernel type Caputo derivative. Adv. Differ. Equ. **2021**(1), 1–18 (2021)
103. Ullah, A., Abdeljawad, T., Ahmad, S., Shah, K.: Study of a fractional-order epidemic model of childhood diseases. J. Funct. Spaces (2020)
104. Mandal, M., Jana, S., Nandi, S.K., Kar, T.K.: Modelling and control of a fractional-order epidemic model with fear effect. Energy, Ecol. Environ. **5**(6), 421–432 (2020)
105. Iyiola, O.S., Zaman, F.D.: A fractional diffusion equation model for cancer tumor. AIP Adv. **4**(10), 107121 (2014)
106. Mahdy, A.M., Almalki, N., Higazy, M.: A general fractional breast cancer model: model graph energy, Caputo-Fabrizio derivative existence and uniqueness plus numerical simulation. Part. Differ. Equ. Appl. Math. **10**, 100723 (2024)
107. Azeem, M., Farman, M., Akgül, A., De la Sen, M.: Fractional order operator for symmetric analysis of cancer model on stem cells with chemotherapy. Symmetry **15**, 533 (2023). https://doi.org/10.3390/sym15020533
108. Ahmad, S., Ullah, A., Abdeljawad, T., Akgül, A., Mlaiki, N.: Analysis of fractal-fractional model of tumor-immune interaction. Results Phys. **25**, 104178 (2021)
109. Salah, E.Y., Sontakke, B., Hamoud, A.A., et al.: A fractal–fractional order modeling approach to understanding stem cell-chemotherapy combinations for cancer. Sci. Rep. **15**, 3465 (2025). https://doi.org/10.1038/s41598-025-87308-w
110. Amilo, D., Kaymakamzade, B., Hincal, E.: A fractional-order mathematical model for lung cancer incorporating integrated therapeutic approaches. Sci. Rep. **13**, 12426 (2023). https://doi.org/10.1038/s41598-023-38814-2
111. Kumar, D., Singh, J., Al Qurashi, M., et al.: A new fractional SIRS-SI malaria disease model with application of vaccines, antimalarial drugs, and spraying. Adv. Differ. Equ. **2019**, 278 (2019). https://doi.org/10.1186/s13662-019-2199-9
112. Gizaw, A.K. and Deressa, C.T., Adel, W.: Analysis of age-structured mathematical model of malaria transmission dynamics via classical and ABC fractional operators. Math. Probl. Eng. Hindawi **2024**, 1–24 (2024)
113. Gunasekar, T., Manikandan, S., Suba, M., Akgül, A.: A fractal-fractional mathematical model for COVID-19 and tuberculosis using Atangana-Baleanu derivative. Math. Comput. Model. Dyn. Syst. **30**(1), 857–881 (2024)
114. Nandi, T.R., Saha, A.K., Roy, S.: Analysis of a fractional order epidemiological model for tuberculosis transmission with vaccination and reinfection. Sci. Rep. **14**, 28290 (2024). https://doi.org/10.1038/s41598-024-73392-x
115. Khan, A., Shah, K., Abdeljawad, T. et al. Fractal fractional model for tuberculosis: existence and numerical solutions. Sci Rep 14, 12211 (2024). https://doi.org/10.1038/s41598-024-62386-4

116. Fatmawati, M.A.K., Bonyah, E., Hammouch, Z., Shaiful, E.M.: A mathematical model of tuberculosis (TB) transmission with children and adults groups: a fractional model. AIMS Math. **5**(4), 2813–2842 (2020). https://doi.org/10.3934/math.2020181
117. Abboubakar, H., Regonne, R.K., Nisar, K.S.: Fractional dynamics of typhoid fever transmission models with mass vaccination perspectives. Fractal Fract. **5**, 149 (2021). https://doi.org/10.3390/fractalfract5040149
118. Dayan, F., Ahmed, N., Ali, A.H., et al.: Numerical investigation of a typhoid disease model in fuzzy environment. Sci. Rep. **13**, 21993 (2023). https://doi.org/10.1038/s41598-023-48405-w
119. Alalhareth, F.K., Alharbi, M.H., Ibrahim, M.A.: Modeling typhoid fever dynamics: stability analysis and periodic solutions in epidemic model with partial susceptibility. Mathematics **11**, 3713 (2023). https://doi.org/10.3390/math11173713
120. Zhao, Y., Li, T., Kang, R., et al.: Dynamical analysis of a novel fractional order SIDARTHE epidemic model of COVID-19 with the Caputo–Fabrizio(CF) derivative. Adv. Cont. Discr. Mod. **2024**, 3 (2024). https://doi.org/10.1186/s13662-024-03798-4
121. Panda, S.K.: Applying fixed point methods and fractional operators in the modelling of novel coronavirus 2019-nCoV/SARS-CoV-2. Results Phys. **19**, 103433 (2020)
122. Partohaghighi, M., Yusuf, A., Karaca, Y., Li, Y., Ibrahim, T., Younis, B., et al.: A new fractal fractional modeling of the computer viruses system. Fractals **30**(5), 2240184 (2022)
123. Atangana, A.: Fractal-fractional differentiation and integration: connecting fractal calculus and fractional calculus to predict complex, system. Chaos, Solitons Fractals **102**, 396–406 (2017)
124. Wardowski, D.: Fixed point theory of a new type of contractive mappings in complete metric spaces. Fixed Point Theory Appl. **2012**, 94 (2012)
125. Panda, S.K., Tassaddiq, A., Agarwal, R.P.: A new approach to the solution of non-linear integral equations via various FBe-contractions. Symmetry **11**, 206 (2019). https://doi.org/10.3390/sym11020206
126. Kumari, P.S., Zoto, K., Panthi, D.: d-Neighborhood system and generalized F-contraction in dislocated metric space. Springerplus **4**, 368 (2015)
127. Kumari, P.S., Panthi, D.: Connecting various types of cyclic contractions and contractive self-mappings with Hardy-Rogers self-mappings. Fixed Point Theory Appl. **2016**, 15 (2016)
128. Kumari, P.S., Panthi, D.: Cyclic compatible contraction and related fixed point theorems. Fixed Point Theory Appl. **2016**, 28 (2016)
129. Sumati Kumari, P., Alqahtani, O., Karapınar, E.: Some fixed-point theorems in B-dislocated metric space and applications. Symmetry **10**, 691 (2018)
130. Rasham, T., Shoaib, A., Hussain, N., Arshad, M., Khan, S.U.: Common fixed point results for new Ciric-type rational multivalued F-contraction with an application. J. Fixed Point Theory Appl. **20**, 45 (2018)
131. Gornicki, J.: Fixed points theorems for F-expanding mappings. Fixed Point Theory Appl. **2017**, 10 (2017)
132. Piri, H., Kumam, P.: Some fixed point theorems concerning F-contraction in complete metric spaces. Fixed Point Theory Appl. **2014**, 210 (2014)
133. Turinici, M.: Wardowski implicit contractions in metric spaces (2013). arXiv:1212.3164v2
134. Wardowski, D.: Solving existence problems via F-contractions. Proc. Am. Math. Soc. **146**(4), 1585–1598 (2018)

Medical Image Processing

8

Ant-Based Algorithms, particularly **"Ant Colony Optimization (ACO)"**, can be described mathematically by modeling the behavior of artificial ants traversing a graph. The graph $G = (V, E)$ represents the problem, where V is the set of vertices (nodes) and E is the set of edges (paths between nodes). Each edge $(i, j) \in E$ has an associated *pheromone level* $\tau_{ij}(t)$ and a heuristic value η_{ij}, which guide the ants in constructing solutions.

At a given step t, an ant located at node i selects the next node j based on a probabilistic rule. The probability $P_{ij}(t)$ of moving from node i to node j is given by:

$$P_{ij}(t) = \begin{cases} \frac{[\tau_{ij}(t)]^\alpha [\eta_{ij}]^\beta}{\sum_{k \in \mathcal{N}_i} [\tau_{ik}(t)]^\alpha [\eta_{ik}]^\beta}, & \text{if } j \in \mathcal{N}_i, \\ 0, & \text{otherwise.} \end{cases}$$

Here:

- $\mathcal{N}_i$ is the set of nodes adjacent to i that the ant has not yet visited.
- α controls the influence of pheromone levels.
- β controls the influence of the heuristic information η_{ij}, often defined as $\eta_{ij} = 1/d_{ij}$, where d_{ij} is the distance between nodes i and j.

Once all ants complete their tours (solutions), the pheromone levels are updated. The pheromone on each edge (i, j) is updated as follows:

$$\tau_{ij}(t + 1) = (1 - \rho)\tau_{ij}(t) + \Delta\tau_{ij},$$

where:

S. K. Panda et al., *Fixed Point Theory from Early Foundations to Contemporary Challenges*, Synthesis Lectures on Mathematics & Statistics,
https://doi.org/10.1007/978-3-032-03585-1_8

- ρ $(0 < \rho < 1)$ is the *evaporation rate*, which reduces the pheromone intensity over time to avoid excessive accumulation and premature convergence.
- $\Delta\tau_{ij}$ is the pheromone deposited by ants during the current iteration, defined as:

$$\Delta\tau_{ij} = \sum_{k=1}^{m} \Delta\tau_{ij}^{k},$$

where m is the number of ants, and $\Delta\tau_{ij}^{k}$ is the pheromone deposited by the k-th ant:

$$\Delta\tau_{ij}^{k} = \begin{cases} Q/L_k, & \text{if ant } k \text{ traverses edge } (i, j), \\ 0, & \text{otherwise.} \end{cases}$$

Here:

- Q is a constant representing the pheromone quantity.
- L_k is the total length (cost) of the tour completed by ant k.

The goal of the algorithm is to minimize (or maximize) a given objective function $f(S)$, where S is the solution constructed by the ants. For example, in the *Traveling Salesman Problem (TSP)*, the objective is to minimize the total path length:

$$f(S) = \sum_{(i,j)\in S} d_{ij},$$

where S is the sequence of edges forming a valid tour.

Over multiple iterations, the pheromone intensities guide the ants toward constructing better solutions. The algorithm converges when the solutions stabilize, and pheromone trails strongly favor optimal or near-optimal paths. This mathematical framework enables Ant-Based Algorithms to efficiently solve a variety of **"discrete optimization problems"** by simulating the collective behavior of ant colonies in nature.

Example 8.1 Let us consider a graph with nodes A, B, C, D, E, F, G, H representing locations, and edges connecting these nodes representing paths between them. Each edge has a weight that represents the distance between nodes. For example, the edge (A, B) has a weight of 1.2, and (C, D) has a weight of 1.7. The goal is to find the shortest path from node A to node H using an Ant-Based Algorithm.

Initially, all edges have an equal amount of pheromone, $\tau_{ij}(0) = 1$. The ants start at node A and move through the graph to construct paths to H. The decision to move to the next node depends on a probability function influenced by two factors: the pheromone level τ_{ij} of an edge and the heuristic information $\eta_{ij} = 1/d_{ij}$, where d_{ij} is the edge's weight. Paths with higher pheromone levels and shorter distances are more attractive to the ants.

Let us say one ant constructs the path $A \rightarrow C \rightarrow D \rightarrow G \rightarrow H$, and its total path cost is calculated as:

$$f(S) = d_{AC} + d_{CD} + d_{DG} + d_{GH} = 2.8 + 1.7 + 3.0 + 2.8 = 10.3.$$

Another ant takes a different path, $A \rightarrow B \rightarrow D \rightarrow E \rightarrow H$, and its total cost is:

$$f(S) = d_{AB} + d_{BD} + d_{DE} + d_{EH} = 1.2 + 1.5 + 2.1 + 2.5 = 7.3.$$

After all ants have completed their paths, the algorithm updates the pheromone levels on the edges. Pheromone evaporates over time to prevent overemphasis on a single path, modeled by:

$$\tau_{ij}(t+1) = (1-\rho)\tau_{ij}(t) + \Delta\tau_{ij}.$$

Here, ρ is the evaporation rate, and $\Delta\tau_{ij}$ is the additional pheromone deposited by the ants that used the edge (i, j). Ants that found shorter paths deposit more pheromone, reinforcing these paths for subsequent iterations.

Over several iterations, ants favor shorter paths due to increased pheromone levels on those edges. Eventually, the algorithm converges to the path $A \rightarrow B \rightarrow D \rightarrow E \rightarrow H$, identified as the optimal path with the least total cost. This path is highlighted in the graphical representation with red edges.

The Fig. 8.1 visually shows how the edges' thickness corresponds to pheromone levels, with the red edges representing the shortest path found by the algorithm. The dynamic interplay between pheromone reinforcement and evaporation ensures that the solution evolves toward optimality over iterations.

8.1 Ant-Based Algorithms in Medical Image Edge Detection

Ant-based algorithms are particularly useful in medical image edge detection due to their ability to identify edges by mimicking the foraging behavior of ants. The primary objective of edge detection in medical imaging is to locate boundaries and structures (e.g., organs, tumors, or blood vessels) for diagnosis, treatment planning, and research. This section explains how these algorithms contribute to edge detection.

8.1.1 Key Principles Applied to Edge Detection

1. **Pheromone Deposition and Intensification**: In medical image analysis, each pixel in the image is treated as a node in a graph. Ants traverse the image and lay down "pheromones" along paths with significant intensity changes, corresponding to edges. Over time, paths along true edges accumulate more pheromones due to consistent gradient or intensity differences.

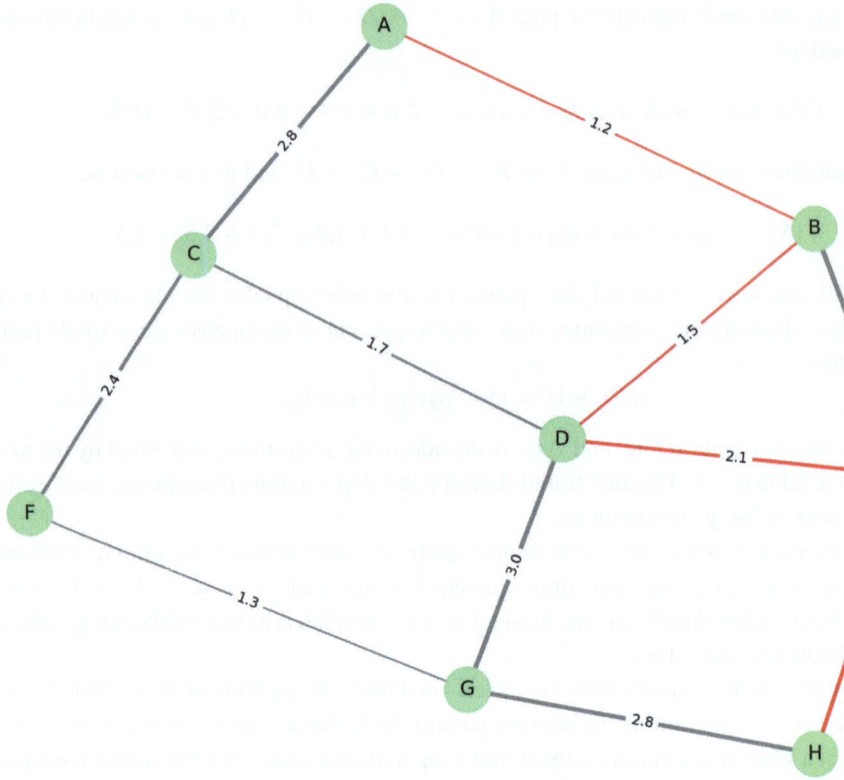

Fig. 8.1 Graphical representation of shortest path by ant-based algorithm

2. **Gradient-Based Heuristics**: The heuristic information η_{ij}, guiding the movement of ants, is defined using the gradient magnitude of the image. Areas with high gradients (sharp changes in intensity) attract ants, leading to the identification of edges.

8.1.2 Algorithm Steps in Medical Edge Detection

The steps of the algorithm for edge detection in medical imaging are as follows:

Step 1: Initialization. Convert the medical image (e.g., MRI, CT, or X-ray) into a grayscale intensity map. Initialize pheromone levels $\tau_{ij}(0)$ uniformly across all pixels.
Step 2: Ant Movement. Ants are deployed on the image and move according to a probabilistic rule:

$$P_{ij} = \frac{[\tau_{ij}]^{\alpha} [\eta_{ij}]^{\beta}}{\sum_{k \in N_i} [\tau_{ik}]^{\alpha} [\eta_{ik}]^{\beta}},$$

where:

- τ_{ij}: This term represents the Pheromone level on edge (i, j).
- η_{ij}: Gradient-based heuristic information.
- α, β: These parameters managing the effect of pheromones and heuristics, correspondingly.

Step 3: Pheromone Update. After ants complete their traversal, pheromones are updated as follows:

$$\tau_{ij}(t+1) = (1-\rho)\tau_{ij}(t) + \Delta\tau_{ij},$$

where:

- ρ: Evaporation rate.
- $\Delta\tau_{ij}$: Additional pheromone deposited by ants that used edge (i, j), proportional to the gradient strength.

Step 4: Edge Extraction. After several iterations, paths with high pheromone intensity correspond to edges. These edges are extracted and visualized.

Example 8.2 Consider a grayscale image resembling a medical MRI scan. Each pixel in the image represents tissue density, and regions with sharp intensity changes correspond to potential edges, such as tumor boundaries. The objective is to use an Ant-Based Algorithm to detect these edges effectively.

Problem Representation

1. **Nodes**: Each pixel in the image is treated as a node. 2. **Edges**: Connections between neighboring pixels represent possible paths for ants to traverse. 3. **Heuristic Information**: The gradient magnitude of the image, computed using a Sobel filter, quantifies the intensity change at each pixel. High gradient values indicate potential edges.

The gradient magnitude map serves as the heuristic information, guiding ants toward high-gradient regions.

The heuristic information $\eta(x, y)$, derived from the gradient magnitude map, is computed for the entire image. Initially, pheromones $\tau(x, y)$ are uniformly distributed across the image, and ants are randomly placed on the image. Each ant probabilistically moves to a neighboring pixel, favoring pixels with higher gradient magnitudes and higher pheromone levels. The probability of moving from pixel i to pixel j is given by:

$$P_{ij} = \frac{[\tau_{ij}]^{\alpha}\,[\eta_{ij}]^{\beta}}{\sum_{k\in N_i}[\tau_{ik}]^{\alpha}\,[\eta_{ik}]^{\beta}},$$

where:

- τ_{ij}: Pheromone level between pixels i and j,
- η_{ij}: Gradient magnitude between pixels i and j,
- $\mathcal{N}_i$: Neighborhood of pixel i,
- α, β: Parameters controlling the influence of pheromones and heuristic information.

After all ants complete their movements, pheromone levels are updated as follows:

$$\tau_{ij}(t+1) = (1-\rho)\tau_{ij}(t) + \Delta\tau_{ij},$$

where:

- ρ: Evaporation rate,
- $\Delta\tau_{ij}$: Additional pheromone deposited by ants traversing edge (i, j), proportional to the gradient.

Over several iterations, ants reinforce paths along high-gradient regions (edges), while low-gradient regions lose pheromones due to evaporation.

The gradient magnitude map highlights regions with sharp intensity changes in the image. As ants traverse the image, they deposit pheromones on these high-gradient regions, eventually converging on paths that correspond to edges. This approach is robust to noise and effectively detects irregular or faint boundaries.

The ants' behavior mimics a collective optimization process, balancing exploration of the image and exploitation of high-gradient paths. This method is especially suited for noisy medical images, where traditional edge detection methods might fail to detect weak edges. Below Fig. 8.2 is an illustration of the simulated MRI scan and its corresponding gradient magnitude map: The gradient map (right) highlights areas with sharp intensity changes, serving as a guide for ants to identify edges.

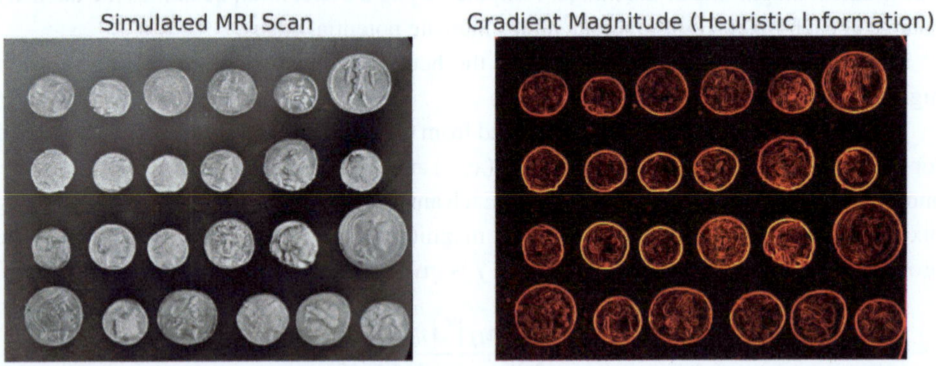

Fig. 8.2 Gradient magnitude (heuristic information)

8.2 The Role of Fixed Points in Image Process

Fixed points represent one of the primary elements of image processing mathematics that guarantee both stability and convergence to iterative algorithms. They play a crucial role wherever iterative optimization is needed, for instance, in image registration, which is aligned with overlapping images. In segmentation, fixed points help to control the contour change so that certain region or object of interest may be well described. The reconstruction techniques that could be used in MRI or CT scans involve using some reference points to repeatedly improve images generated by the use of robust or noisy data. It forms the basis of the methods that result in strong, fast, and accurate computation for different uses of images.

8.2.1 Digital Image Process

A fixed point is essential to several branches of mathematics. Numerous topological and scrutiny distinctions can benefit from a comprehension of the existence of fixed points. It can also be used in almost every area of computer science, including mathematical morphology, computer graphics, image processing, and so forth. If an equation has a solution, it is appropriate to catch a way out. It is possible to consistently frame a number of mathematical challenges that originate from innumerable mathematical outlets as fixed point problems.

The effects of digital images are balanced by digital topology. Rosenfield first made people familiar with it in [1]. In 1986, he created the appearance of digital continuity in photographs [2]. Boxer [3, 4] expanded the digital types of several perspectives, including digital continuous functions, in 1994. A few years ago, the concept of digital continuity was expanded upon [5]. Ege and Karaca [12] initially provided the fixed point attributes and fixed point theory for digital images, which they called the ***"Lefschetz fixed point theorem"***. In order to count fixed points, they resolved certain applications of the ***"Lefschetz fixed point theorem"*** and the ***"Nielsen fixed point theorem"*** in digital images [6–8].

Additionally, they have developed almost novel markings that combine fixed point theory and digital homotopy [7]. The authors of [9] investigated the Banach fixed point theorem for digital images in 2015. Recently, digitally continuous functions' approximated fixed points and ***"approximated fixed point property"*** (AFPP) have been introduced in [10]. In [11], the fixed point theorems of Banach and Eldestein were also presented.

To put it more sharply, in 2013 the authors of [12] concentrated on discrete variants of classical topological fixed point theorems, especially, ***"Lefschetz fixed point theorem"***. Using the algebraic devices of homology groups and simplifier complexes, they investigate the fixed point properties of digital image and formulate the ***"Lefschetz number"*** for digital spaces in a complete way. They illustrate the relationship between algebraic invariants (notably the Euler characteristic) and the existence of fixed points in *discrete topological structures*, providing many examples to corroborate their theoretical results. Building on the connectivity between the two homological realms, this work opens many new pathways for studies in image processing and computer vision.

The authors of [12] introduced the ***"Lefschetz fixed point theorem"*** for digital images, a result derived from algebraic topology, applied to finite digital simplifier complexes. They defined a digital image (X, κ) as possessing the fixed point property if every (κ, κ)-continuous function $\mathbb{T} : (X, \kappa) \to (X, \kappa)$ admits at least one fixed point $x \in X$ such that $\mathbb{T}(x) = x$. The Lefschetz number $\lambda(\mathbb{T})$, a critical invariant, was introduced as

$$\lambda(\mathbb{T}) = \sum_{i=0}^{n} (-1)^i \operatorname{tr}(\mathbb{T}_*),$$

where $\mathbb{T}_* : H_i^{\kappa}(X) \to H_i^{\kappa}(X)$ is the homomorphism induced by $\mathbb{T}$ on the digital simplicial homology groups $H_i^{\kappa}(X)$.

The authors of [12] rigorously established that if (X, κ) is a finite digital simplicial complex, or a retract thereof, and $\lambda(\mathbb{T}) \neq 0$, then f necessarily has a fixed point. This theorem parallels its classical counterpart but adapts to the adjacency relations and discrete structures inherent to digital images. They demonstrated that the Lefschetz number encapsulates the essential combinatorial and topological information of $\mathbb{T}$, making it a robust tool for analyzing digital transformations.

An illustrative example was provided in [12] by using the two-dimensional Brouwer fixed point theorem for a digital square $X = \{(0, 0), (1, 0), (0, 1), (1, 1)\} \subset \mathbb{Z}^2$ with 4-adjacency. The induced map $\mathbb{T}_*$ on the homology group $H_0^4(X) \cong \mathbb{Z}$ was shown to be the identity map, yielding $\lambda(\mathbb{T}) = 1$. As $\lambda(\mathbb{T}) \neq 0$, the fixed point property was guaranteed for $\mathbb{T}$.

Further, the authors of [12] showed that the fixed point property is a topological invariant in the digital setting, preserved under digital homotopy and digital isomorphisms. For example, they demonstrated that if $X \approx (A, \kappa')$, and X has the fixed point property, then so does A.

Above stated article highlighted the adaptability of algebraic topology to discrete settings, providing tools such as the Lefschetz number for analyzing fixed points in digital images and advancing the theoretical framework for applications in digital image processing and computational topology.

In 2015, the authors of [7] *"Banach Fixed Point Theorem for Digital Images"* rigorously investigated the adaptation of the Banach fixed point theorem within the realm of digital topology, a mathematical framework dedicated to analyzing the structural and connectivity properties of discretized Euclidean spaces, particularly in dimensions two and three. The study commenced by establishing formal definitions for κ-adjacency relations, κ-neighborhoods, and digital κ-paths, which collectively define the algebraic and topological underpinnings of digital images. By constructing digital metric spaces (X, d, κ), where d represents the Euclidean metric and κ encodes adjacency constraints, the authors developed a discrete analog of classical metric spaces to extend fixed point theory to finite and lattice-based spaces.

The key contents in [7] are:

- Constructed the ***"contraction mapping theorem"*** for digital metric spaces by introducing the concept of a digital contraction map $\mathbb{T} : (X, d, \kappa) \to (X, d, \kappa)$, characterized by the inequality
$$d(\mathbb{T}(x), \mathbb{T}(y)) \leq \lambda d(x, y),$$
for some contraction coefficient $\lambda \in (0, 1)$.
- Associated theorem satisfied the convergence of iterated sequences $\{x_n\}$, where $x_{n+1} = \mathbb{T}(x_n)$, to the fixed point c as $n \to \infty$, ensuring convergence within the digital framework.

By analyzing the sequence of distances between iterates and invoking properties of monotonic and convergent sequences, the authors established that $\mathbb{T}$ not only possesses a unique fixed point but also guarantees the convergence of any iterated sequence $\{\mathbb{T}^n(x)\}$ to this fixed point for all $x \in X$.

The application of the Banach fixed point theorem to digital image processing underscored its practical significance, particularly in the domain of image compression. The authors modeled digital images as sets in digital metric spaces and iteratively transformed these sets using contractive mappings. By demonstrating the convergence of a sequence of transformed images $\{F_n\}$ to a fixed digital image F_2, they showcased the utility of fixed points in achieving compression while preserving critical structural features. The limiting fixed image F_2 provided a mathematically optimal representation that could substitute higher-resolution images like F_5 in computational tasks, ensuring efficiency and fidelity within the constraints of digital topology.

In 2023, the authors of [13] present a novel approach for digital image compression by extending fixed-point theorems on complete symmetric G-metric spaces. They introduce the concept of an extended G-contraction mapping, leveraging a comparison function $\phi : [0, \infty) \to [0, \infty)$ that satisfies specific monotonicity and convergence criteria. Using this framework, they prove that a self-mapping $\mathbb{T} : X \to X$ in a G-metric space has a unique fixed point if there exist constants $\alpha, \beta, \gamma \in [0, 1)$ such that the contractive condition:

$$\begin{aligned}\phi(G(\mathbb{T}x, \mathbb{T}y, Tz)) \leq& \alpha\phi(G(x, y, z)) + \beta\phi(\max\{G(x, \mathbb{T}x, \mathbb{T}x), G(x, y, z)\}) \\ &+ \gamma\phi\left(\frac{G(x, y, z)(1 + \sqrt{G(x, \mathbb{T}x, \mathbb{T}x)G(x, y, z)})^2}{(1 + G(x, y, z))^2}\right)\end{aligned}$$

is satisfied, where $0 \leq L = \alpha + \beta + \gamma < 1$. They apply this theoretical foundation to compress grayscale digital images, representing images as matrices and reducing their size via G-contraction mappings. By dividing an image into blocks and applying the mapping to each block, the authors achieve significant size reduction while maintaining image quality. The results demonstrate the effectiveness of this mathematical framework in digital image compression, showcasing high-quality contracted images with minimal distortion. Moreover, the presented application in [13] is particularly compelling as it rigorously applies the extended G-metric framework to achieve efficient compression of grayscale images, ensuring minimal information loss while adhering to strict mathematical principles. This condition guarantees that the mapping $\mathbb{T}$ is a strict contraction, leading to a unique fixed point for the iterative compression process. By partitioning images into $n \times n$ blocks and applying the contraction iteratively, the approach achieves significant reductions in image size, up to 99%, while maintaining topological and geometric fidelity. The Frobenius norm, employed to compute the matrix distance function, quantitatively demonstrates that deviations between the original and compressed images are negligible under the G-metric. This innovative application underscores the robustness and scalability of the G-metric-based compression framework, positioning it as a mathematically rigorous and computationally efficient solution for digital image processing.

8.2.2 Medical Image Process

Fixed point theory has emerged as a critical mathematical framework in medical image processing, offering robust solutions for problems requiring iterative convergence and stability. Most of these application areas include image registration or segmentation and image enhancements where the existence and uniqueness of solutions are guaranteed by fixed points. For example, iterative algorithms that are applied to medical image registration make use of fixed points in order to align a number of medical images initiated with CT, MRI or PET scans to provide maximum accuracy in comparative diagnostics.

Also, fixed point methods are useful in deformable image registration, where complex mapping functions in anatomical structures are described. These methods guarantee that several iterations all lead to the same fixed alignment of features. In segmentation tasks, the fixed point iteration forms a delineation of tissues or identification of additional regions of interest, in this case, tumors or lesion areas. There are other segmentation techniques which are based on level set form and active contours; both employ ***"fixed point theory"*** to ensure the ***"stabilization of dynamic contours"***.

In medical image enhancement, fixed points form the basis of iterative denoising and reconstruction techniques especially in computed tomography or magnetic resonance imaging. These elements guarantee fixity and convergence in iterative schemes that recover effective images from noisy or partial information. More simply, fixed point theory remains a foundation for medical image processing because it promises stable, efficient, and mathematically rigorous algorithms that will underpin crucial functions in diagnostics and treatment.

Iterative methods—particularly those based on fixed point principles—play a pivotal role in applying mathematical techniques to real-world challenges in areas such as acoustic signal processing, optical imaging, and medical tomography. In the realm of medical image reconstruction, many underlying mathematical models and computational algorithms are formulated using nonexpansive or nonexpansive-type operators. Consequently, solving the associated fixed point equations often relies on carefully designed fixed point iterative schemes (see, for instance, the monograph [14]).

A recent study [15] aimed to improve edge detection outcomes in medical imaging through ant-based algorithms by employing test functions that involve admissible perturbations of specific demicontractive operators. This class of mappings, which generalizes nonexpansive mappings with fixed points and includes quasi-nonexpansive mappings, offers favorable damping characteristics for edge detection tasks. The effectiveness of demicontractive mappings in this context was supported by numerical results presented in [15].

Beyond image processing, demicontractive mappings have proven highly effective in addressing a broad spectrum of problems in applied mathematics. These include solving nonlinear functional equations [16], fixed point problems [17–19], multiple-set split feasibility problems [20], equilibrium models, and variational inequality problems [21–23], highlighting their versatility and broad applicability.

In [24], the authors formulate a new Mathematical Framework called theory of admissible perturbations for the purpose of studying iterative approximation methods for the fixed points of a nonlinear operator. They define a **demicontractive operator** [17, 19] as following:

Let $\mathbb{T} : C \to C$ such that there exists a constant $k < 1$ satisfying:

$$\|\mathbb{T}x - p\|^2 \leq \|x - p\|^2 + k\|x - \mathbb{T}x\|^2,$$

where p is a fixed point of $\mathbb{T}$. The work emphasizes the importance of this inequality in ensuring convergence. An example provided in [24] is the mapping:

$$\mathbb{T}(x) = \begin{cases} \frac{2}{3}x \sin\left(\frac{1}{x}\right), & \text{if } x \neq 0, \\ 0, & \text{if } x = 0. \end{cases}$$

Further, the authors of [25] defined an **admissible mapping** $G : X \times X \to X$ satisfying two conditions:

1. $G(x, x) = x$ for all $x \in X$,
2. $G(x, y) = x$ implies $y = x$.

Using these mappings, the authors describe **admissible perturbations** of an operator $\mathbb{T} : X \to X$ as $\mathbb{T}_G(x) = G(x, \mathbb{T}(x))$. This formulation creates modified operators while preserving certain essential properties of the original operator. Two key examples in [24] include the **Krasnoselskij admissible mapping** [26]:

$$G(x, y) = (1 - \lambda)x + \lambda y,$$

and the **(χ)-admissible mapping**:

$$G(x, y) = (1 - \chi(x, y))x + \chi(x, y)y,$$

where $\chi(x, y) = \frac{x^2y^2}{(1+x^2)(1+y^2)}$. These mappings are further employed to derive iterative algorithms for fixed-point approximation.

In [24], the authors apply Ant Colony Optimization (ACO) to edge detection in medical images. ACO uses artificial ants that mimic natural behavior by constructing a **pheromone matrix** representing the edge structure of an image. The ants' movements are guided by a **transition probability**:

$$p_{IJ} = \frac{(\tau_{IJ})^{\alpha} (\eta_{IJ})^{\beta}}{\sum_{J' \in \Omega_I} (\tau_{IJ'})^{\alpha} (\eta_{IJ'})^{\beta}},$$

where τ_{IJ} is the pheromone value, η_{IJ} is the heuristic information, and α, β control their respective influences. The heuristic η_{IJ} is computed using variation functions, such as:

Sin Operator:

$$\mathbb{T}(x) = \begin{cases} \sin\left(\frac{\pi x}{2\lambda}\right), & 0 \leq x \leq \lambda, \\ 0, & \text{otherwise.} \end{cases}$$

KH Operator (based on Krasnoselskij perturbation):

$$\mathbb{T}(x) = (1 - \lambda)x + \lambda\mathbb{T}(x),$$

where $y(x) = \frac{2}{3}x \sin\left(\frac{1}{x}\right)$.

(χ)-Operator:

$$\mathbb{T}(x) = (1 - \chi(x, \mathbb{T}(x)))x + \chi(x, \mathbb{T}(x))\mathbb{T}(x).$$

The pheromone matrix is updated locally after each ant's movement using:

$$\tau_{IJ} \leftarrow (1 - \rho)\tau_{IJ} + \rho\Delta_{IJ},$$

and globally after all ants' movements:

$$\tau_{IJ} \leftarrow (1 - \psi)\tau_{IJ} + \psi\tau_0.$$

These updates reinforce paths that correspond to edges in the image. For more information, the reader can refer [15, 27, 28].

In [24], the authors conduct a case study by applying the proposed ACO-based algorithm to detect edges in medical images, including CT scans and X-rays. The ants start at random positions and explore the image grid, with their movements dictated by the heuristic values computed from the proposed operators. The algorithm involves three main steps:

1. **Initialization**: Parameters such as the number of ants, movement steps, and initial pheromone values are defined.
2. **Construction**: Ants iteratively move, guided by transition probabilities, to create the pheromone matrix.

3. **Decision Process**: A threshold is applied to the final pheromone matrix to produce a binary edge map.

The authors validate the approach using medical images of brain CT and hand X-rays, demonstrating the ability of the algorithm to capture edge details effectively. In [24], the authors present experimental results showing that the proposed KH and χ operators improve the accuracy of edge detection in medical images. The Sin operator serves as a benchmark, while the new operators demonstrate higher precision in identifying edge pixels. Statistical analysis reveals:

- The χ operator has the smallest standard deviation, indicating consistent performance.
- The KH operator exhibits stability across all tests.
- Post-processing with a denoising convolutional neural network (DnCNN) further enhances the edge quality.

The results confirm the robustness of the proposed operators. The authors highlight the operators' mathematical properties, which enable better heuristic computation and guide the ants to regions of significant intensity variation. These improvements suggest that the methodology can be extended to other complex image processing tasks.

In [15], the authors explore the enhancement of nonlinear mappings through the concept of admissible perturbations. They define a mapping $G : X \times X \to X$ as admissible if it satisfies two core properties: $G(x, x) = x$ for all $x \in X$, and $G(x, y) = x$ implies $y = x$. Using this framework, a perturbed operator $\mathbb{T}_G(x) = G(x, \mathbb{T}(x))$ is constructed, retaining key characteristics of the original operator $\mathbb{T}$ while introducing controlled modifications to improve performance.

The authors of [15] focus on demicontractive operators, which are particularly relevant for iterative processes. A mapping $\mathbb{T} : C \to C$, where C is a subset of a Hilbert space, is considered demicontractive if there exists a constant $k < 1$ such that:

$$\|\mathbb{T}(x) - p\|^2 \leq \|x - p\|^2 + k\|x - \mathbb{T}(x)\|^2,$$

for all $x \in C$ and any fixed point p of $\mathbb{T}$. An illustrative example provided is $\mathbb{T}(x) = \frac{2}{3}x \sin\left(\frac{1}{x}\right)$ for $x \neq 0$, and $\mathbb{T}(0) = 0$, which satisfies the criteria for demicontractiveness. Furthermore, the Krasnoselskij perturbation is introduced as a significant enhancement. Defined as $\mathbb{T}_\lambda(x) = (1 - \lambda)x + \lambda\mathbb{T}(x)$ for $\lambda \in (0, 1)$, this perturbation ensures asymptotic regularity, which is the property that the sequence $\{\mathbb{T}_\lambda^n(x)\}$ converges for any $x \in C$. This regularity is beneficial in iterative methods for fixed-point computations.

In [15], the authors investigate the application of admissible perturbations of demicontractive operators as test functions in ant colony optimization (ACO) algorithms for edge detection. The primary focus is on leveraging these operators to enhance the algorithm's capability to identify edges in medical images accurately.

The study involves several perturbed operators constructed from a base demicontractive operator T. The operators tested include the following:

1. **Krasnoselskij Perturbation**:

$$\mathbb{T}_\lambda(x) = (1-\lambda)x + \lambda\mathbb{T}(x),$$

   where λ is a parameter within (0, 1).
2. χ**-Admissible Perturbation**:

$$\mathbb{T}_\chi(x) = (1-\chi(x, \mathbb{T}(x)))x + \chi(x, \mathbb{T}(x))\mathbb{T}(x),$$

   where χ is a function defined as:

$$\chi(x, \mathbb{T}(x)) = \frac{x^2(\mathbb{T}(x))^2}{(1+x^2)(1+(\mathbb{T}(x))^2)}.$$

3. **Sin-Based Operator**:

$$\mathbb{T}_{\sin}(x) = \sin\left(\frac{\pi x}{2\alpha}\right),$$

   where α is a parameter adjusting the operator's sensitivity.

These operators were used as test functions in the ACO algorithm to optimize edge detection. The iterative process involves ants exploring the image, guided by pheromone levels and heuristic values derived from the test functions. The ants' movements are probabilistic, governed by the transition probability:

$$p_{ij} = \frac{(\tau_{ij})^\alpha (\eta_{ij})^\beta}{\sum_{k\in\Omega_i} (\tau_{ik})^\alpha (\eta_{ik})^\beta},$$

where:

- τ_{ij}: pheromone value between pixel i and j,
- η_{ij}: heuristic value derived from the operator,
- α, β: parameters controlling the influence of pheromones and heuristics,
- Ω_i: neighborhood of pixel i.

The authors of [24] conducted experiments using medical images, including Head CT, Brain CT, and Hand X-rays. The perturbed operators significantly improved edge detection, with sharper and more defined edges observed in the results. Among the operators tested, the χ and Krasnoselskij perturbations yielded superior results compared to the base operator T, especially when the parameter λ was set to smaller values.

Additionally, the study revealed that the extracted edges contained more pixels as the perturbation parameter λ decreased, making the edge representation more compact and precise. These findings underscore the advantage of using admissible perturbations in ACO algorithms for edge detection tasks. In their study, each image is represented as a matrix. The values in the matrices for every image shown in Fig. 8.3 are integers with inclusive values ranging from 0 to 255. Figure 8.4 showcases the detected edges for the hand X-ray, head CT and brain CT.

The authors of [15] highlighted that the use of non-demicontractive functions resulted in inferior edge detection quality. This confirms the critical role of demicontractive mappings and their perturbations in enhancing algorithmic performance.

This chapter also includes one example that satisfies demicontractive and their admissible perturbations. Next, we present some relevant numerical tests to demonstrate how the admissible perturbations of demicontractive mappings improve ant-based algorithms for medical image edge detection when used as test functions. Our study's edge detection method takes into account medical images from hand X-ray, Brain MRI, and Head MRI. The algorithm's performance was empirically assessed through parameter evaluation and visually tested using a variety of medical images.

Example 8.3 Let $H = \mathbb{R}$ (the real line) and $C = \left[-\frac{\pi}{4}, \frac{\pi}{4}\right]$. Define the mapping $\mathbb{T} : C \longrightarrow C$ by

$$\mathbb{T}x = \frac{1}{2}\tan x, \text{ for } x \in C.$$

Then, $Fix(\mathbb{T}) = \{0\}$, then $\mathbb{T}$ is demicontractive.

For $x \in C$ and $y = 0$,

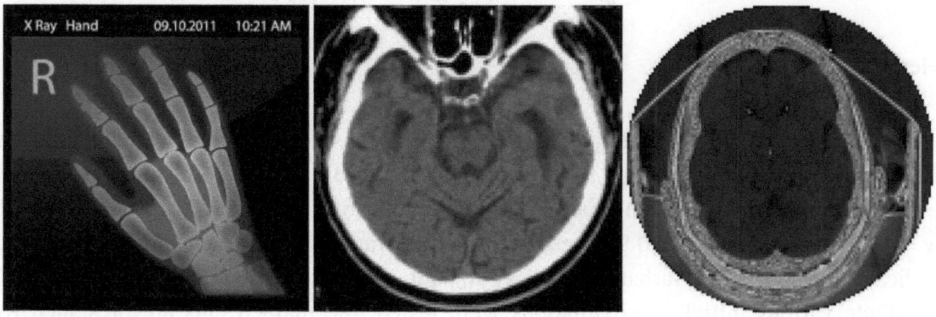

Fig. 8.3 Medical images of hand X-Ray [29], head CT and brain CT [30]

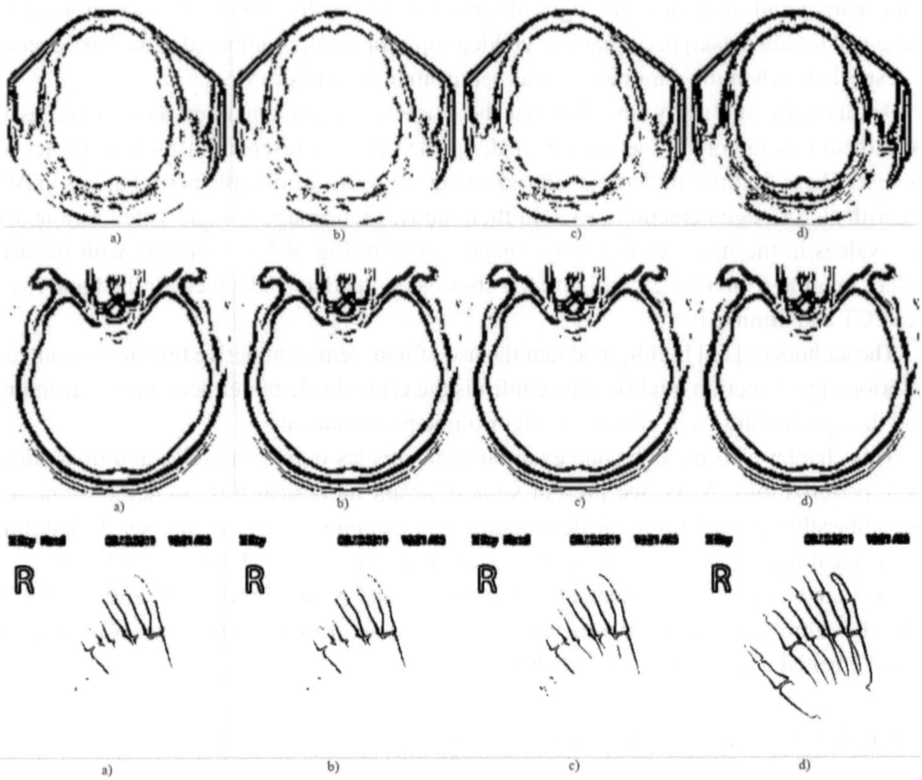

Fig. 8.4 Detected edges for brain CT, head CT and hand X-ray with test function [24]

$$|\mathbb{T}x - 0|^2 = |\mathbb{T}x|^2 = \left|\frac{1}{2}\tan x\right|^2 = \frac{1}{4}\tan^2 x$$
$$\leqslant |x - 0|^2 + k\left|\frac{1}{2}\tan x - x\right|^2 \quad \text{for any } k < 1.$$

Hence $\mathbb{T}$ is demicontractive.

The operator defined in Example 8.3, denoted $\mathbb{T}$, is considered throughout the subsequent discussions. To approximate its fixed points, both Picard and Krasnoselskij iterations are employed, applying admissible perturbations for different values of λ, namely $\lambda = 0.1$, $\lambda = 0.2$, $\lambda = 0.5$, and $\lambda = 0.9$. These perturbed operators are represented as $\mathbb{T}_\lambda$, where λ indicates the specific perturbation, such as $\mathbb{T}_{0.1}$ for $\lambda = 0.1$, and so on.

While various initial approximations were tested, the main results are presented for an initial value of $x_0 = 0.6$. Successive approximations, x_n, were calculated for iterations up to 60, considering $\mathbb{T}$, $\mathbb{T}_{0.1}$, $\mathbb{T}_{0.2}$, $\mathbb{T}_{0.5}$, and $\mathbb{T}_{0.9}$. Table 8.1 summarizes the findings. After

100 iterations, all the tested functions exhibited a convergence rate where the difference between successive approximations did not exceed 10^{-6}, meaning that $|x_{n+1} - x_n| \leq 10^{-6}$. The above Fig. 8.5 illustrates the behavior of the successive approximations for the function $\mathbb{T}(x) = \frac{1}{2}\tan(x)$ and its Krasnoselskij perturbations $\mathbb{T}_\lambda(x)$ with varying values of λ, namely $\lambda = 0.1$, $\lambda = 0.2$, $\lambda = 0.5$, and $\lambda = 0.9$. The iterative process starts with an initial approximation of $x_0 = 0.6$, and the figure shows the values of x_n over 100 iterations. Each iteration value is plotted on the Y-axis, with the number of iterations on the X-axis. The successive approximations show how each function converges towards a fixed point as the iterations progress.

The plot clearly demonstrates the convergence behavior of the original function $\mathbb{T}(x)$ and its Krasnoselskij perturbations. As the number of iterations increases, all the functions approach a stable value, with the convergence being more pronounced for higher values of λ. In particular, the perturbations with larger λ values exhibit a faster convergence rate compared to the original function. This indicates that the perturbations are effective in improving the convergence behavior of the iterative process.

The impact of the Krasnoselskij perturbations is evident in the figure, as they significantly enhance the convergence rate, especially for smaller values of λ. The iterative process for $\mathbb{T}_\lambda(x)$ with lower values of λ converges more quickly to the fixed point compared

Table 8.1 Successive approximation results

n	$\mathbb{T}$	$\mathbb{T}_{0.1}$	$\mathbb{T}_{0.2}$	$\mathbb{T}_{0.5}$	$\mathbb{T}_{0.9}$
1	0.342068404	0.57420684	0.548413681	0.471034202	0.367861564
5	0.022568445	0.478551913	0.373696664	0.157569457	0.035577732
10	0.000705423	0.377081712	0.224839861	0.03761737	0.001791456
15	2.20445E-05	0.294967047	0.133637604	0.008929808	9.01613E-05
20	6.8889E-07	0.229728621	0.079092038	0.002119126	4.53768E-06
25	2.15278E-08	0.17845246	0.046740254	0.000502879	2.28374E-07
30	6.72744E-10	0.138404941	0.027607316	0.000119336	1.14937E-08
35	2.10233E-11	0.107244437	0.016303422	2.83189E-05	5.78461E-10
40	6.56977E-13	0.083052949	0.009627332	6.7202E-06	2.9113E-11
45	2.05305E-14	0.064296889	0.00568491	1.59474E-06	1.46521E-12
50	6.41579E-16	0.049766583	0.003356897	3.78438E-07	7.37419E-14
55	2.00493E-17	0.038515327	0.001982217	8.98052E-08	3.71131E-15
60	6.26542E-19	0.02980562	0.00117048	2.13112E-08	1.86785E-16
70	6.11857E-22	0.017847556	0.000408121	1.20011E-09	4.73117E-19
80	5.97517E-25	0.010686384	0.000142303	6.75823E-11	1.19838E-21
90	5.83513E-28	0.006398417	4.9618E-05	3.8058E-12	3.03544E-24
100	5.69837E-31	0.003830987	1.73007E-05	2.14318E-13	7.68863E-27

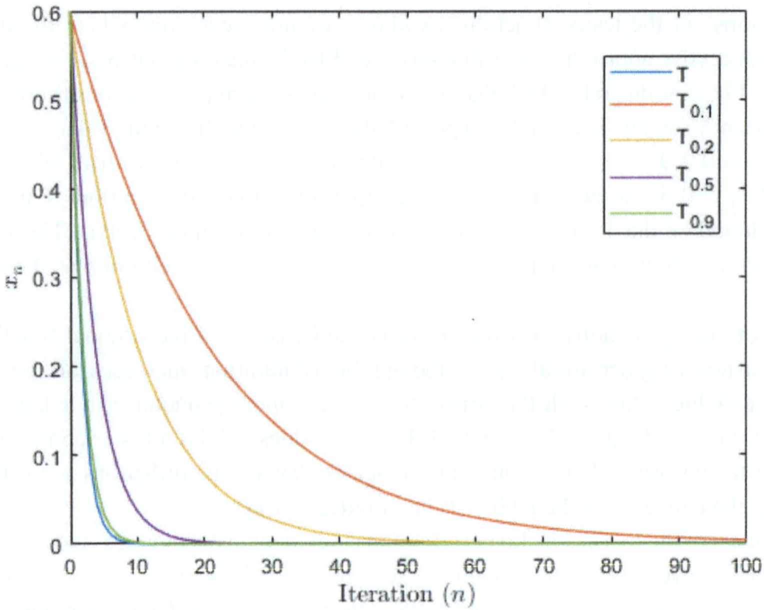

Fig. 8.5 Successive approximations for $\mathbb{T}$ and $\mathbb{T}_\lambda$ (Initial $x_0 = 0.6$)

to the unperturbed function $\mathbb{T}(x)$. The figure illustrates how modifying the operator with perturbations can accelerate the process of approximation in iterative methods.

Particularly, we present and examines the impact of varying the parameter λ, which plays a crucial role in the admissible perturbation of the demicontractive mapping $\mathbb{T}$. By adjusting λ, we analyze its influence on edge detection performance.

Each image in our study is represented as a matrix, where pixel values range from 0 to 255. Specifically, for the images shown in Figs. 8.6, 8.7 and 8.8 the matrix elements are integers within this range. The matrices corresponding to the detected edges contain only two possible values: 0 or 255. A value of 0 represents an edge, whereas 255 corresponds to the white (non-edge) region.

To determine the most effective edge detection outcome, we identify the matrix that contains the highest number of zero-valued elements, as this indicates the most prominent edge representation. Furthermore, we establish both the most similar and the least similar detected edges based on comparative analysis.

The next thing we performed was use MATLAB software to turn Figs. 8.6, 8.7 and 8.8 into grayscale images, as seen in Figs. 8.9, 8.10 and 8.11.

An ant colony optimization algorithm for image edge detection was then carried out. Figs. 8.12, 8.13 and 8.14 illustrate the detected edges for Hand X-ray images, Brain-MRI and Head-MRI respectively. In our notation, D represents the edge matrix generated by the operator $\mathbb{T}$, while $\mathbb{T}_{0.9}$, $\mathbb{T}_{0.5}$, $\mathbb{T}_{0.1}$, and $\mathbb{T}_{0.06}$ correspond to edge matrices derived from $\mathbb{T}_{0.9}$, $\mathbb{T}_{0.5}$, $\mathbb{T}_{0.1}$, and $\mathbb{T}_{0.06}$, respectively.

For quantitative assessment, we counted the number of pixels forming the detected edges. The results of this pixel count are summarized in Table 8.2, providing a comparative evaluation of edge detection efficacy across different parameter values. We demonstrate both theoretically and empirically that employing acceptable perturbations of demicontractive mappings as test functions can enhance ant-based methods for image edge detection. We refer to the classes of so-called enriched mappings, which are actually admissible perturbations of different contractive operators and were recently examined by the authors of [34–40], as an additional potential method of identifying more selective test functions that can enhance ant-based algorithms for image edge detection. Moreover, we believe that the ant colony method of edge detection is equivalent to other approaches such as, Prewitt edge detection, Sobel edge detection, and Roberts edge detection. We provide Fig. 8.15 to support this claim.

Fig. 8.6 Test data: hand X-Ray [29, 31]

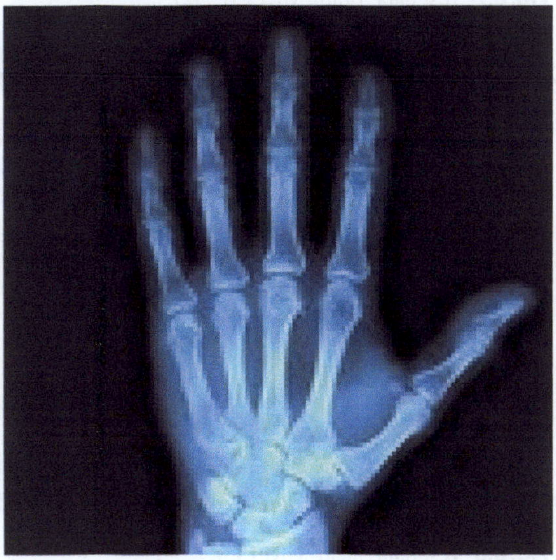

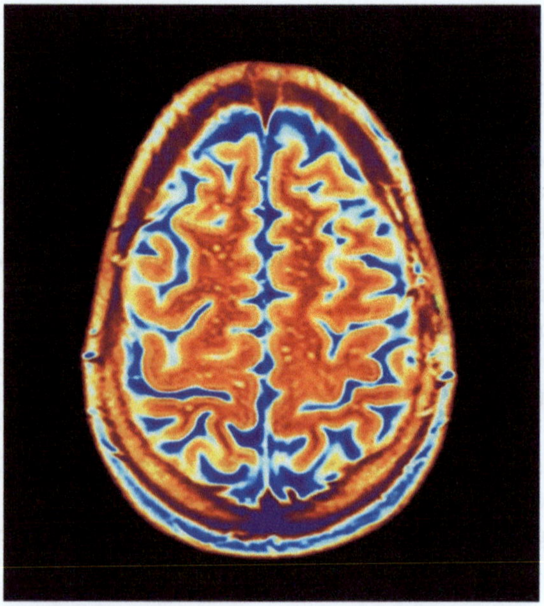

Fig. 8.7 Test data: brain-MRI [31, 32]

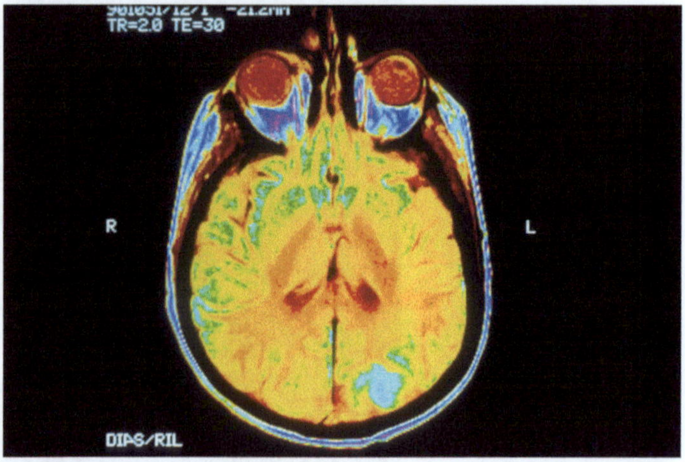

Fig. 8.8 Test data: head-MRI [33].

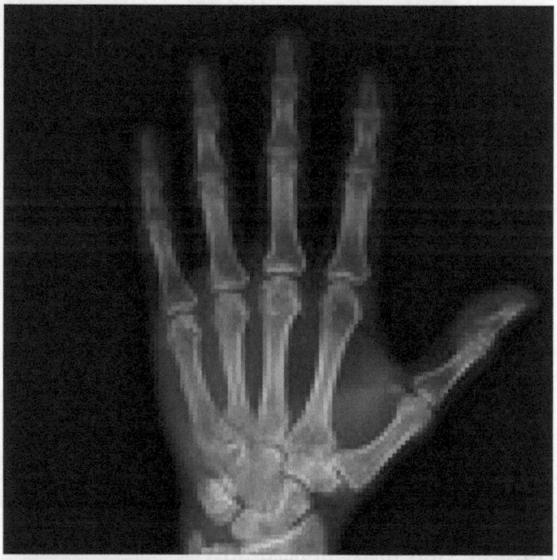

Fig. 8.9 Gray image of hand X-Ray

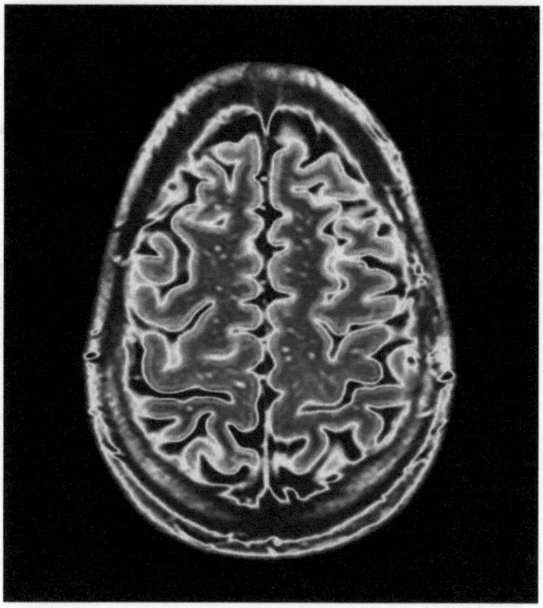

Fig. 8.10 Gray image of brain-MRI

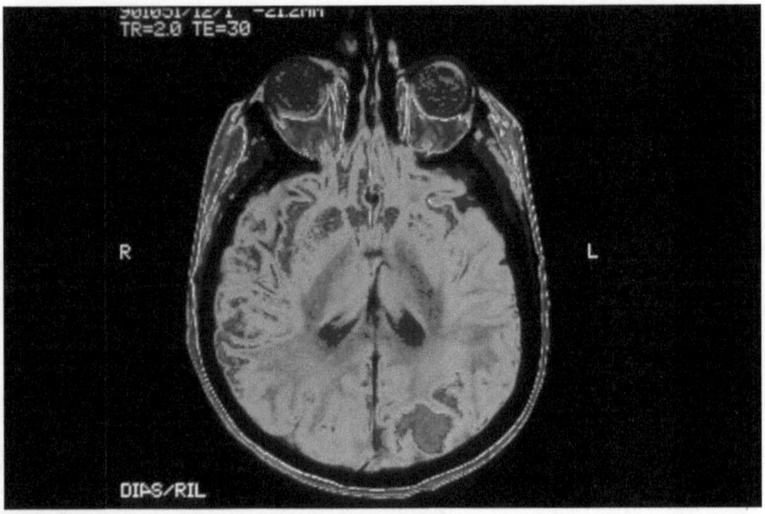

Fig. 8.11 Gray image of head-MRI

Open Questions

8.1 Is it possible to generalize the concept of admissible perturbations to spaces other than Hilbert spaces, e.g. to topological spaces that are not locally convex or even not metric, and still have convergence and stability properties?

8.2 How should one choose the perturbation parameter lambda, lambda in Krasnoselskij-type perturbations to achieve a good compromise between speed of convergence and quality of edge detection? Is this something that can be learnt adaptively through iteration?

8.3 When is the admissible perturbation applicable to not only fixed point set but also the rate of convergence or the stability basin of the iterative process?

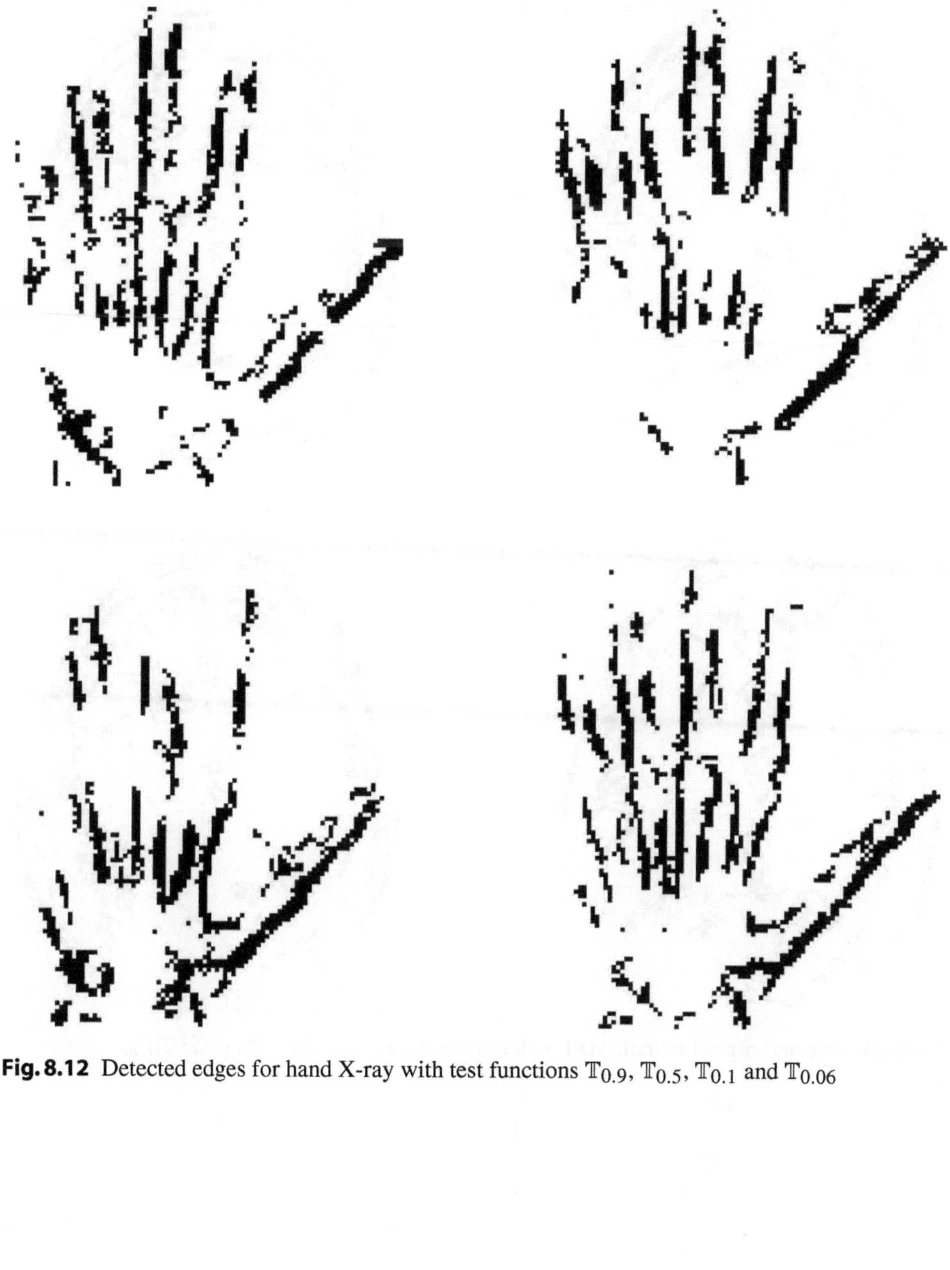

Fig. 8.12 Detected edges for hand X-ray with test functions $\mathbb{T}_{0.9}$, $\mathbb{T}_{0.5}$, $\mathbb{T}_{0.1}$ and $\mathbb{T}_{0.06}$

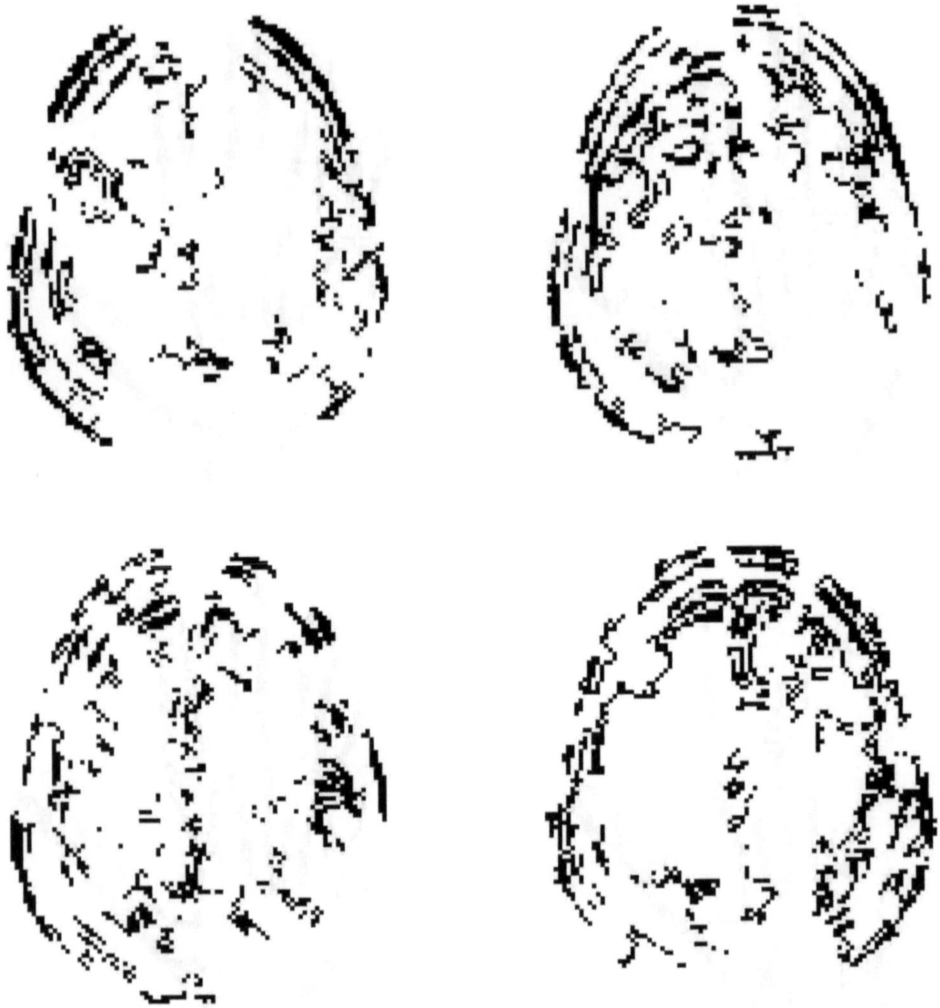

Fig. 8.13 Detected edges for brain-MRI with test functions $\mathbb{T}_{0.9}$, $\mathbb{T}_{0.5}$, $\mathbb{T}_{0.1}$ and $\mathbb{T}_{0.06}$

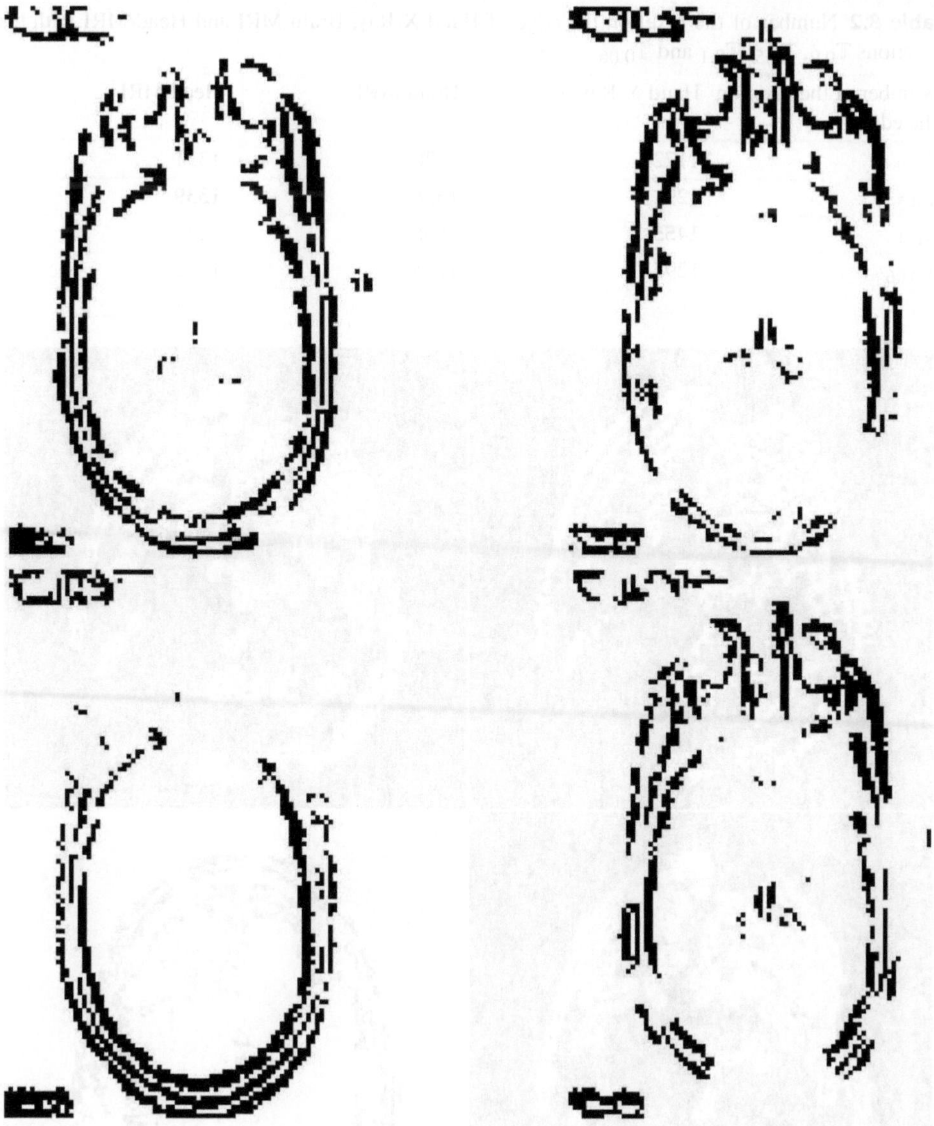

Fig. 8.14 Detected edges for head-MRI with test functions $\mathbb{T}_{0.9}$, $\mathbb{T}_{0.5}$, $\mathbb{T}_{0.1}$ and $\mathbb{T}_{0.06}$

Table 8.2 Number of the pixel on the edge of Hand X-Ray, Brain MRI and Head-MRI with test functions $\mathbb{T}_{0.9}$, $\mathbb{T}_{0.5}$, $\mathbb{T}_{0.1}$ and $\mathbb{T}_{0.06}$

Number of the pixel on the edge	Hand X-Ray	Brain MRI	Head MRI
$\mathbb{T}_{0.9}$	1172	1520	1331
$\mathbb{T}_{0.5}$	1291	**1658**	**1339**
$\mathbb{T}_{0.1}$	**1455**	885	1247
$\mathbb{T}_{0.06}$	1295	1195	1222

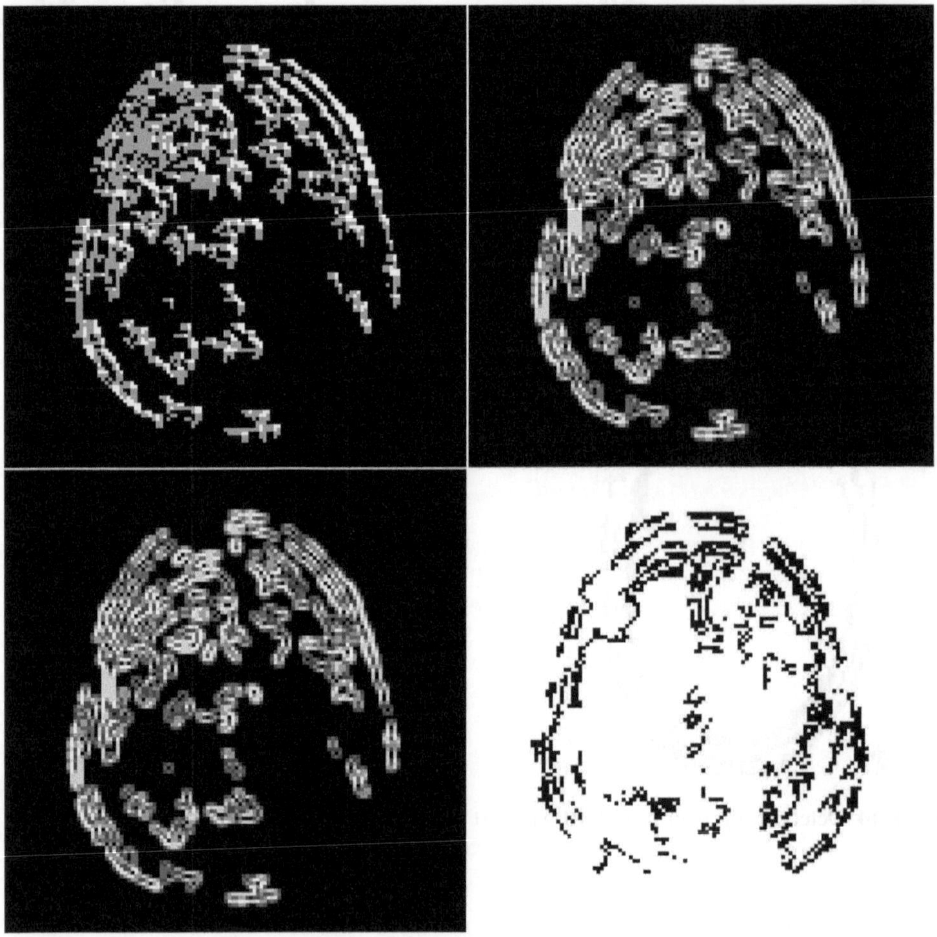

Fig. 8.15 Extracted edges for brain-MRI with different extraction methods, such as prewitt, sobel, roberts and ACO with test function

8.4 Is it possible to learn the optimal form of admissible perturbation $G(x, \mathbb{T}x)$ with techniques of deep learning or reinforcement learning, by optimizing edge detection performance on a dataset of medical images?

8.5 What are the admissible perturbations that can optimise several edge detection criteria simultaneously, e.g. compactness, continuity and contrast, perhaps with the aid of multi-criteria decision theory?

References

1. Rosenfeld, A.: Digital topology. Amer. Math. Soc. **86**, 621–630 (1979)
2. Rosenfeld, A.: Continuous functions on digital pictures. Pattern Recognit. Lett. **4**, 177–184 (1986)
3. Boxer, L.: Digitally continuous function. Pattern Recognit. Lett. **15**, 833–839 (1994)
4. Boxer, L., Ege, O., Karaca, I., Lopez, J., Louwsma, J.: Digital fixed points, approximate fixed points, and universal functions. Appl. Gen. Topol. **17**(2), 159–172 (2016)
5. Han, S.E.: Non-product property of the digital fundamental group. Inform. Sci. **171**, 73–91 (2005)
6. Ege, O., Karaca, I.: Applications of the Lefschetz number to digital images. Bull. Belg. Math. Soc.-Simon Stevin **21**(5), 823–839 (2014)
7. Ege, O., Karaca, I.: Digital homotopy fixed point theory. Comptes Rendus Mathematique **353**(11), 1029–1033 (2015)
8. Ege, O., Karaca, I.: Nielsen fixed point theory for digital images. J. Comput. Anal. Appl. **22**(5), 874–880 (2017)
9. Ege, O., Karaca, I.: Banach fixed point theorem for digital images. J. Nonlinear. Sci. Appl. **8**, 237–245 (2015)
10. Boxer, L., Ege, O., Karaca, I., Lopez, J., Louwsma, J.: Digital fixed points, approximate fixed points, and universal functions. Appl. Gen. Topol. **17**(2), 159–172 (2016)
11. Hossain, A., Ferdausi, R., Mondal, S., Rashid, H.: Banach and Edelstein fixed point theorems for digital images. J. Math. Sci. Appl. **5**(2), 36–39 (2017)
12. Ege, O., Karaca, I.: Lefschetz fixed point theorem for digital images. Fixed Point Theory Appl. **2013**, 253 (2013)
13. Anna, T.R., Thalapathiraj, S.: A novel approach for digital image compression in some fixed point results on complete g-metric space using comparison function. Int. J. Anal. Appl. **21**, 110 (2023)
14. Byrne, C.L.: Applied Iterative Methods. A K Peters Ltd: Wellesley. MA, USA (2008)
15. Ticala, C., Zelina, I., Pintea, C.M.: Admissible perturbation of demicontractive operators within ant algorithms for medical images edge detection. Mathematics **8**, 1040 (2020)
16. Maruster, S.: Sur le calcul des zéros d'un opérateur discontinu par itération. Canad. Math. Bull. **16**, 541–544 (1973)
17. Hicks, T.L., Kubicek, J.D.: On the Mann iteration process in a Hilbert space. J. Math. Anal. Appl. **59**, 498–504 (1977)
18. Maingé, P.E.: A hybrid extragradient-viscosity method for monotone operators and fixed point problems. SIAM J. Control Optim. **47**, 1499–1515 (2008)

19. Maruster, S.: The solution by iteration of nonlinear equations in Hilbert spaces. Proc. Am. Math. Soc. **63**, 69–73 (1977)
20. Qin, L.-J., Wang, G.: Multiple-set split feasibility problems for a finite family of demicontractive mappings in Hilbert spaces. Math. Inequal. Appl. **16**, 1151–1157 (2013)
21. Suantai, S., Phuengrattana, W.: A hybrid shrinking projection method for common fixed points of a finite family of demicontractive mappings with variational inequality problems. Banach J. Math. Anal. **11**, 661–675 (2017)
22. Thong, D.V., Hieu, D.V.: Modified subgradient extragradient algorithms for variational inequality problems and fixed point problems. Optimization **67**, 83–102 (2018)
23. Vuong, P.T., Strodiot, J.J., Nguyen, V.H.: On extragradient-viscosity methods for solving equilibrium and fixed point problems in a Hilbert space. Optimization **64**, 429–451 (2015)
24. Berinde, V., icală, C.: Enhancing ant-based algorithms for medical image edge detection by admissible perturbations of demicontractive mappings. Symmetry **13**, 885 (2021). https://doi.org/10.3390/sym13050885
25. Rus, I.A.: An abstract point of view on iterative approximation of fixed points: impact on the theory of fixed point equations. Fixed Point Theory **13**, 179–192 (2012)
26. Krasnosel'skii, M.A.: Two remarks about the method of successive approximations. Uspehi Mat. Nauk **10**, 123–127 (1955)
27. Tian, J., Yu, W., Xie, S.: An ant colony optimization algorithm for image edge detection. In: Proceedings of the IEEE Congress on Evolutionary Computation (IEEE World Congress on Computational Intelligence), Hong Kong, China, 1–6 June 2008, pp. 751–756
28. Ticala, C., Zelina, I.: New ant colony optimization algorithm in medical images edge detection. Creat. Math. Inform. **29**, 101–108 (2020)
29. X-ray Hand. Vista Medical Pack. License: Free for Non Commercial Use. Id, 236487. https://www.iconspedia.com/. Accessed 1 May 2020
30. Head, C.T.: Online Medical Free Image. http://www.libpng.org/pub/png/pngvrml/ct2.9-128x128.png. Accessed 1 May 2020
31. Hand X-Ray, Brain-MRI and Head-MRI: Online Medical Free Images. https://hslguides.med.nyu.edu/medicalimages
32. The Radiology Assistant: Multiple Sclerosis 2.0. 1 December 2021. https://radiologyassistant.nl/neuroradiology/multiple-sclerosis/diagnosis-and-differential-diagnosis-3#mri-protocol-ms-brain-protocol. Accessed 22 May 2023
33. http://commons.wikimedia.org/wiki/File:Brain_MRI.jpg
34. Berinde, V.: Weak and strong convergence theorems for the Krasnoselskij iterative algorithm in the class of enriched strictly pseudocontractive operators. Analele Universiătii de Vest din Timisoara: Seria Matematică-Informatică, vol. 56, pp. 13–27 (2018)
35. Berinde, V.: Approximating fixed points of enriched nonexpansive mappings by Krasnoselskij iteration in Hilbert spaces. Carpathian J. Math. **35**, 293–304 (2019)
36. Berinde, V.: Approximating fixed points of enriched nonexpansive mappings in Banach spaces by using a retraction-displacement condition. Carpathian J. Math. **36**, 27–34 (2020)
37. Berinde, V., Pacurar, M.: Approximating fixed points of enriched contractions in Banach spaces. J. Fixed Point Theory Appl. **22**, 1–10 (2020)
38. Berinde, V., Pacurar, M.: Kannan's fixed point approximation for solving split feasibility and variational inequality problems. J. Comput. Appl. Math. **386**, 113217 (2021)
39. Berinde, V., Pacurar, M.: Existence and approximation of fixed points of enriched contractions and enriched J-contractions. Symmetry **13**, 498 (2021)
40. Berinde, V., Pacurar, M.: Fixed points theorems for unsaturated and saturated classes of contractive mappings in Banach spaces. Symmetry **13**, 713 (2021)

MIX
Papier aus verantwortungsvollen Quellen
Paper from responsible sources
FSC® C105338

If you have any concerns about our products,
you can contact us on
ProductSafety@springernature.com

In case Publisher is established outside the EU,
the EU authorized representative is:
Springer Nature Customer Service Center GmbH
Europaplatz 3, 69115 Heidelberg, Germany

Printed by Libri Plureos GmbH
in Hamburg, Germany